Haptic Interfaces for Accessibility, Health, and Enhanced Quality of Life

Troy McDaniel • Sethuraman Panchanathan
Editors

Haptic Interfaces for Accessibility, Health, and Enhanced Quality of Life

 Springer

Editors
Troy McDaniel
The Polytechnic School
Arizona State University
Mesa, AZ, USA

Sethuraman Panchanathan
Arizona State University
Tempe, AZ, USA

ISBN 978-3-030-34232-6   ISBN 978-3-030-34230-2   (eBook)
https://doi.org/10.1007/978-3-030-34230-2

This Springer imprint is published by the registered company Springer Nature Switzerland AG.
The registered company address is: Gewerbestrasse 11, 6330 Cham, Switzerland

Preface

The focus of this book on Haptic Interfaces for Accessibility, Health, and Enhanced Quality of Life is to serve as a principal resource that provides an in-depth coverage of haptic (touch-based) technologies in the realm of assistive, rehabilitative, and health-related applications. Application topics are grouped into thematic areas spanning haptic devices for sensory impairments, health and wellbeing, and physical impairments. A diverse group of experts in the field were invited to contribute complementary and multidisciplinary perspectives. Unlike other books on haptics, Haptic Interfaces for Accessibility, Health, and Enhanced Quality of Life takes an application-oriented approach to present a comprehensive view of how the field of haptics has advanced with respect to important and impactful thematic focuses.

The editors thank Dr. Ramin Tadayon for his assistance with editing. The editors also acknowledge the funding support of Arizona State University and the National Science Foundation (Grant No. 1828010), which made this book project possible.

Mesa, AZ, USA   Troy McDaniel
Tempe, AZ, USA   Sethuraman Panchanathan

Contents

Part I   Haptics for Sensory Impairments

1   Enabling Learning Experiences for Visually Impaired
     Children by Interaction Design 3
     Florian Güldenpfennig, Armin Wagner, Peter Fikar, Georg Kaindl,
     and Roman Ganhör

2   Haptically-Assisted Interfaces for Persons with Visual
     Impairments... 35
     Yeongmi Kim and Matthias Harders

3   Maps as Ability Amplifiers: Using Graphical Tactile Displays
     to Enhance Spatial Skills in People Who Are Visually Impaired...... 65
     Fabrizio Leo, Elena Cocchi, Elisabetta Ferrari, and Luca Brayda

4   Haptics for Sensory Substitution 89
     Bijan Fakhri and Sethuraman Panchanathan

Part II   Haptics for Health and Wellbeing

5   Haptics in Rehabilitation, Exergames and Health 119
     Mohamad Hoda, Abdulmotaleb El Saddik, Philippe Phan,
     and Eugene Wai

6   Therapeutic Haptics for Mental Health and Wellbeing 149
     Troy McDaniel and Sethuraman Panchanathan

7   Applications of Haptics in Medicine 183
     Angel R. Licona, Fei Liu, David Pinzon, Ali Torabi, Pierre
     Boulanger, Arnaud Lelevé, Richard Moreau, Minh Tu Pham,
     and Mahdi Tavakoli

Part III   Haptics for Physical Impairments

8   **Assistive Soft Exoskeletons with Pneumatic Artificial Muscles** 217
     Yuichi Kurita, Chetan Thakur, and Swagata Das

9   **Haptics for Accessibility in Hardware for Rehabilitation** 243
     Ramin Tadayon

10   **Intelligent Robotics and Immersive Displays for Enhancing
     Haptic Interaction in Physical Rehabilitation Environments** 265
     Jason Fong, Renz Ocampo, and Mahdi Tavakoli

Contributors

Pierre Boulanger Department of Computing Science, University of Alberta, Edmonton, AB, Canada

Luca Brayda Robotics, Brain and Cognitive Sciences Department, Fondazione Istituto Italiano di Tecnologia, Genoa, Italy

Elena Cocchi Istituto David Chiossone, Genoa, Italy

Swagata Das Graduate School of Engineering, Hiroshima University, Hiroshima, Japan

Bijan Fakhri Center for Cognitive Ubiquitous Computing, Arizona State University, Tempe, AZ, USA

Elisabetta Ferrari Robotics, Brain and Cognitive Sciences Department, Fondazione Istituto Italiano di Tecnologia, Genoa, Italy

Peter Fikar Multidisciplinary Design & User Research (TU Wien), Vienna, Austria

Jason Fong Department of Electrical and Computer Engineering, University of Alberta, Edmonton, AB, Canada

Roman Ganhör Multidisciplinary Design & User Research (TU Wien), Vienna, Austria

Florian Güldenpfennig New Design University (NDU), St. Pölten, Austria

Matthias Harders Department of Computer Science, University of Innsbruck, Innsbruck, Austria

Mohamad Hoda University of Ottawa, Ottawa, ON, Canada

Georg Kaindl Independent Researchers, Vienna, Austria

Yeongmi Kim Department of Mechatronics, MCI, Innsbruck, Austria

Yuichi Kurita Graduate School of Engineering, Hiroshima University, Hiroshima, Japan

Fabrizio Leo Robotics, Brain and Cognitive Sciences Department, Fondazione Istituto Italiano di Tecnologia, Genoa, Italy

Arnaud Lelevé Laboratoire Ampère (UMR 5005), INSA Lyon, University of Lyon, Lyon, France

Angel R. Licona Laboratoire Ampère (UMR 5005), INSA Lyon, University of Lyon, Lyon, France

Fei Liu Laboratoire Ampère (UMR 5005), INSA Lyon, University of Lyon, Lyon, France

Troy McDaniel The Polytechnic School, Arizona State University, Mesa, AZ, USA

Richard Moreau Laboratoire Ampère (UMR 5005), INSA Lyon, University of Lyon, Lyon, France

Renz Ocampo Department of Electrical and Computer Engineering, University of Alberta, Edmonton, AB, Canada

Sethuraman Panchanathan Arizona State University, Tempe, AZ, USA

Minh Tu Pham Laboratoire Ampère (UMR 5005), INSA Lyon, University of Lyon, Lyon, France

Philippe Phan Division of Orthopaedic Surgery, University of Ottawa, Ottawa, ON, Canada

David Pinzon Department of Computing Science, University of Alberta, Edmonton, AB, Canada

Abdulmotaleb El Saddik University of Ottawa, Ottawa, ON, Canada

Ramin Tadayon Center for Cognitive Ubiquitous Computing, Arizona State University, Tempe, AZ, USA

Mahdi Tavakoli Department of Electrical and Computer Engineering, University of Alberta, Edmonton, AB, Canada

Chetan Thakur Graduate School of Engineering, Hiroshima University, Hiroshima, Japan

Ali Torabi Department of Electrical and Computer Engineering, University of Alberta, Edmonton, AB, Canada

Eugene Wai The Ottawa General Hospital, Ottawa, ON, Canada

Armin Wagner Independent Researchers, Vienna, Austria

Part I
Haptics for Sensory Impairments

Chapter 1
Enabling Learning Experiences for Visually Impaired Children by Interaction Design

Florian Güldenpfennig, Armin Wagner, Peter Fikar, Georg Kaindl, and Roman Ganhör

Abstract Interaction design and tangible computing offer rich opportunities for supporting children with impairments by means of enhanced therapeutic toys and educational materials. In order to explore how technology can be utilized to meet special requirements in the education of visually impaired children (and teenagers), we set up a practice-based research project at a special health center and school for the blind. Drawing on a number of design experiments involving educators and affected children, we came up with design proposals that enabled instructive (sensory) experiences despite their impairments in the sensory system. We describe two interactive prototypes in detail – a tangible color-picker toy, that we named *The Cuebe*, and an *Audio-Tactile Map* designed for e-learning – and show how they can support children in building new skills by augmenting physical properties and affordances. In both prototypes, tactility, haptics, and interactivity were crucial features, since all experiences originated at the fingertips and then unfolded higher-level sensory and cognitive processes. Moreover, the prototypes were also characterized by a high degree of *open-endedness* and *customizability* in their design, allowing educators to incorporate them in flexible ways to meet the needs of the children.

F. Güldenpfennig (✉)
New Design University (NDU), St. Pölten, Austria
e-mail: florian.gueldenpfennig@ndu.ac.at

A. Wagner · G. Kaindl
Independent Researchers, Vienna, Austria
e-mail: wagner@wiretouch.net; kaindl@wiretouch.net

P. Fikar · R. Ganhör
Multidisciplinary Design & User Research (TU Wien), Vienna, Austria
e-mail: peter.fikar@tuwien.ac.at; roman.ganhoer@tuwien.ac.at

© Springer Nature Switzerland AG 2020
T. McDaniel, S. Panchanathan (eds.), *Haptic Interfaces for Accessibility, Health, and Enhanced Quality of Life*, https://doi.org/10.1007/978-3-030-34230-2_1

1   Enabling Experiences

We had the opportunity to work with visually impaired children and their educators in order to explore the design of interactive and tangible artifacts to be used as therapeutic toys and educational materials. There had been no detailed design brief except that we should make use of interactive technology to enable novel learning experiences for the children. This open-ended approach posed both a great creative opportunity to us, as well as a challenge, because the design space was vast and hard to navigate for designers without disability-related training. Therefore, we decided to take a practice-based approach to research (e.g., [35]) with participatory design elements (e.g., [52]). That is, we engaged in extensive field work to learn about the requirements of special-needs educators and visually impaired children, and we created a series of design experiments aiming to meet these requirements.

From an artifact- or interaction-design perspective, two features were particularly important. (1) For one thing, our artifacts were *open-ended* in their design and did not prescribe what educators and therapists should do with them. Rather, they were built in a way that allowed flexible adaptation according to current needs. (2) For another, our prototypes drew on tangible interaction and haptics to trigger or unfold deeper learning experiences. Literally speaking, these experiences started at the children's fingertips and then addressed additional parts of the sensory and cognitive system. Our designs amplified, augmented, and added stimuli. Put differently, they made previously inaccessible information perceivable to visually impaired children and teenagers.

The artifacts presented in this book chapter have been created together with children and educators from the *Federal Institute for the Blind* in Vienna/Austria[1] and its affiliated association for supporting children with visual impairments named *Contrast*.[2] We focus on two prototypical devices incorporating the features described above, but designed for different purposes: *The Cuebe* is a small tangible toy which can be used for 'picking colors' and is intended to support young children in exercising motor skills and in training their visual perception (*sensomotorical/procedural knowledge*). *Audio-Tactile Map*, on the other hand, has been designed for teaching at a school for the blind. The prototypes are aimed at infants and preschool children (*The Cuebe*), as well as students (*Audio-Tactile Map*) with severe visual impairments.

Both prototypes were informed by in-depth field work. We engaged in passive observations during class sessions as well as during therapeutic sessions, open-ended interviews, and group workshops to learn from our participants and to collaboratively develop design proposals. During these participatory activities, we started implementing early working prototypes to be deployed in the field for collecting feedback.

---

[1]https://www.bbi.at/ – last accessed 27 Mar 2019.
[2]http://www.contrast.or.at – last accessed 27 Mar 2019.

Backed by novel rapid prototyping methods in material design (e.g. 3D-printing) and electronics (e.g., the Arduino platform [6]), we were able to present fully-interactive devices early in the process, which underwent multiple design iterations. We argue that both exposing participants to functional prototypes and creating multiple iterations was important to our endeavor. The high degree of interactivity did not allow substituting system logic with *Wizard of Oz components* (e.g., [12]).

In this chapter, we report what we have learned from this interaction design-driven or practice-based research process [35]. We offer a detailed description of the designed artifacts, which play an important role in practice-based research. They were used to simultaneously explore the problem and solution space, since prototypes are often understood as carriers of embodied design knowledge (e.g., [36, 47, 65]). We follow this line of argumentation and draw on *The Cuebe* and *Audio-Tactile Map* in order to illustrate how we used interaction design to enable experiences for visually impaired children and to support them in their education (see Sect. 3 for more details).

1.1   Chapter Structure

The chapter is structured as follows. First of all, we contextualize the practice-based endeavor and provide relevant background information. This entails a short description of the physical/psychological condition of the affected children and related work from HCI. Next, we detail our research/design philosophy and explain the underlying methodology. This explanation is followed by the main part of this chapter's contribution, the presentation of the design research artifacts *The Cuebe* and *Audio-Tactile Map*. Here, our focus is on describing the artifacts' underlying design process, the design decisions, and the design features. Finally, we discuss the results of our design explorations and reflect about the potential of our approach from a broader perspective.

2   Background

In this section, we provide background information about visual impairments from a physical/psychological perspective as well as related design work from HCI. It is divided into three parts: (1) prior work from HCI about supporting children with (visual) impairments, (2) tangible computing and interaction design of haptic technology for children, and (3) tactile maps for the blind featuring (multi-)touch technology. Before we continue, we will also briefly present background information about the research context of our projects.

2.1 Project Background

As our background is in computer science and interaction design, we closely collaborated with special needs therapists, educators, and teachers in order to bring relevant competencies in understanding children with disabilities into the design project. In particular, we collaborated closely with four specialists for Early Intervention and one teacher for blind children. Those were the experts who, together with affected children, engaged in the co-design activities that formed the foundation for all design and research efforts.

From an administrative perspective, the present work was divided into two separate projects with individual funding. Despite this formal separation, our collaborators, as introduced in the above paragraph, all belonged to the same special needs center, the *Federal Institute for the Blind* in Vienna/Austria and its affiliated association for supporting children with visual impairments named *Contrast*.

The first project, with *The Cuebe* as the final artifact outcome, was commissioned to design novel special needs toys for young children (0–6 years old preschool children). Our initial idea was that these toys should draw on interactive technology in order to make them more appropriate as well as more interesting to play with in the context of Early Intervention programs. We describe this kind of intervention in more detail below. For now, it is evident that Early Intervention seeks to build up various competencies in young children with disabilities. For example, they are often supported by exposure to bright lights and high-contrast images in order to stimulate visual neurons and their growth. In addition, therapists aim to train motor skills, as those are often affected by visual impairment. Children also benefit from Early Intervention by growing their self-confidence, social skills, and by boosting their general development. The most important and ultimate design specification for this project was that the novel toys should fit well into this context.

The second project, with the *Audio-Tactile Map* as the result, aimed at developing an audio-tactile tool for e-learning to be used in the school for the blind. Other than *The Cuebe* project, it was targeted at older (14+ years) visually impaired students. Accordingly, it also addressed 'higher-level' activities, such as learning facts and studying content for school instead of supporting basic skills and competencies.

Having provided some information about the background of the design projects, we go on to explain the physical and cognitive conditions of the children for whom the novel technologies were designed and hence whose needs had to be met.

2.2 Physical and Cognitive Situation of the Target Group

Due to the complexity of the visual system, there are numerous causes of visual impairments. In general, people suffering from such conditions may show a diverse set of symptoms. From the perspective of our research project, those symptoms and manifestations were more important than the causes for the conditions, because our work or contribution was not framed in a medical sense where the cause of a disease had to be cured. Rather, we started with certain symptoms and then searched for appropriate responses to meet the needs of the people with our design work. While the educators and therapists engage with the children and seek ways of strengthening their abilities, we as designers sought to support the educators in their efforts. Thus, the symptoms of the visual impairment were of prime interest to us as designers.

2.2.1 Blindness and Cortical Visual Impairment (CVI)

In 2015, a global meta-analysis estimated that of 7.33 billion people alive, 36.0 million were blind [7]. For their analysis, the authors defined individuals with a visual acuity of worse than $(3/60)^3$ in the better eye as being blind. Another 217.0 million people suffered from moderate or severe vision impairment (visual acuity worse than 6/18, but 3/60 or better) [7]. Degeneration of the macula and posterior pole, glaucoma, diabetic retinopathy, and cataracts were the most common causes for blindness [10, 19].

The children that we worked with during our first project *The Cuebe* were sighted, at least to some extent, and the majority of them had been diagnosed with Cortical Visual Impairment (CVI). This condition is characterized by visual dysfunction caused by damages in the brain and visual pathway during early perinatal development. CVI describes a complex clinical picture. Depending on the location of the affected structure in the brain, visual impairments show different symptoms. Common impairments affect visual processing and attention, which again can cause problems in learning, development, and in the children's independence [41]. Moreover, CVI is often associated with additional comorbidities, caused by lesions of the brain resulting in complex and diverse impairments of different severity (e.g., such as cerebral palsy) [39].

2.2.2 Professional Support of Children and Students with Visual Impairments

Since visual impairments and related conditions can cause special needs for the affected children and their families, Early Intervention Services [50] as provided by healthcare systems around the world [25] support preschool children with neurological issues. This kind of support typically starts at the first few weeks after birth and continues its efforts until about the age of six years [39]. The ultimate goal of Early Intervention Service is to prevent or minimize developmental delays and corresponding negative long-term effects [39]. To this end, Early Intervention introduces affected infants and young children to stimulating environments and

---

[3] That is, while a healthy subject can read a letter from 60 m distance, the visually impaired person can only do so from 3 m.

activities. Often, therapists make use of technology in the shape of aids and (special need) toys in order to provide such stimulation and appropriate cues [39]. In cases where specific skills cannot be acquired, therapists seek to teach compensation techniques to enable the children to later live independently.

The project partner *Contrast* is such a service provider for Early Intervention with an additional specialization in supporting children with CVI. *Contrast* shares the same building with the *Federal Institute for the Blind* in Vienna, our partner for the second project. While *Contrast* supports preschoolers, this federal institute offers schooling for students affected by blindness and severe visual impairment. Their curriculum for visual impaired students [4] covers, among other things, the following disability related skills:

- training of the senses, in particular active tactile perception and auditory perception
- training of body perception
- training of strategies for orientation and mobility
- use of Assistive Technologies [11] for individuals with visual impairments
- braille reading
- training of communication and social skills

The above description of the different conditions and health services should help the reader to understand the nature of the design space and why *The Cuebe* and *Audio-Tactile Map* eventually emerged as design artifacts. In the next section, we present work that is related to these two prototypes and the overall project.

2.3 Related Work from Interaction Design and HCI

There is a growing body of work about children and disability in HCI. In this section, we focus on research that is either concerned explicitly with *visually* impaired children or with *Tangible Computing* for children with some kind of disability. In addition, we briefly review work on tactile maps for the blind. Hence, there are three areas of prior work that were of particular importance to our research: HCI for visually impaired children, Tangible Computing for impaired children (visually or in some other sense that can be addressed by tangibles), and efforts in creating tactile maps featuring (multi-)touch technology, etc., for building touch based interfaces.

2.3.1 HCI and Visually Impaired Children

HCI researchers have investigated various ways of supporting children with visual impairments including, for example, tools for navigation and orientation [51] or educational tools (e.g., for teaching programming [58]). Recently, there has also been a strengthening of HCI research into the inclusive education of visually impaired children that is looking for ways of integrating impaired children into regular schooling through interactive technologies (e.g., [42, 43, 58]).

Since children usually love toys and much of their knowledge is acquired through playing, some design interventions were actually toys or (serious) games that could be played without visuals. Linehan et al. [38] and Waddington et al. [60] created a video game in order to exercise the vision of children with CVI. The authors used participatory design methods in order to find appropriate visual cues to be used in their gaming-based exercises. These efforts represent rather rare examples of HCI research explicitly designing for children with CVI.

In recent years, our own research added some examples of therapeutic toys for CVI theraphy to the literature. In addition to the *The Cuebe* [17] (presented in Sect. 4), we created *Boost Beans* [23, 24] and a virtual ball run [16]. Our first-mentioned prototype comprised four different remote-controlled actuators that we named 'beans' with the size of approximately a piece of soap. By pressing the remote, children could either make some noise, vibrate, light up or ventilate some air. These four actuators were created to be integrated during exercise by therapists in order to stimulate the children's senses on various levels. The virtual ball run [16] made use of a computer and monitor to emulate a conventional ball run in order to make it perceivable for children with CVI. Due to high contrasts and a modulated pace, affected children could focus on tracking the virtual ball. Moreover, the ball run included computerized tangible elements to also exercise the children's motor skills. Clearly, these prototypes could be classified as instances of Tangible Computing. In the subsequent section, we present further related prototypes with emphasis on their tangible design.

2.3.2 Tangible Computing and (Visually) Impaired Children

There is a great potential of tangibles for educational purposes (e.g., [48]) and for supporting children with disabilities, as evident from many research projects. Especially in recent years, since physical prototyping has become more feasible, an increasing number of researchers turned to designing and investigating Tangible Computing (see, e.g., [29]) applications for impaired children. By means of embedded controllers and rapid prototyping techniques, it became possible for even small teams of researchers and designers to create interactive, often toy-like artifacts, suitable for use by children and addressing their particular needs. ChillFish by Sonne and Jensen [55], for example, was a tangible application for exploring video-game based biofeedback for children suffering from attention deficit hyperactivity disorder (ADHD). The game was operated by the ChillFish mouthpiece which could sense the players' respiratory rate. The children used their breath to control a character in the accompanying ChillFish video-game (a side-scroller or jump'n'run game). The game was designed to calm children with ADHD down, since it constituted a relaxing breathing-exercise that should make them more focused and alert. Antle et al. [2] introduced PhonoBlocks, a tangible user interface consisting of smart and illuminated letters. The prototype's concept was to support children with dyslexia in decoding letters by utilizing sound- and color-cues. Jadan-Guerrero et al. [30] proposed Kiteracy, another tangible application, this time with the aim to support children with Down Syndrome in improving their literacy.

As we will see in the description of the following related projects, novel interactive technology has often also been utilized to create tangibles that could be *adapted or customized* to fit the needs of their users better.

Polipo was a hand-held device by Tam et al. [57] for improving the motor skills of children's hands. It employed interaction design to motivate children during exercise, that is, to make training fine motor skills fun and engaging. To accomplish this, they designed Polipo to allow for customization by the therapists, as an example, by mounting different knobs or handles to it. The therapists could also record audio messages for the Polipo. When the child successfully completed a motor task, these messages were played out loud, accompanied by different light animations.

Moraiti et al. [44] also drew on tangibles and the concept of customization, but to an even larger extent. They created a Do-It-Yourself (DIY) toolkit for occupational therapists, empowering them to design and create their own tailor-made Assistive Technologies to be handed to their clients. The toolkit enabled the therapists to turn everyday objects, such as soft pillows, into tangible user interfaces and game controllers. This enabled the therapists to adapt materials and equipment to the needs of their clients.

Verhaegh, Fontijn and Hoonhout [59] and Garzotto and Gonella [21] designed tangible and tile-based (board) games for children to support their learning. Similar to the research described above, the researchers incorporated high degrees of adaptability into their tangible applications. While Verhaegh, Fontijn and Hoonhout addressed a broad range of skills and difficulties (e.g., fine motor skills, cognitive and social skills), Garzotto and Gonella created a completely open system, allowing the users to reprogram or customize all tangible elements.

The work described above represents only a selection of prior research. In this work, we focus on some work for visually impaired children, tangibles, and also prototype systems that incorporated the concept of *customizability* and *open-endness* into their design, as this has been important for both *The Cuebe* and *Audio-Tactile Map*. We go on to report additional related work for the latter prototype.

2.3.3 Tactile Maps for the Blind

The basic concept of the *Audio-Tactile Map* is not new. Enhancing tactile geographical information with audible information has been explored since the mid 1980s [45].

One of the earlier prototypes tested at the Federal School for the Blind in Vienna was implemented by Seisenbacher et al. [54] in 2005. It featured a video camera mounted above a surface for placing the model, for example, a tactile map or picture. By means of computer vision, the authors inferred which part of the map the user is pointing to. An accompanying software could then play back audio files according to the user's tracked interactions with the map.

A more recent prototype created by Albouys-Perrois et al. [1] also employed visual tracking, but added overhead projection. It also introduced an innovative map construction mode, which allowed users to create their own content by arranging tangible materials.

The design of technologically augmented tactile maps is still an area of ongoing research. Ducasse, Brock, and Jouffrais provide a comprehensive and topical overview [14].

Audio-Tactile Map is clearly in line with the research efforts described. However, it follows yet another approach to tracking the users' interactions, not only from a technical, but also from a design-strategical standpoint, as we will show in the following pages.

3 Research Philosophy and Methods

HCI is a highly dynamic field and has seen many changes in its research orientation during its relatively short history (see, e.g., [5, 26]). As a consequence, the community has also adapted its repertoire of research methods to be able to address current topics (e.g., desktop computers, or the workplace, or the situated use of ubiquitous technological artifacts). Today, HCI researchers can draw on a broad variety of analytic tools, ranging from classical lab experiments to ethnographic methods and participatory design.

With the present work, we participate in one of the more recent approaches in HCI, which is informed by the work of the design research community. This approach attributes *design* and the *designerly way* of working (or knowing) a special role in the acquisition of knowledge in HCI research. One of the classic texts that has paved the way for this epistemological stance was written by Fallman [15], who contrasted the conservative and widely spread *design-as-engineering* (every step in the design process can be prescribed) with design as interpreted by, for example, product design (the process is dependent on the designers and their interpretation of the situation). We embrace this understanding of design and employed a practice-based research strategy in exploring the design space of enabling instructive experiences for visually impaired children and teenagers.

Such practice-based approaches have been described with different labels in HCI, most prominently, Research through Design [22, 65]. Lately, Koskinen et al. [35] coined the term of Constructive Design Research to stress that the *construction* of products (e.g, system, space, or media) plays the most important role in the *construction of knowledge* when engaging in this kind of design research.

Our projects featuring *The Cuebe* and *Audio-Tactile Map* were also based on hands-on and constructive explorations. Through our engagements in the process, we learned from reflective experiences, from manipulating different parameters in the design work, and from observing the effects, including responses from the participants (cf. Schön's seminal investigations of such reflective processes of professionals during their daily work and practices [53]).

Practice-based design research is suitable for addressing ill-defined design problems including, as an example, the design of therapeutic toys for visually impaired children. Rittel [49] famously emphasized that often user requirements/social systems are not stable or predictable, and therefore solutions must be found by trying out different ideas and framings of the problem. Conventional techniques of science (e.g., attempting to control and explain a limited set of variables) and engineering (e.g., attempting to model *design-as-engineering* as a linear process, starting with a limited set of requirements and ending at some determinable point) are less appropriate for this problem space [56], because of the plethora of different variables involved. Moreover, we were interested in a possible future technology and not so much in the description and analysis of the status quo.

On the other hand, the long tradition of science and engineering has established accepted means for elaborating knowledge, for example, by proposing theories and their verification/falsification by empirical evidence. Practice-based research like Research through Design or Constructive Design Research cannot look back in a similar way on centuries of negotiations about epistemology, etc.. However, we can make clear statements about the nature and scope of the knowledge that we seek to contribute through our practice-based approach.

3.1 Design Knowledge in Practice-Based Research

With the present design explorations, we seek to contribute *design knowledge*, that is, our prototypes, which are grounded in the needs of the participants, are offered as design exemplars in order to inspire novel design artifacts. We do not aim to provide prescriptive design rules or guidelines as project results. Rather, we believe that in design, knowledge can be accumulated across different prototypes that embody the design decisions made and the knowledge that has been revealed in the process (i.e., the design rationale [27]). Fellow designers can then identify patterns across different exemplars and apply this knowledge in their own work [64].

To little surprise, text has been described as problematic in the dissemination of practice-based design research outcomes [27], because artifacts in HCI are made from material and/or they behave interactively. For this reason, the community has developed alternative means for conveying embodied knowledge, for example, by creating annotated portfolios and thereby enabling the comparison between different designs [8, p.71]. However, annotated portfolios, photo essays, or similar means for dissemination haven't reached full acceptance in academia yet, and the regular paper format is not optimal for holding this kind of information [31]. In the present paper, we therefore chose to stick to the traditional form of presentation (text and some pictures). Still, we focus on aspects that were relevant from a *designerly* perspective in conceiving and building the prototypes. In addition, as detailed in the discussion section, we made publishing the blueprints and providing open-source access to our design work an important part of our dissemination strategy.

3.2   Data Collection, Participants, and Data Analysis

We incorporated strong elements of participatory design (e.g., [52]) and qualitative research into our projects. Our motivation was twofold. On the one hand, we had no expertise in working with visually impaired children, as our background is interaction design and computer science. Therefore, we informed our design work by co-design activities and the valuable feedback of four Early Intervention specialists (all female, specialized in CVI with many years of experience), several teachers for the blind (multiple years of experience each) and from many affected children to guide our design decisions (pre-school and school children).

On the other hand, we wanted to make systematic observations and analyses, so we employed proven qualitative methods. In more detail, we made field observations during Early Intervention therapy in the homes of the children and during classes at the school of the blind. We also conducted design workshops, where professionals and affected children could try out different materials (e.g., experimenting with different types of translucent acrylic glass) and bring in their own ideas. For analysis, we made use of methods like Thematic Analysis [9] for synthesizing collected interview transcripts, field notes, photos, sketches by the participants, etc., into meaningful units of information, that is, clusters of related information or *themes*. One of our most comprehensive analyses described the overall practice of Early Intervention specialized in visual impairments with focus on the therapeutic toys used [18]. This analysis constituted the foundation of our subsequent constructive research efforts around *The Cuebe*, which we describe in the next section.

4   Supporting Visually Impaired Children and Teenagers with The Cuebe and Audio-Tactile Map

In this section, we describe our two design artifacts, which enable experiences for visually impaired children by incorporating interactive technology. These experiences would not have been possible or perceivable without augmentations due to the limitations in vision. We first present *The Cuebe*, targeted at preschool children with CVI, who receive Early Intervention therapy. Then we report our research on *Audio-Tactile Map*, which we designed for visually impaired school children.

4.1 The Cuebe: A Therapeutic Toy for Preschool Children

As we knew little about visual impairments, we started the project with four in-depth interviews with Early Intervention specialists (and additional informal interviews), lasting 1–2 h each. This prepared us for the subsequent intensive fieldwork that we engaged in. In detail, we made 18 in-situ observations of Early Intervention sessions at twelve children's homes lasting 1 h each. This experience helped us to understand the requirements of the daily work of the therapists, and it gave us an idea of the children's needs. A detailed qualitative analysis of the use of therapeutic toys in Early Intervention is beyond the scope of this chapter, but can be found in prior work [18].

We learned that exercising is important for young children with CVI, because it can enable them to build up new abilities and skills.

For example, the participating Early Intervention specialists exposed the visual system of affected children to bright or **high-contrast optical stimuli**, because in early age, this can lead to the growth of new neuronal synapses. Moreover, by exercising, children can develop **alternative strategies in seeing** to compensate for missing abilities (e.g., they can exercise fixating and tracking moving objects by holding their head in a certain position that enables them to keep the objects in their intact field of vision). As the visual system is closely connected to the motor system, children can also benefit from exercising by **improving their motor skills** (e.g. hand-eye coordination). Moreover, it is also important to engage with affected children, because it can teach them overall social skills, **foster their general development**, give them valuable experiences of self-efficacy, and so on.

Hence, there are many good reasons for carefully supporting affected children with particular exercises. Unfortunately, as we learned, this **exercise is often boring**, because it involves monotonous tasks like looking at high-contrast patterns or following illuminated objects with the eyes over and over again, several times a week. For this reason, many therapists in Early Intervention are keen to expand their repertoire with new therapeutic toys and objects, since they seek to motivate the children by **embedding the exercises in play activities**. Precisely this turned into the objective of our project: To produce novel interactive artifacts that can be used to engage children in often boring exercises and that simultaneously deliver appropriate stimuli for addressing their senses.

4.1.1 Goals of The Cuebe

The Cuebe results from the explorations described above and embodies what we have learned from our participants. It is an *open-ended tangible* aiming to support the goals of Early Intervention specialists and fits with the needs of children with visual (and often also multiple) disabilities.

Accordingly, the major goals of *The Cuebe* were:

- to deliver visual and haptic stimuli that are appropriate for therapeutic exercise in the context of Early Intervention
- to support Early Intervention specialists with an open-ended toy to inspire and enable the creation of play activities
- to motivate affected children to engage in their often boring exercises

The principal concept of *The Cuebe* is to detect color when pushed against a colored surface and to replicate this color by means of an array of RGB LED-lights, as illustrated in Fig. 1.1. In analogy to a *reading stone* that magnifies text, *The Cuebe* amplifies the color it reads. Hence, the device turns the color of the physical surface into a **visually perceivable stimuli** to support children with low vision who can hardly perceive the original color.

Moreover, we carefully designed *The Cuebe's* **physical properties to afford appropriate haptic interactions** for exercising the children's motor skills. The latest working prototype is a nearly cubic hand-held device, measuring $6 \times 6 \times 7$ centimeters and fitting well the into hands of the target group (preschool children). The lower part of *The Cuebe* is a movable platform mounted on springs and can be pushed inwards (2 mm). It functions as a button, and the children can use it by

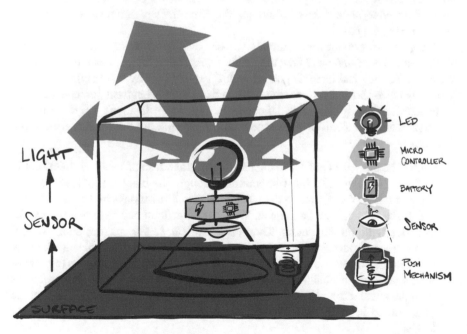

Fig. 1.1 Schematic illustration of *The Cuebe*. The downwards facing sensor detects the color of the surface below. The detected color is then reproduced by means of bright *RGB LEDs*, lightening up the semi-translucent upper part of *The Cuebe*. The bottom part/platform of the device is movable and mounted on springs. In order to 'pick' a color, the children therefore need to place the device on a surface, and gently push it against this surface

placing their hands on top of *The Cuebe*, pushing it, and thereby *updating* the color. The color will be read continuously, until the child stops pushing it, and it keeps emitting this same color until it is pushed against another surface.

From a technical perspective, the top of *The Cuebe* houses an array of RGB LEDs and is made from translucent material. We experimented with cutting acrylic glass and 3D-printing transparent filament in order to accomplish a homogeneous light effect. This was important for the children with low vision to be able to perceive and *understand* the shape. As requested by the therapists, we implemented an additional mode for displaying the light – blinking light mode – because for some children who have CVI it is easier to visually perceive a stimuli when it alternates between bright and low lights.

4.1.2   Field Deployment of The Cuebe

We observed six Early Intervention sessions (each lasting approximately 1 h) in the homes of children with CVI where the therapists used *The Cuebe* for exercising. In order not to bias these observations, we provided no instructions about how the device should be used. It was left to the therapist to decide how to incorporate it into their exercises. We put no pressure on them; that is, we encouraged them to *stop* using the device, should the children not like *The Cuebe* or should any other issues arise with the exercise.

We go on to provide four exemplary cases, which illustrate how the therapists made use of *The Cuebe* and how the children responded to it. Like all of our young participants, they had been diagnosed with CVI; that is, they were sighted to some extent, but also showed various symptoms and difficulties in visual perception, along with additional complementing deficits (e.g., motor coordination). We describe our observations in the form of short scenarios for drawing out the situated use of *The Cuebe*:

- One of the therapists put a blue blanket on the ground, so that she and John[4] could sit down (see Fig. 1.2). Most importantly though, the blanket provided a strong contrast to *The Cuebe* and to the wooden tiles and toy blocks in different colors, which she also put on the blanket. With her help, John had to match toy blocks and tiles with similar colors. They used *The Cuebe* for making comparisons. While some of the colors provided a very strong contrast against the blue blanket (e.g. orange), others were of similar colors. This allowed the therapists to observe John's abilities in handling these different contrasts. Moreover, picking colors from the smaller toy blocks was much harder than pushing the device against the tiles, and so John also had to handle motor exercises at two different difficulty levels.

---

[4]The names of the children were altered for anonymity.

Fig. 1.2 *The Cuebe* prototype in action. One of the therapists laid down a soft blanket on the ground and put tiles and small wooden cubes in different colors on it. She then made up a game where the child was asked to pick certain colors and move along the tiles, similar to playing a board game

- A therapist wanted to exercise the coordination/orientation and autonomy of Kyle. He should walk in a room independently and solve certain color-related tasks. Hence, this exercise was about coordination and independence, but it also involved a cognitive component and provided a strong stimulus for the eyes through *The Cuebe*. The therapists therefore asked: "Will you pick something green for me?" Kyle then carefully navigated in the room, closely examining different colors. Finally, he decided for a green plastic watering can, placed *The Cuebe* on it and proudly returned the device to the therapist, which was now illuminated in green color.
- Tim was asked to pick colors from colored wooden disks that were placed inside a relatively big and empty metal can (formerly for food). The reflective steel mirrored the emitted light from *The Cuebe* and the can lit up strongly. Tim was putting his head close to the opening of the can, curiously observing the joyful 'dancing' of the colors. He picked colors, pulled *The Cuebe* out of the can, proudly showed it to the therapist, to finally put it back into the can to repeat his play all over.
- An Early Intervention specialist used *The Cuebe* and other common materials in therapy to set up a board game-like arrangement. On a flat surface, she spread out colored disks and planes. Mario was asked to throw a dice, with each side in a different color, and then to find the disk in the corresponding color. To learn the different colors, the child had to say a "magic spell" including the color's

name and then pick it with *The Cuebe* to complete the task. The "magic spell" was introduced by the therapist, to make this exercise more engaging and fun for Mario.

The above scenarios illustrated how *The Cuebe* was integrated into the routines and practices of Early Intervention of preschool children with CVI. The device could be used flexibly as appropriate for the current situation due to its undefined and open-ended design. It addressed the sensory system both on the visual and on the tactile level, as the children had to manipulate *The Cuebe* with their hands in order to change its bright colors. Next, we describe our second prototype, *Audio-Tactile Map*, which was designed to support children with visual impairments in learning at school. That is, we move on to children of another age group, who were in school and already too old for Early Intervention.

4.2 The Audio-Tactile Map: An E-Learning Toolset for Visually Impaired Students

While *The Cuebe* was built to support Early Intervention specialists, aiming at reducing or compensating developmental delays in early childhood (e.g., in vision, motor skills, basic cognition, social skills; in the age of 0–6 years), the *Audio-Tactile Map* was designed for students at least 14 year old and to be used in educational activities during school.

Similar to *The Cuebe*, this device drew on technology to enable rich interactive experiences for visually impaired children, which again sought to foster their development. However, this time we did not focus on therapeutic exercises, but on group and independent study activities. We started by conducting a contextual inquiry, followed up by several iterations of prototyping. The purpose of the first constructed prototypes was mainly to facilitate the co-design process and to help develop good design ideas [20]. As the project progressed, questions of resource constraints and related design trade-offs gained importance. The last prototype therefore tried to manifest, in the sense of what Korsgaard et al. described as a *computational alternative*, illustrations of different possible directions that can be pursued [34] while taking infrastructural actualities into account.

4.2.1 Contextual Inquiry

We initiated a contextual inquiry of the *Federal Institute for the Blind* in Vienna by exploring their facilities, guided by a visually impaired teacher, and by attending regular class sessions. Moreover, we conducted open-ended interviews and group workshops held together with mobility trainers, teachers and students at the age of 14–17 years.

These activities gave us valuable insights into the processes and the requirements of blind children's education in general, and the daily routines at this school in particular. It also allowed us to take a closer look at the materials and tools used during class, including drawing pads and rulers for tactile sketching, braille output devices, and various forms of tactile teaching objects, such as reliefs, models, sculptures and taxidermic recreations. Common to all were high levels of portability, usability, affordability, durability, and the existence of an adequate long-term support service [61]. Some of the tools have been in use for several decades already.

Abandoned educational materials and tools lacking those qualities could be found in a storage room, where we also discovered the remains of earlier research studies. This discovery motivated us to carefully consider issues around the acceptance and abandonment of Assistive Technology in general and functional prototypes in particular.

Considering the robustness of the educational materials used in class – highly durable tools that stood the test of time – as well as the obvious need for long-term support services, challenged our usual design approach. Most notably, the use of "hacked" proprietary off-the-shelf products as components of a functional prototype now appeared to be inadequate, if longevity had to be taken into account. While the rather short life-span and support of electronic mainstream consumer products hardly matches the requirements of the educational context already, manipulating its hardware and software would most probably lower it even more.

Another problem raised by the teachers was the **lack of usable educational content available**. It appeared that designing educational tools would require more than building a prototypical device. What was needed was a whole system that addresses content production and exchange as well.

On those grounds, it soon became evident that the teachers or participants were concerned if our academic research could actually have a direct impact on their actual daily practice. That is, **despite their general enthusiasm for cooperative design, they had reservations about the direct value of the project and academic research in general**. (As these concerns corresponded with a re-occurring debate within the field of design research, we will further reflect on them in Sect. 5.4.)

In summary, it became clear that a single design research project could not fulfill all these requirements. Creating and implementing a fully fledged production and maintenance strategy clearly exceeded the available resources. What a design study could provide, however, would be the first steps within a larger initiative, wherein every design decision takes into account how it impacts subsequent ones. Our prototypical tools and services should be built not as insular solutions, but in a way that could be expanded on later – in subsequent studies and non-academic initiatives alike.

We followed this strategy by **assessing the existing infrastructure** and production processes on site, investigating what was available and what already worked. The mobility trainers showed us **self-made schematic maps of nearby shops and streets**. These maps were created by drawing directly on *swell paper*. Using a *fusing* *machine*, located near the teachers' lounge, the swell paper was then transformed into reliefs, to be later used in preparatory mobility teaching sessions.

Other teachers showed us how they created tactile diagrams and illustrations out of paper and cardboard. Although carefully crafted, these tactile illustrations were not self-explanatory – and weren't meant to be. Teachers used these tactile materials as a teaching supplement in class. The students themselves were not involved in the creation of tactile illustrations. However, it appeared that students enjoyed showing each other specific details on tangible teaching objects. The exploration of these teaching materials could be a guided, but also an open and playful exercise. Eventually, we developed the idea to **augment these educational materials as well as the means of production using contemporary information technology**. We started to explore the design space with two initial ideas in mind: the *Audio Box* and the *Audio-Tactile Map*.

4.2.2 Towards the Audio-Tactile Map: Initial Explorations with the Audio Box

Early prototypes of the *Audio Box* were portable hand-sized containers, capable of holding one or several small objects and of recording, as well as playing back, one single audio memo (see Fig. 1.3). The initial idea was to connect haptically appealing materials to sounds or audio descriptions. Each student would use one or

Fig. 1.3 The Audio Box was able to record and play back one single audio recording. It could also function as a container for small and haptically appealing objects

more *Audio Boxes* to collect items during a field trip to re-examine them later on. However, after our first informal tests, we decided that these portable devices could only be part of a more comprehensive educational program. The students should start from the familiar, then slowly expand the known space step by step. Hence, we postponed the refinement of the *Audio Box* and started to focus on the immediate surrounding of the students: the school building. Here the *Audio-Tactile Map* would find its use. Only after learning the basics of cartographic abstraction should the students switch to lesser known areas: the immediate surroundings of the school building at first, followed by the destinations of field trips.

4.2.3   Audio-Tactile Map Prototype 1: Visual Tracking

In contrast to some earlier studies on audio-tactile maps, our device should take the actual local educational processes and needs into account – as well as the individual explorative and expressive interests and abilities of the various stakeholders. From the beginning we were careful not to disrupt existing 'conversational' practices. Supplementing the teacher with a mere information retrieval device was not the goal and considered to be an "anti-pattern". The device should neither hinder proven educational practices, nor collaborative activities. Instead, we aimed at building a tool which would allow teachers and students alike to actively enrich available teaching materials and extend their interaction beyond class sessions.

To address the lack of tactile content, we decided to define an open file format, which combined digital illustrations with links to networked audio recordings. Teachers should be able to share maps and illustrations easily, so that others could reproduce them by downloading and then printing them on swell paper. Storing the linked audio recordings on the Web would also allow the creator to remotely add and update the audible information layer at a later stage. This turned the audio layer from a static collection of recordings into something that could potentially grow with its user base. It also shifted the *Audio-Tactile Map* from a static teaching supplement towards a *communication medium*. While our focus remained on the direct teacher-student relationship, this opened a wide range of new potential use cases.

To download and play back the recorded sounds, we used an embedded computer (BeagleBone) based on open-hardware. Open hardware, to our understanding, would increase the maintainability and adaptability of the system as a whole.

To identify the tactile maps, we connected an RFID (radio-frequency identification) reader. RFID tags, attached to the back side of the swell paper, allowed the computer to recognize the map, download the map file, and prepare the audio files for instant playback. A student would place and fixate the tactile map on the device. The device would download the necessary data. Then the student would trigger the playback by simply touching an area of interest.

For our first tests, we used a visual tracking mechanism relying on a top-mounted camera and optical markers – an approach that had already been explored by an earlier feasibility study [54]. However, this system was only able to track the *position* of outstretched fingers, and not the act of *touching* certain areas. This limited the possibilities of interaction considerably. Hence, we decided to investigate the possibilities of pen-based interaction as an alternative.

4.2.4   Audio-Tactile Map Prototype 2: Pen-Based Interaction

We based our second prototype on a commercially available pen-based input device (see Fig. 1.4). In a first test, a visually impaired teacher lifted the paper with the left hand to "scan" the map with his right hand, which was placed flat on the surface. After putting the map back on the tracking surface, he used the fingertips of the left hand to explore the tactile map in more detail, while using the right hand to hold the pen. Using the pen as a pointing device allowed him to select very small areas (down to 1×1 mm). But holding the pen hindered two-handed exploratory procedures. Moreover, the system was not able to support the user in the search for audible information. The only way to set audible-enhanced areas apart from the rest were the tactile properties of the map itself. This influenced the design of the tactile surface and reduced the flexibility of the whole setup considerably.

For a second test, we invited visually impaired students to actively engage in the creation of their own augmented map. Guided by a visually impaired teacher, the students explored the school building and collected audio recordings – field recordings or short interviews with people working in their offices. These recordings

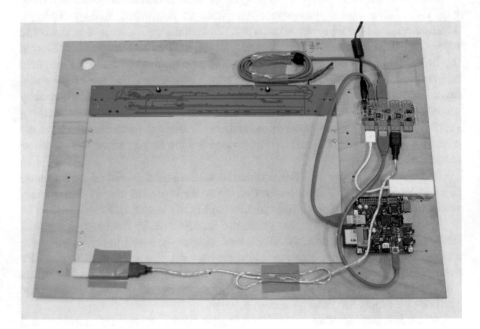

Fig. 1.4 The backside of prototype 2, featuring an open-hardware embedded computer, connected to an off-the-shelf graphic tablet and an RFID reader

were then positioned on a tactile map consisting of several A4-sized sheets of swell paper. The rooms represented on this map were delimited by raised lines resembling walls. The test subject used both hands to explore the map, often using one hand to continuously "walk" from one room to the next, while using the other hand to find the exits. The pen was put aside during these exploratory movements.

We concluded that a preferable tracking mechanism should be able to continuously track both hands when in contact with the tactile surface. This would not only provide uninterrupted exploration, but also allow the system to give audible feedback when an area of interest was within reach. We started to investigate available technical options.

4.2.5   Audio-Tactile Map Prototype 3: Open Projected Capacitive Touch Tracking

Tests with commercially available tablets that supported multi-touch input seemed promising at first. However, as the tablets were based on closed hardware and software, we weren't able to fully customize their behavior and properties; in particular, the limited surface size did not match the format of the maps used in school. Moreover, we weren't able to solve tracking problems that appeared on seemingly random occasions, as the highly specialized fingertip tracking technology wasn't always reacting in a predictable way when multiple parts of the hands rested on the swell paper. Furthermore, as already explained in Sect. 4.2.1, relying on off-the-shelf products as integral parts of the system would influence the longevity of the system as a whole.

We therefore decided to build our own tracking device based on carefully selected widely-available and basic electronics components (see Fig. 1.5). Although this would take up considerable project resources, we hoped for beneficial long-term effects, as we designed for repairability and adaptability. The tracking device should be able to be used with a wide variety of tactile maps placed on top of the sensor panel, including maps based on swell paper. At the same time it should be able to track two-handed explorative movements. The technological approach which seemed most adequate to us was *projected capacitance touch tracking*. By implementing our own solution, we were able to fine-tune all relevant parameters in correspondence with the material in use at the school.

By releasing our tracking system under a free license,[5] and by building it in a modular and highly customizable way, we also hoped for synergy effects. Researchers and hobbyists, not necessarily interested in Assistive Technology but working on touch-based interaction, should be able to use as well as contribute to the design of the tracking device.

To expand the explorable areas on the Audio-Tactile Map and to address the challenge of **content creation**, we decided to pursue two approaches in parallel.

---

[5] A detailed description of the prototype can be found in a separate publication [62].

Fig. 1.5 The third prototype of the *Audio-Tactile Map* featured multi-touch and multi-hand tracking based on open-sourced projected capacitance touch technology. A tactile map was placed on the sensor panel. Audio recordings were triggered when the users double tapped locations with their fingertips

On the one hand, we initiated the design and prototypical implementation of a sophisticated map making tool based on OpenStreetMap. On the other hand, we built an accessible web interface that allowed rapid creation of black and white renderings of Google Maps images. While the first approach aimed at a long-term solution, the second one was geared at a "quick win". Completely dependent on an external proprietary service, it served as an intermediary solution providing immediate value. We also helped the school to acquire and install a special printer, which facilitated the process of transforming the created images directly into embossed graphics on normal paper sheets, thereby reducing production costs. To add sounds to geographical maps, we recorded exemplary audio files ourselves. We also tested the integration of web-based field recording platforms in order to benefit from already existing crowd-sourced services. Students should be able to explore the city using several audio information layers, one of them providing "city sounds".

After our first tests, we decided to implement a rudimentary sonification service, providing subtle stereophonic sounds which informed the user about the relative position of the touched area to nearby areas of interest. The sounds changed in volume, pitch and balance during exploratory movements. To trigger the playback of a recording, once the selectable area was found, the user could double tap it.

After we successfully built a *usable* user interface, we concluded that the designed system and resulting *user experience* spearhead a promising direction, but would benefit considerably from research and further design iterations. We were not able to reach a level of product quality that fully corresponded to the requirements of teaching tools used at the school, but we were able to design a toolkit hopefully *resilient* enough to serve as a platform for further research and refinement (see also future work Sect. 5.4).

5   Discussion

As outlined above, there are many ways of conducting research in HCI. We took a practice-based approach (e.g., [35]) to exploring the potential of haptics and related modalities in the education of visually impaired children. Here we discuss some of the consequences of this decision.

5.1   Reflections on the Design Process

In our project, we combined fieldwork with a practice-based Research through Design approach (see, e.g., [32] for an overview of the design research discipline). In retrospect, we think this choice was appropriate, since throughout the process, there were many design choices to be made which could not be inferred from a static requirements specification or "desktop research" [63] (cf. Dorst's work for a comparison of different approaches to *designing* [13]). That is, while the project certainly featured conventional engineering problems that could be tackled by *algorithmic* thinking and procedures (e.g., arranging all components within a shell and manufacturing this shell in a reasonable as well as economical way), it also demanded choices made through the experience of design (both as an artifact and as a reflective process [53]).

For example, we used samples of different translucent materials and shapes to decide the final appearance of the *The Cuebe*. Both look and feel had to be experienced, and the components could not be investigated in isolation. The same argument applies to the push mechanism: one needed to grip the device and pick a color by pushing it against a surface to actually experience this kind of interaction. The designer should experience this with their own senses as the design is eventually intended to address the children's senses, and their therapeutic experience (or outcome, even) relies on that feel. At this point, we would like to emphasize that we regarded all people involved in the project – the children, teachers, therapists, and authors of this article – as designers or *co-designers*, because each one of these groups brought in their own particular expertise, which the others did not have, that were necessary for the project to come into being.

In our estimation, it was also necessary to successively work towards the final design using interactive prototypes. Due to the nature of the project, we could not expect the participants to anticipate interacting with therapeutic toys or educational tools based on simple mock-ups. This required us, relatively early in the process, to use platforms like *Arduino* for 'sketching in hardware', as opposed to sketching on paper or creating mock-ups. Especially in the beginning, we carefully sketched the interactive behavior, while at the same time not paying too much attention to the "polish" or physical appearance of the prototypes.

5.2 Open-Endedness and Customizability of the Design Artifacts

Throughout the co-design process, *open-endedness* and *customizability* emerged as important themes in the design of *The Cuebe* as well as *Audio-Tactile Map*. Both the Early Intervention specialists and the educators working at the Federal Institute for the Blind stressed that the technological artifacts should allow appropriations according to local practices and the incorporation of custom content, for example, by making up their own games with *The Cuebe* or creating their own maps with *Audio-Tactile Map*. Hence, the resulting technology is not attempting to "fix" disabilities by prescribing specific actions or therapeutic plans. Rather, our prototypes offer technological opportunities that can be taken and appropriated to facilitate developmental growth.

5.3 Dissemination of Knowledge

As mentioned in the methods section, practice-based design research is different to conventional science with regard to knowledge dissemination. The research goal is not to describe, explain or predict phenomena, but to understand how artifacts can be constructed. Design exemplars are recognized as important carriers of knowledge as they embody the findings and design decisions the designers made during the process [36].

On these grounds, we described our prototypes *The Cuebe* and *Audio-Tactile Map* in detail, including the underlying design processes, design decisions, and illustrative use scenarios.

Admittedly, scientific contributions in the shape of such design archetypes [63] are problematic in that the embodied knowledge can only fully unfold or be experienced when the user interacts with them directly. Hence, to make this knowledge accessible on an additional level, we decided to publish essential parts of the design of our artifacts under an open-source license. Build instructions for *The Cuebe* can be found online (www.guelden.info/cuebe.php). The blueprints for the tracking device of the *Audio-Tactile Map* have been published on an Open Science platform [62]. Through this strategy, we also hope to increase some of the sustainable effects of our research and to respond to some of its inherent challenges, as elaborated in the next section.

5.4 Reflections on the Sustainability of Our Design Process

As described in Sect. 4, much of our thinking during the design process centered around respecting the participants and their local needs, sustainability, product resilience, and the problem of device abandonment.

These issues have also been investigated by at least two research communities. Academic literature on Assistive Technology defines an abandoned assistive device as a product that gets purchased and put aside before the expected lifetime of usage has been reached, and that is not replaced with a related updated device [46] – a fate that befalls, according to Riemer-Reiss and Wacker [37], at least one third of all assistive products. This not only poses a problem in terms of time and money being wasted in acquisition and training, but can also lead, as Martin and McCormack [40] have noted, to disillusionment about specific categories of Assistive Technologies or Assistive Technologies in general. This dynamic can have far-reaching consequences, not only on the Assistive Technology market in the long term, but also on the intended user, who misses out on the untapped potential of emerging technologies.

To involve all relevant stakeholders, foremost the users themselves, in the acquisition process – to let them decide what assistive product should be purchased or not – is believed to reduce the chance of rejection. Similarly, Kane and colleagues [33] recently highlighted the importance of integrating affected individuals closely into the design process to increase the user acceptance and product success – a proposition which aligns well with the principles of user-centered design.

Such an approach was also favourably noted by our participants. However, although they expressed their general enthusiasm for cooperative design, they also put forward their concerns towards the frequently underwhelming direct impact of academic research on their actual daily practice (see also Sect. 3). These concerns are reflected by a re-occurring debate within the field of design research, in particular within the participatory design movement: It is generally agreed upon that giving the intended users a voice in the design process increases the chance that the designed tool fits into the environment of the co-designers. However, it does not guarantee that the end result is actually able to fulfill the set expectations. The quality of a functional prototype depends on the resources available, and maintenance and long-term support typically exceed the scope of a single research project. Simply facilitating discourse and reflecting on it is clearly insufficient when direct local impact is part of the desired project outcome. As Asaro emphasizes [3], participation should be *realized*; that is, the participants' voices should be echoed by actual implementations in order to make a difference in daily practices.

5.5 Future Work

As next steps, we intend to conduct additional user studies in order to understand how people integrate our technologies into therapeutic/educational settings and ultimately into their life. This will support us in generating intermediate design knowledge [28] about designing technologies for and with visually impaired children and students. Moreover, we want to improve our prototypes and iterate on their technical implementations.

From a broader perspective and connecting to Sect. 5.4, we would also like to spark a discussion about the opportunities and challenges of academic practice-based design with regards to effective local change. We consciously addressed mid- and long-term goals in our design-decisions. At the same time, we attempted to achieve "quick wins" with direct and immediate value for the participants, whenever possible. However, it is not clear how long these quick improvements will last and by which mechanisms long-term goals can be reached after a project comes to an end. It remains to be seen how well our approach fits into the current academic milieu, what structural inhibitors can be identified, and what can be done about them – in order to support future long-term design research where actual local impact is not an afterthought.

5.6 Limitations

The present work collated our experiences from two practice-based research projects in the context of the education of visually impaired children. The resulting prototypes (*The Cuebe* and *Audio-Tactile Map*) were driven by field-work and co-design activities with professional therapists and affected children. In this article, we do not aim at providing generalizable findings or discovering some sort of "truth" in the design of such tools for visually impaired children. Rather, the purpose of our work is to describe the design process, to explain design decisions, and to illustrate how the prototypes can be used. Further research is needed to evaluate possible positive effects of *The Cuebe* and *Audio-Tactile Map* on the education and well-being of children with visual disabilities.

6 Conclusion

Our objective in this research was to design interactive technology for enabling stimulating and educational experiences for visually impaired children of two different age groups. For the preschool children, we wanted to create a therapeutic toy that could be used by Early Intervention Specialists to motivate children with different manifestations of CVI to engage them in exercising. As a response to this challenge, we proposed a tangible color-picker toy, *The Cuebe*. In our ongoing field study, we found that the toy was well received and appreciated by the participants (children and therapists). For the school children, we aimed at conceiving an e-learning tool, which would fit into existing practices and enrich them. This design process led to the development of *Audio-tactile Map*, which could be used with custom content made by students or teachers, thereby facilitating haptic interaction in an educational setting.

We decided for the dissemination under an *open-source* license. In this way, we wanted to ensure the sustainability of the project.

Acknowledgements This research has been funded by *"Gemeinsame Gesundheitsziele aus dem Rahmen-Pharmavertrag, eine Kooperation von österreichischer Pharmawirtschaft und Sozialversicherung"* and by the Sparkling Science program, led by the Austrian Federal Ministry of Science and Research. The authors also thank all involved participants – the preschool children, their Early Intervention therapists, the students, and their teachers as well as the Institute "Integrated Study" for their support.

References

1. Albouys-Perrois, J., Laviole, J., Briant, C., Brock, A.M.: Towards a multisensory augmented reality map for blind and low vision people: A participatory design approach. In: Proceedings of the 2018 CHI Conference on Human Factors in Computing Systems. pp. 629:1–629:14. CHI '18, ACM, New York, NY, USA (2018), https://doi.org/10.1145/3173574.3174203
2. Antle, A.N., Fan, M., Cramer, E.S.: Phonoblocks: A tangible system for supporting dyslexic children learning to read. In: Proceedings of the Ninth International Conference on Tangible, Embedded, and Embodied Interaction. pp. 533–538. TEI '15, ACM, New York, NY, USA (2015), https://doi.org/10.1145/2677199.2687897
3. Asaro, P.M.: Transforming society by transforming technology: the science and politics of participatory design. Accounting, Management and Information Technologies 10(4), 257–290 (2000), http://www.sciencedirect.com/science/article/B6VFY-40X8FS2-1/2/ac8ff34bf4812794b471d535068bea6a
4. Austrian Federal Ministry of Education, Science and Research: Curriculum for Blind School Children – Lehrplan der Sonderschule für blinde Kinder. Report, Austrian Government (2008), https://www.cisonline.at/fileadmin/kategorien/BGBl_II__Nr_137_Anlage_C_3.pdf
5. Bannon, L.: Reimagining HCI: Toward a More Human-centered Perspective. Interactions 18(4), 50–57 (Jul 2011), https://doi.org/10.1145/1978822.1978833
6. Banzi, M., Cuartielles, D.: *Arduino*. Open-source Electronics Platform [Cross-platform] (2005), https://www.arduino.cc/. Accessed 27 March 2019.
7. Bourne, R.R.A., Flaxman, S.R., Braithwaite, T., Cicinelli, M.V., Das, A., Jonas, J.B., Keeffe, J., Kempen, J.H., Leasher, J., Limburg, H., Naidoo, K., Pesudovs, K., Resnikoff, S., Silvester, A., Stevens, G.A., Tahhan, N., Wong, T.Y., Taylor, H.R.: Magnitude, temporal trends, and projections of the global prevalence of blindness and distance and near vision impairment: a systematic review and meta-analysis. The Lancet Global Health 5(9), e888–e897 (2017), https://doi.org/10.1016/S2214-109X(17)30293-0
8. Bowers, J.: The logic of annotated portfolios: Communicating the value of 'research through design'. In: Proceedings of the Designing Interactive Systems Conference. pp. 68–77. DIS '12, ACM, New York, NY, USA (2012), https://doi.org/10.1145/2317956.2317968
9. Braun, V., Clarke, V.: Using thematic analysis in psychology. Qualitative Research in Psychology 3(2), 77–101 (2006), http://www.tandfonline.com/doi/abs/10.1191/1478088706qp063oa

10. Bunce, C., Wormald, R.: Leading causes of certification for blindness and partial sight in England & Wales. BMC Public Health 6, 58–58 (2006), https://www.ncbi.nlm.nih. gov/pubmed/16524463 https://www.ncbi.nlm.nih.gov/pmc/PMC1420283/, 16524463[pmid] PMC1420283[pmcid] 1471-2458-6-58[PII] BMC Public Health

11. Cook, A.M., Hussey, S.: Assistive Technologies: Principles and Practice (2nd Edition). Mosby, 2 edn. (Dec 2001), http://www.worldcat.org/isbn/0323006434

12. Dahlbäck, N., Jönsson, A., Ahrenberg, L.: Wizard of Oz studies – why and how. Knowledge-Based Systems 6(4), 258–266 (1993), http://www.sciencedirect.com/science/ article/pii/095070519390017N, Special Issue: Intelligent User Interfaces

13. Dorst, C.: Describing Design – A comparison of paradigms. TU Delft, Delft, Netherlands (1997)

14. Ducasse, J., Brock, A.M., Jouffrais, C.: Accessible interactive maps for visually impaired users. In: Pissaloux, E., Velazquez, R. (eds.) Mobility of Visually Impaired People: Fundamentals and ICT Assistive Technologies, pp. 537–584. Springer International Publishing, Cham (2018), https://doi.org/10.1007/978-3-319-54446-5_17

15. Fallman, D.: Design-oriented human-computer interaction. In: Proceedings of the SIGCHI Conference on Human Factors in Computing Systems. pp. 225–232. CHI '03, ACM, New York, NY, USA (2003), https://doi.org/10.1145/642611.642652

16. Fikar, P., Güldenpfennig, F., Ganhör, R.: Pick, place, and follow: A ball run for visually impaired children. In: Proceedings of the 2018 ACM Conference Companion Publication on Designing Interactive Systems. pp. 165–169. DIS '18 Companion, ACM, New York, NY, USA (2018), https://doi.org/10.1145/3197391.3205430

17. Fikar, P., Güldenpfennig, F., Ganhör, R.: The Cuebe: Facilitating Playful Early Intervention for the Visually Impaired. In: Proceedings of the Twelfth International Conference on Tangible, Embedded, and Embodied Interaction. pp. 35–41. TEI '18, ACM, New York, NY, USA (2018), https://doi.org/10.1145/3173225.3173263

18. Fikar, P., Güldenpfennig, F., Ganhör, R.: The use(fulness) of therapeutic toys: Practice-derived design lenses for toy design. In: Proceedings of the 2018 Designing Interactive Systems Conference. pp. 289–300. DIS '18, ACM, New York, NY, USA (2018), https://doi.org/10. 1145/3196709.3196721

19. Flaxman, S.R., Bourne, R.R.A., Resnikoff, S., Ackland, P., Braithwaite, T., Cicinelli, M.V., Das, A., Jonas, J.B., Keeffe, J., Kempen, J.H., Leasher, J., Limburg, H., Naidoo, K., Pesudovs, K., Silvester, A., Stevens, G.A., Tahhan, N., Wong, T.Y., Taylor, H.R.: Global causes of blindness and distance vision impairment 1990-2020: a systematic review and meta-analysis. The Lancet Global Health 5(12), e1221–e1234 (2017), https://doi.org/10.1016/ S2214-109X(17)30393-5

20. Floyd, C.: A systematic look at prototyping. In: Budde, R., Kuhlenkamp, K., Mathiassen, L., Züllinghoven, H. (eds.) Approaches to Prototyping, pp. 1–18. Springer, Berlin, Heidelberg (1984)

21. Garzotto, F., Gonella, R.: An open-ended tangible environment for disabled children's learning. In: Proceedings of the 10th International Conference on Interaction Design and Children. pp. 52–61. IDC '11, ACM, New York, NY, USA (2011), https://doi.org/10.1145/1999030.1999037

22. Gaver, W.: What should we expect from research through design? In: Proceedings of the SIGCHI Conference on Human Factors in Computing Systems. pp. 937–946. CHI '12, ACM, New York, NY, USA (2012), https://doi.org/10.1145/2207676.2208538

23. Güldenpfennig, F., Fikar, P., Ganhör, R.: Designing interactive and motivating stimuli for children with visual impairments. In: Proceedings of the 31st British Computer Society Human Computer Interaction Conference. pp. 64:1–64:4. HCI '17, BCS Learning & Development Ltd., Swindon, UK (2017), https://doi.org/10.14236/ewic/HCI2017.64

24. Güldenpfennig, F., Fikar, P., Ganhör, R.: Interactive and open-ended sensory toys: Designing with therapists and children for tangible and visual interaction. In: Proceedings of the Twelfth International Conference on Tangible, Embedded, and Embodied Interaction. pp. 451–459. TEI '18, ACM, New York, NY, USA (2018), https://doi.org/10.1145/3173225.3173247

25. Guralnick, M.J.: The System of Early Intervention for Children with Developmental Disabilities. In: Jacobson, J.W., Mulick, J.A., Rojahn, J. (eds.) Handbook of Intellectual and Developmental Disabilities, pp. 465–480. Springer US, Boston, MA (2007), https://doi.org/10.1007/0-387-32931-5_24

26. Harrison, S., Sengers, P., Tatar, D.: Three paradigms in HCI. In: Proceedings of the SIGCHI Conference on Human Factors in Computing Systems. CHI '07, ACM, New York, NY, USA (2007)

27. Hengeveld, B., Frens, J., Deckers, E.: Artefact matters. The Design Journal 19(2), 323–337 (2016), https://doi.org/10.1080/14606925.2016.1129175

28. Höök, K., Löwgren, J.: Strong Concepts: Intermediate-level Knowledge in Interaction Design Research. ACM Trans. Comput.-Hum. Interact. 19(3), 23:1–23:18 (Oct 2012), https://doi.org/10.1145/2362364.2362371

29. Hornecker, E., Buur, J.: Getting a grip on tangible interaction: A framework on physical space and social interaction. In: Proceedings of the SIGCHI Conference on Human Factors in Computing Systems. pp. 437–446. CHI '06, ACM, New York, NY, USA (2006), https://doi.org/10.1145/1124772.1124838

30. Jadan-Guerrero, J., Jaen, J., Carpio, M.A., Guerrero, L.A.: Kiteracy: A kit of tangible objects to strengthen literacy skills in children with down syndrome. In: Proceedings of the 14th International Conference on Interaction Design and Children. pp. 315–318. IDC '15, ACM, New York, NY, USA (2015), https://doi.org/10.1145/2771839.2771905

31. Jarvis, N., Cameron, D., Boucher, A.: Attention to detail: Annotations of a design process. In: Proceedings of the 7th Nordic Conference on Human-Computer Interaction: Making Sense Through Design. pp. 11–20. NordiCHI '12, ACM, New York, NY, USA (2012), https://doi.org/10.1145/2399016.2399019

32. Joost, G., Bredies, K., Christensen, M., Conradi, F., Unteidig, A.: Design as Research. Birkhäuser De Gruyter, Basel, Switzerland (2016)

33. Kane, S.K., Hurst, A., Buehler, E., Carrington, P.A., Williams, M.A.: Collaboratively Designing Assistive Technology. Interactions 21(2), 78–81 (Mar 2014), https://doi.org/10.1145/2566462

34. Korsgaard, H., Klokmose, C.N., Bødker, S.: Computational alternatives in participatory design: Putting the T back in socio-technical research. In: Bossen, C., Smith, R.C., Kanstrup, A.M., McDonnell, J., Teli, M., Bødker, K. (eds.) Proceedings of the 14th Participatory Design Conference: Full Papers – Volume 1. vol. 1, pp. 71–79. ACM, New York, NY, USA (2016)

35. Koskinen, I., Zimmerman, J., Binder, T., Redström, J., Wensveen, S.: Design Research Through Practice: From the Lab, Field, and Showroom. Morgan Kaufmann (2011)

36. Koskinen, I., Frens, J.: Research prototypes. Archives of Design Research 30(3), 17–26 (8 2017)

37. L. Riemer-Reiss, M., Wacker, R.: Factors associated with assistive technology discontinuance among individuals with disabilities. Journal of Rehabilitation 66(3), 44–50 (07 2000)

38. Linehan, C., Waddington, J., Hodgson, T.L., Hicks, K., Banks, R.: Designing Games for the Rehabilitation of Functional Vision for Children with Cerebral Visual Impairment. In: CHI '14 Extended Abstracts on Human Factors in Computing Systems. pp. 1207–1212. CHI EA '14, ACM, New York, NY, USA (2014), https://doi.org/10.1145/2559206.2581219

39. Majnemer, A.: Benefits of Early Intervention for children with developmental disabilities. Seminars in Pediatric Neurology 5(1), 62–69 (1998), http://www.sciencedirect.com/science/article/pii/S107190919880020X, topics in Developmental Delay

40. Martin, B., McCormack, L.: Issues surrounding Assistive Technology use and abandonment in an emerging technological culture. In: in Proceedings of Association for the Advancement of Assistive Technology in Europe (AAATE) Conference. pp. 413–417. IOS Press, Düsseldorf, Germany (1999)

41. Martin, M.B.C., Santos-Lozano, A., Martin-Hernandez, J., Lopez-Miguel, A., Maldonado, M., Baladron, C., Bauer, C.M., Merabet, L.B.: Cerebral versus Ocular Visual Impairment: The Impact on Developmental Neuroplasticity. Frontiers in Psychology 7, 1958 (2016), https://www.frontiersin.org/article/10.3389/fpsyg.2016.01958

42. Metatla, O., Cullen, C.: "Bursting the Assistance Bubble": Designing Inclusive Technology with Children with Mixed Visual Abilities. In: Proceedings of the 2018 CHI Conference on Human Factors in Computing Systems. pp. 346:1–346:14. CHI '18, ACM, New York, NY, USA (2018), https://doi.org/10.1145/3173574.3173920
43. Metatla, O., Thieme, A., Brulé, E., Bennett, C., Serrano, M., Jouffrais, C.: Toward classroom experiences inclusive of students with disabilities. Interactions 26(1), 40–45 (Dec 2018), https://doi.org/10.1145/3289485
44. Moraiti, A., Vanden Abeele, V., Vanroye, E., Geurts, L.: Empowering Occupational Therapists with a DIY-toolkit for Smart Soft Objects. In: Proceedings of the Ninth International Conference on Tangible, Embedded, and Embodied Interaction. pp. 387–394. TEI '15, ACM, New York, NY, USA (2015), https://doi.org/10.1145/2677199.2680598
45. Parkes, D.: Nomad: an audio-tactile tool for the acquisition, use and management of spatially distributed information by visually impaired people. In: Tatham, A.F., Dodds, A.G. (eds.) Proceedings of the second International Conference on Maps and Graphics for Visually Impaired People, pp. 24–29. International Cartographic Association Commission VII (Tactile and Low Vision Mapping) and Royal National Institute for the Blind, London, UK (1988)
46. Phillips, B., Zhao, H.: Predictors of Assistive Technology Abandonment. Assistive Technology 5(1), 36–45 (1993), https://doi.org/10.1080/10400435.1993.10132205, pMID: 10171664
47. Pierce, J.: On the Presentation and Production of Design Research Artifacts in HCI. In: Proceedings of the 2014 Conference on Designing Interactive Systems. pp. 735–744. DIS '14, ACM, New York, NY, USA (2014), https://doi.org/10.1145/2598510.2598525
48. Reitberger, W., Güldenpfennig, F., Fitzpatrick, G.: Persuasive Technology Considered Harmful? An Exploration of Design Concerns Through the TV Companion. In: Proceedings of the 7th International Conference on Persuasive Technology: Design for Health and Safety. pp. 239–250. PERSUASIVE'12, Springer-Verlag, Berlin, Heidelberg (2012), https://doi.org/10.1007/978-3-642-31037-9_21
49. Rittel, H.W.J., Webber, M.M.: Dilemmas in a general theory of planning. Policy Sciences 4(2), 155–169 (1973), https://doi.org/10.1007/BF01405730
50. Rosenberg, S.A., Zhang, D., Robinson, C.C.: Prevalence of Developmental Delays and Participation in Early Intervention Services for Young Children. Pediatrics 121(6), e1503–e1509 (2008), https://pediatrics.aappublications.org/content/121/6/e1503
51. Sanchez, J., Tadres, A., Pascual-Leone, A., Merabet, L.: Blind children navigation through gaming and associated brain plasticity. In: 2009 Virtual Rehabilitation International Conference. pp. 29–36 (June 2009)
52. Sanders, E.B.N., Stappers, P.J.: Co-creation and the new landscapes of design. CoDesign 4(1), 5–18 (2008), https://doi.org/10.1080/15710880701875068
53. Schön, Donald A.: The reflective practitioner: how professionals think in action. Temple Smith, London (1983)
54. Seisenbacher, G., Mayer, P., Panek, P., Zagler, W.: 3D-Finger – System for Auditory Support of Haptic Exploration in the Education of Blind and Visually Impaired Students – Idea and Feasibility Study. In: Assistive Technology: From Virtuality to Reality. pp. 73–77. IOS Press, Amsterdam (09 2005)
55. Sonne, T., Jensen, M.M.: Chillfish: A respiration game for children with ADHD. In: Proceedings of the TEI '16: Tenth International Conference on Tangible, Embedded, and Embodied Interaction. pp. 271–278. TEI '16, ACM, New York, NY, USA (2016), https://doi.org/10.1145/2839462.2839480
56. Stolterman, E.: The nature of design practice and implications for interaction design research. International Journal of Design 2(1), 55–65 (2008), http://www.ijdesign.org/ojs/index.php/IJDesign/article/view/240
57. Tam, V., Gelsomini, M., Garzotto, F.: Polipo: A tangible toy for children with neurodevelopmental disorders. In: Proceedings of the Eleventh International Conference on Tangible, Embedded, and Embodied Interaction. pp. 11–20. TEI '17, ACM, New York, NY, USA (2017), https://doi.org/10.1145/3024969.3025006

58. Thieme, A., Morrison, C., Villar, N., Grayson, M., Lindley, S.: Enabling collaboration in learning computer programing inclusive of children with vision impairments. In: Proceedings of the 2017 Conference on Designing Interactive Systems. pp. 739–752. DIS '17, ACM, New York, NY, USA (2017), https://doi.org/10.1145/3064663.3064689
59. Verhaegh, J., Fontijn, W., Hoonhout, J.: Tagtiles: Optimal challenge in educational electronics. In: Proceedings of the 1st International Conference on Tangible and Embedded Interaction. pp. 187–190. TEI '07, ACM, New York, NY, USA (2007), https://doi.org/10.1145/1226969.1227008
60. Waddington, J., Linehan, C., Gerling, K., Hicks, K., Hodgson, T.L.: Participatory design of therapeutic video games for young people with neurological vision impairment. In: Proceedings of the 33rd Annual ACM Conference on Human Factors in Computing Systems. pp. 3533–3542. CHI '15, ACM, New York, NY, USA (2015), https://doi.org/10.1145/2702123.2702261
61. Wagner, A.: Collaboratively generated content on the audio-tactile map. In: Miesenberger, K., Klaus, J., Zagler, W., Karshmer, A. (eds.) Computers Helping People with Special Needs. pp. 78–80. Springer Berlin Heidelberg (2010)
62. Wagner, A., Kaindl, G.: WireTouch: An Open Multi-Touch Tracker based on Mutual Capacitance Sensing (September 2016), https://doi.org/10.5281/zenodo.61461
63. Wensveen, S., Matthews, B.: Prototypes and prototyping in design research. In: Rodgers, P.A., Yee, J. (eds.) The Routledge Companion to Design Research. pp. 262–276. Routledge (2014), https://www.routledgehandbooks.com/doi/10.4324/9781315758466.ch21
64. Zimmerman, J., Forlizzi, J.: The role of design artifacts in design theory construction. Artifact 2(1), 41–45 (2008), https://doi.org/10.1080/17493460802276893
65. Zimmerman, J., Forlizzi, J., Evenson, S.: Research Through Design As a Method for Interaction Design Research in HCI. In: Proceedings of the SIGCHI Conference on Human Factors in Computing Systems. pp. 493–502. CHI '07, ACM, New York, NY, USA (2007), https://doi.org/10.1145/1240624.1240704

Chapter 2
Haptically-Assisted Interfaces for Persons with Visual Impairments

Yeongmi Kim and Matthias Harders

Abstract Persons with visual impairments may encounter difficulties in their activities of daily living. To alleviate the situation, the use of assistive devices has been proposed in the past. In the development of such devices, haptic feedback has received increasing attention over the years. Notably, touch stimuli do not block or distort the auditory signals coming from the environment. In this chapter, we first will provide a brief overview of related work on haptically-assisted solutions for helping persons with visual impairments. Next, we address electronic travel aids that employ haptic feedback to support safe and independent ambulation and navigation of these users. A newly developed system and several related studies with end users will be discussed. Finally, haptically-assisted solutions to provide access to visual media on mobile devices will be addressed. A number of prototype systems, combining shape displays with surface haptics will be covered.

1 Introduction

Worldwide, several hundreds of millions of individuals suffer from visual impairments, as reported by the World Health Organization (WHO Global Data on Blindness and Vision Impairment 2018). Due to a partial impediment of or complete absence of vision, persons with such impairments often face difficulties in their activities of daily living. To alleviate the situation, numerous assistive devices have been developed in the past, supporting access to visual information. Several of these solutions are being successfully employed in everyday life,

Y. Kim
Department of Mechatronics, MCI, Innsbruck, Austria
e-mail: yeongmi.kim@mci.edu

M. Harders (✉)
Department of Computer Science, University of Innsbruck, Innsbruck, Austria
e-mail: matthias.harders@uibk.ac.at

© Springer Nature Switzerland AG 2020
T. McDaniel, S. Panchanathan (eds.), *Haptic Interfaces for Accessibility, Health, and Enhanced Quality of Life*, https://doi.org/10.1007/978-3-030-34230-2_2

such as refreshable Braille displays, screen readers, audiobooks, or smartphones with voiceover/talkback function; commercially-available solutions being the *ALVA Braille Controller*, the *Esys40* Braille display, the *VoiceOver* screen reader, or *Audible* audiobooks, to name a few available systems.

The sense of touch and haptically-assisted interfaces play a key role in the current development of novel assistive devices. A key advantage of using haptic feedback, in comparison to audio feedback, is that it neither blocks nor distorts the sounds coming from the environment, which represent an essential input for blind users. Numerous research projects have been carried out in recent years to advance the available technology. In this chapter, we focus on two concepts of technological solutions for assisting persons with visual impairments. In the first, we will address work focusing on electronic travel aids that employ haptic feedback to support safe and independent ambulation and navigation of visually impaired users. In the second, we will discuss haptically-assisted solutions to provide access to visual media on mobile devices, also aimed at this user group. In addition, we also provide a review of the state-of-the-art, covering early and recent developments.

2 State of the Art

2.1 Early Developments of Assistive Devices

The first attempts to encode visual data into the haptic modality date back over a century. It has been reported that already in 1881, a system called *Anoculoscope* has been envisioned [1], which would project an image onto an 8×8 array of selenium cells; however, the system had never been realized. In the 1920s, Gault described a device to transform speech into vibrational stimuli applied to the skin [2]. The method enabled subjects to distinguish colloquial sentences and certain vowels. His *Teletactor*, a multi-vibration unit delivering stimuli to five fingers, enabled subjects with an auditory disability to perceive speech, music, and other sounds. This denotes a typical example of sensory substitution, where one sensory stimulus is replaced by another, e.g. here sound to tactile substitution. In the 1960s the *Optacon* (Optical to Tactile Converter) was developed to enable visually impaired users to access tactile representations of black-and-white images captured by the system [3]. The device comprised of a photoelectric sensor and a 24×6 array of piezo-electrically driven tactile pins. While the Optacon presented tactile images of small areas, the *Tactile Vision* Sensory Substitution system developed by Bach-y-Rita et al. provided a wider range of tactile images [4]. For these, solenoids were employed that provided stimuli according to the visual image of a video camera.

Next, we will overview work addressing haptically-assistive solutions, both for supporting mobility of visually-impaired users, as well as for exploration of visual data on mobile devices.

2.2 Mobility Assistance

2.2.1 White Cane Usage and Electronic Travel Aids

The traditional *white cane* (also called *long cane*) is the primary mobility tool to support persons with visual impairments. It is mostly divided into three sections: a handle, a cane tip, and a long and hollow (usually white) cylinder shaft, made of fiberglass or aluminum. Haptic properties of obstacles (stiffness, viscosity) or ground surface information (texture) is sensed in the hand holding the handle, transferred via the cane tip and hollow cylinder part from the ground. Due to the limited length of the cane, only the swept surrounding at a distance of approximately 1.2 m can be detected by the traditional white cane [5]. In addition to the limited detection range, it is not possible to detect obstacles located at the upper body or head level. According to the survey conducted by Manduchi and Kurniawan (2011), more than 90% of persons with visual impairments had experienced a head level incident [6]. Moreover, 23% of these resulted in medical consequences. Extending the physical length and vertical range of the traditional white cane could be a solution to address this issue, but such a cane would become cumbersome and too heavy to carry as a daily mobility tool. Instead of physically extending a white cane, electronic travel aids (ETAs) have been developed, incorporating various sensing technologies to increase sensing ranges. Such devices generally comprise a sensing element for obstacle detection, as well as a displaying element to deliver information through other sensory channels. To substitute vision, auditory or haptic feedback is often adopted in ETAs to communicate sensor readings. A few decades after the white cane was introduced, the first ETA was marketed in the 1960s [7]. Since then assorted ETAs with different sensing and displaying components have been developed and commercialized.

2.2.2 Sensing Elements in ETAs

The most frequently employed sensors in ETAs rely on ultrasound signals – a low cost and lightweight solution providing robust obstacle detection. These sensors emit ultrasonic waves, based on which the time until the echo pulse returns to the receiver is measured, yielding the distance to the reflecting surface. The beam angle of the sensor is about 50–60°, and usually the closest object within the workspace is detected [8]. Due to the employed measurement principle, obstacle detection can fail when the surface of an obstacle is angled with respect to the sensor [9]. In contrast to this, infrared (IR) sensors have a relatively narrow beam angle. IR sensors measure the distance of an object depending on the magnitude of reflection of emitted IR light [10–12]. A key disadvantage of IR sensors is their difficulty with detecting transparent objects such as glass doors. Still, the aforementioned two approaches are the most widely used sensing technologies for ETAs. Other options to obtain proximity information are video cameras, combined with image processing techniques [13], laser range finders [14], depth detection systems such as the Kinect [15], LIDAR sensors [16], and time-of-flight (ToF) cameras [17, 18].

2.2.3 Display Elements

Among the five main human senses, hearing and touch are typically employed in ETAs as the display media to compensate for sight loss. Auditory feedback ranges from simple ON/OFF binary beeps to speech output announcing obstacles or providing directions for navigation. Allowing for the variation of frequencies and/or amplitudes of auditory feedback increases the information transfer (e.g. encoding distances to obstacles) [19, 20]. A larger number of systems have been developed utilizing the sense of touch, comprising of various configurations and display methods. Unlike auditory feedback, haptic stimuli (especially tactile cues) can be displayed to diverse locations on a user's body. Wearable haptic displays in ETAs have been developed for usage on the forehead, forearm, foot, and/or torso [21]. In addition, light-weight and small-sized sensors and actuators have also been integrated into handheld ETAs (e.g. the *miniguide* system). Alternatively, some ETAs are combined with a traditional white cane, to maintain the basic functionality of the latter, while also providing additional information to the user [12, 22]. The majority of ETAs (i.e. industrial as well as research prototypes) employ small-sized coin-type or cylinder-type vibration motors. As an alternative, electrotactile displays present electric currents to body parts, such as the forehead, hands or tongue, via electrodes [21]. Moreover, Gallo et al. [12] developed a mechanical shock generation mechanism to mimic haptic sensations of a traditional white cane when hitting an obstacle. For this, a spinning wheel and a solenoid break were combined. Recently, Spiers and Dollar [23] introduced a handheld shape display that provides two degrees of freedom of haptic sensations and tested the device both with visually impaired and sighted users [24]. Extension and rotation of a physically-held prop represent navigation commands (i.e. direction and distance to a target). Related to this, in the *haptic torch* system, the rotation of a bump fixed on a rotational disk indicates the distance to an obstacle [25]. Finally, a pin display driven by piezoelectric ultrasonic actuators or tendons connected to a small DC motor has been outlined in [26, 27]; however, the applications were not specific to ETAs.

2.2.4 Feedback Coding Methods

ETAs mostly support either general navigation or obstacle avoidance and provide only occasionally both functionalities. In the case of an electronic navigation (or orientation) aid, directional commands for reaching a destination are presented. In contrast, an electronic mobility device provides feedback about the area in front of a cane user [10, 22]. Details about the distance and height of an obstacle can be delivered via auditory and haptic feedback interfaces. Analogous to a car parking system, the *Teletact* device generates 28 tones (ranging from 131 to

1974 Hz), each representing different distance intervals (ranging from 12–16 to 2000–3000 cm) [19]. ETAs with a single vibration motor often vary the intensity level according to the distance to an obstacle (e.g. the $iGlasses^{TM}$ Ultrasonic Mobility Aid). Gallo et al. [12] employed the tactile apparent movement phenomenon to change the velocity of tactile flow on three vibration motors according to the distance to an obstacle. Multiple vibration motors were also used to indicate discrete distance levels as well as the vertical location of obstacles (e.g. detecting obstacles at head or knee level) [10]. Recent work finally investigated optimal tactile coding methods via vibrotactile feedback for determining several sensing distances [28].

2.3 Haptically-Assisted Presentation of Visual Data

2.3.1 Display Using Kinesthetic Feedback

An attempt to present visual data to persons with visual impairments using a peripheral haptic device was described by Moustakas et al. [29]. They generated force fields which permitted users to explore images through a force feedback device. The system could process 3D maps as well as conventional 2D maps. Using various off-the-shelf haptic devices, such as the Phantom or the CyberGrasp, the calculated force fields could be displayed. The system was not portable, however, since visual and haptic rendering was processed on a desktop computer. Similarly, Sjöström and Jönsson [30] developed a system to render 2D functional graphs employing the Phantom haptic device. The functions were haptically presented as a groove or ridge on a flat surface. Also, Fritz and Barner [31] describe a setup to display different forms of lines and surfaces to blind users through the Phantom device. They employed the notion of *virtual fixtures* to let a user trace lines in 3D space.

2.3.2 Display Using Tactile or Mixed Feedback

Raza et al. outlined ideas to convey abstract graphical and non-textual content, such as bar and pie charts, via tactile feedback to blind users, as outlined in an early draft in [32]. An electrostatic friction display was employed to present individual colors in an image as perceptually distinct tactile stimuli. Memeo et al. reported on a haptic system that enabled blind users to navigate between two locations in virtual environments [33]. They combined two haptic devices to transmit allocentric and egocentric data independently. The allocentric perspective is displayed through a 3DOF tactile mouse [34] via the fingerpad. The directional cues are rendered to a vibrotactile head-mounted display. In further work, Sinclair et al. introduced a 1-DoF mechanism, actuated in the (vertical) Z-direction [35]. It has been combined with vibrotactile actuators acting in the (horizontal) XY-plane. The combined stimuli enhanced the haptic feedback; however, no rotational movement was included.

In the developments described below, we made use of the Tactile Pattern Display (TPaD) device, provided in the scope of an open source project that has been launched at the Northwestern University, as described by Winfield et al. [36]. Tactile feedback is delivered to a user through varying friction on a flat smooth surface, achieved by vibrating the latter. In the initial version, the variable friction is applied to the whole screen, by actuating piezoelements attached to a glass plate. Using this interface, various textures can be rendered. A later version has also been outlined in [37]. In general, the concept of friction-altering surfaces is based on the work of Salbu and Wiesendanger, who both investigated the so-called air squeeze-film effect [38, 39].

3   Haptic Feedback in Electronic Travel Aids

Among our common daily activities, an important one is ambulation, which can be challenging for visually impaired persons. To this date, the traditional white cane is the most widely used mobility tool for the latter, while a smaller number relies on the assistance of a guide dog (about 10% vs. 0.5%, respectively) [40]. Although various ETAs have been proposed in the past decades, they have not yet been adopted by the majority of visually impaired users. It can be argued that the drawbacks and limitations of ETAs still outweigh their advantages. To increase the acceptance, the following characteristics should be considered when designing an ETA.

- **Maintaining key functionality:** Most pedestrians with visual impairments have been trained in orientation and mobility using a conventional white cane. It is thus beneficial to maintain the key functionality of the instrument: detection of obstacles and/or drop-offs ahead, detecting and following tactile ground surface indicators, delivering any other general texture cues of the ground surface to the handle, as well as additionally also the indication of the impairment to the surroundings. Since pedestrians with visual impairments are accustomed to walking with a white cane, maintaining these features would ensure the same level of safety, while reducing training time with a new interface.
- **Ease of use:** The straightforward way of utilizing a conventional white cane for safe and independent ambulation is likely the most important reason for its success and acceptance by users. ETAs could potentially provide various new functionality, such as vicinity information via several sensing and display elements. However, this may render the system too complicated, especially with continuous delivery of copious information, possibly confusing users. Such information overload may decrease the overall efficiency of walking (e.g. slowing of walking speed, discontinuity in walking, etc.); therefore, mechanisms requiring too complicated manipulation during walking should be avoided.
- **Efficient enhancement of original functionality:** ETAs should complement the standard functionality of a white cane. Any additionally provided information would presumably be displayed via the auditory or haptic sensory channel. Such

augmented and enhanced feedback should thus be easily perceivable, and not hinder the basic functionality of the white cane.

With these recommendations in mind, an option to reduce the cognitive workload when providing detail information about obstacles via an ETA is to switch between two modes – a *sweeping* mode and a *scanning/pointing* mode. This has been explored, for instance, for displaying different levels of obstacle information in [10, 12, 18]. This strategy permits not only nearly seamless walking but also delivers detailed information only when the cane user requests/desires it. Below we will first characterize the use of a white cane in sweeping mode, and then outline several studies carried out with visually impaired persons, using a newly developed ETA.

3.1 Sweeping Mode Characteristics

The mentioned sweeping mode refers to the standard mobility technique, in which a traditional white cane is swept side to side. In normal white cane usage, physical contact between the cane and an obstacle is mediated by the cane shaft. In contrast in ETAs, an enhanced cane, virtually extended through proximity sensors, would generate information about possible collisions via auditory or haptic feedback. Following the previous discussion, it is recommended to only present e.g. simple binary obstacle notification feedback or distance information to the nearest obstacle. When an obstacle is detected during the sweeping mode, a white cane user could switch manually to the scanning mode, e.g. by pressing a button. However, it would also be possible to effect this switch automatically, based on monitoring angular velocity or sweep angles of an ETA. Therefore, we have obtained data on normal usage of a white cane in several studies.

We have measured 3-DoF angular velocity and acceleration using an inertial measurement unit (IMU) [41] attached to a white cane shaft. In addition, also video data synchronized with the above are obtained via a small ($50 \times 20 \times 29$ mm) portable video camera (TrendMicro, BRAUN) fixed to the shaft. These data were acquired while a white cane user was walking with this sensorized cane. The obtained data enable us to estimate the ambulation center line as well as the maxima of the sweeping motion. As depicted in Fig. 2.1, both acceleration as well as angular velocity vary from the start of the motion and are periodic in all dimensions. Based on these raw data, the angular velocity around the Z-axis (i.e. the yaw angle) could be used to determine the times of switching between the two modes. Only the angular rate of the yaw angle of the cane is shown separately in Fig. 2.2. Note that the highest and lowest sweeping angles reached correspond to angular rates of about $0\,°\,s^{-1}$, whereas the angular rate maxima and minima are observed at the sweep center line. As mentioned above, cane users could switch to the scanning mode, in case they are interested in obtaining further information about an obstacle, e.g. shape, height, color, etc. This switch could be done manually or activated e.g. by a reduction in angular rate.

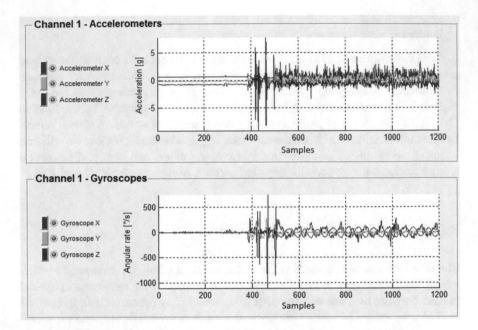

Fig. 2.1 3-axis acceleration and angular velocity during sweeping a white cane

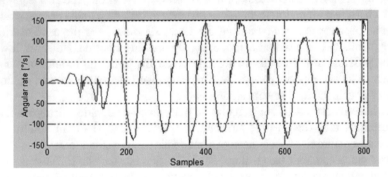

Fig. 2.2 Angular velocity of the white cane in z-axis (yaw angle) when using the sweeping technique

In the following, three user studies will be outlined, performed together with visually impaired white cane users. The goal of the experiments was to examine the usability and effectiveness of a new electronic travel aid developed in our research work. We will also examine how the systems align with the proposed ETA guidelines.

All experiments were carried out with the *Eyecane* ETA [18]. It comprises of a ToF sensor to detect distances of obstacles and further depth properties of obstacles. As a feedback interface, four eccentric rotating mass vibration motors (ERMs) were mounted in the handle, as shown in Fig. 2.3. Each ERM is enclosed by a rubber ring

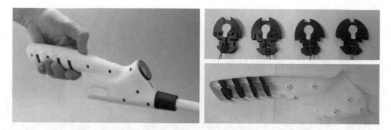

Fig. 2.3 Eyecane handle (left) and vibrotactile actuators mounted into cane handle (right). Each ERM is placed in the middle of a rubber ring (top right). A portion of the rings protrudes outward, where the fingers contact it on the surface (right bottom)

to reduce the transfer of vibrations to the whole handle. This permits presenting localized vibration feedback to individual fingers. The outer portion of the rubber rings is touched by the index, middle, ring, and little finger, respectively. The shaft of a conventional white cane is finally also attached to the enhanced handle.

3.2   Ground Surface Identification

The conventional white cane plays a crucial role in identifying ground surfaces or structures ahead of a cane user [42–44]. Such surfaces or structures could, for instance, be a curb, uneven pavement, a drop-off, tactile surface indicators, slippery floor, or grass. Tactile paving, also called *tactile ground surface indicators* (TGSI) or *tactile walking surface indicators* (TWSI), are installed on pedestrian walkways and passages to assist persons with visual impairments in safe ambulation [42]. Although the appearance of TGSI differs between countries, usually two main types exist: directional TGSI assist in wayfinding, by indicating a path or direction, while warning TGSI, also called attention TGSI, inform and warn about hazards ahead. Any additional haptic feedback displayed by an ETA should not interfere with detecting and discriminating such different ground surfaces. Therefore, a user study with the newly developed ETA was conducted.

Seven panels of ground surface samples, 570 × 570 mm in size, were prepared, equipped with caster wheels for easy exchange between trials. As depicted in Fig. 2.4a–g, the selected ground surfaces were: a sheet of polyvinyl chloride (PVC), wood, artificial turf, tiles, cobblestone paving, directional TGSI, and warning TGSI (comprised of asphalt, embossed with 4 mm tactile paving lines). The experimental task was to identify the sample panels, while the Eyecane was displaying a randomly selected vibrotactile signal. For the latter, signals designed to indicate the distance to an obstacle were used, as developed in [28]. Nine legally blind white cane users (seven male, two female) participated in the study. Participants were sitting comfortably on a chair and wearing a headphone to listen to white noise. In addition, two legally blind participants, who still had residual vision, were asked to wear a

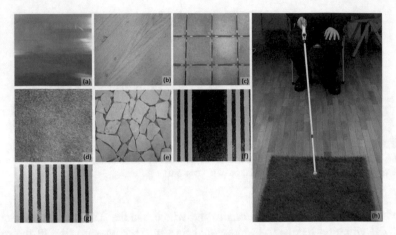

Fig. 2.4 Seven ground surfaces for the identification test. (**a**) Polyvinyl chloride sheet, (**b**) wood, (**c**) artificial turf, (**d**) tiles, (**e**) cobblestone paving, (**f**) directional TGSI, (**g**) warning TGSI, (**h**) experiment setup for the ground surface identification

Table 2.1 Confusion matrix of surface sample identification task

Stim. Resp.	PVC	Wood	Turf	Tiles	Cobbles	Direction	Warning
PVC	99.2%	1.1%	0%	0%	0%	0%	0%
Wood	0.8%	98.9%	0%	0.1%	0%	0%	0%
Turf	0%	0%	100%	0%	0%	0%	0%
Tiles	0%	0%	0%	98.3%	1.2%	0%	0%
Cobbles	0%	0%	0%	1.6%	98.8%	0%	0%
Direction	0%	0%	0%	0%	0%	100%	0%
Warning	0%	0%	0%	0%	0%	0%	100%

blindfold to mask any visual input from the ground surfaces. Each ground surface was presented 20 times, in a pseudo-randomized order. Participants were asked to sweep the ground surface with the Eyecane and to discriminate between the given surfaces, as illustrated in Fig. 2.4h. A confusion matrix compiling the results of the experiment is provided in Table 2.1. As can be seen, the mean correct identification rate was 99.3%. False responses were found mainly in two cases – discriminating between the PVC sheet and the wooden panel, as well as between tiles and cobblestone paving. Nevertheless, incorrect identification rates varied only between 0.1% and 1.6%.

Following the perceptional study, additional quantitative measurements were carried out, to characterize the tactile signals obtained from the surface textures. To this end, a 3-axis accelerometer (ADXL335, Analog Device) was mounted at the middle of the cane shaft. It was connected to a data acquisition device (NI-USB 6008, National Instruments) and acceleration data were obtained via a LabVIEW interface at 1000 Hz. The same subjects as before participated in the measurements.

Fig. 2.5 Surface texture measurement setup

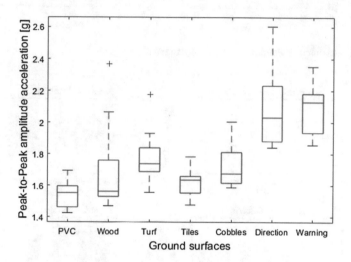

Fig. 2.6 Peak-to-peak amplitudes of acceleration for the seven ground surfaces

Neither a blindfold nor a headphone was used for this study. All subjects were asked to stand up in front of a ground surface panel, and then walk in place, while sweeping the Eyecane across the surface sample, as they normally would employ their white cane (see Fig. 2.5). Measurements for each ground surface lasted about 10 s. Figure 2.6 depicts box plots of peak-to-peak acceleration amplitudes obtained from the nine participants when sweeping on the seven ground surfaces. Smooth ground textures, such as PVC and wood panel show lower amplitudes. In contrast, both TGSI types yielded higher acceleration amplitudes. Figure 2.7 depicts normalized acceleration of two different example surfaces. As can be seen, amplitude profiles differ apparently between a smooth surface and rough surface.

Finally, the propagation of vibrations, generated by the ERMs to the housing, was also studied. For this, a small-sized accelerometer (ADXL325, Analog Device) was attached to the rubber ring enclosing the vibration motor, at the point where the fingers would be in contact on the handle surface. In addition, another small-sized accelerometer was attached to the surface of the handle housing. This setup permits validating the proposed design using the rubber rings as dampers. Vibrotactile feedback was rendered for four distance levels, as described in [28]. Figure 2.8 visualizes the vibration amplitudes measured with the sensors. Note that Level 1

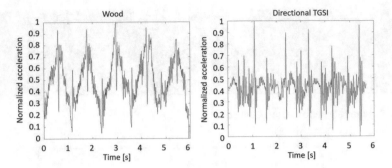

Fig. 2.7 Normalized acceleration when sweeping the cane

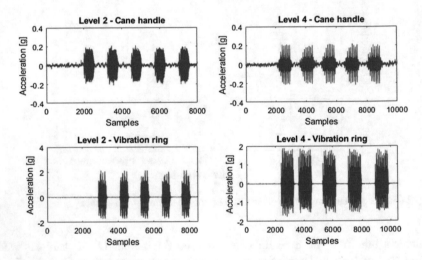

Fig. 2.8 Vibration amplitude measurements on the cane handle and the vibration rubber ring for two different distance levels

indicates the smallest distance to an obstacle and Level 4 the highest, respectively. As can be seen, the vibration amplitude decreases by about an order of magnitude, from the rubber piece to the handle.

3.3 Detection of Tactile Distance Signals During Sweeping

In previous work, we have compared three different modes of rendering tactile signals in ETAs, each encoding a distance range of obstacles [28]. In the first rendering method, signals are generated using only temporal variation between individual stimuli. In the second, both temporal and spatial variation are employed. The third method comprises spatial, temporal, and intensity variation. In our previous work, we found that the first rendering technique resulted in lower identification rates as compared to the other two. In addition to this, we also examined grip types. From the data, we could conclude that a four finger grip configuration provided significantly higher accuracy in detecting distance levels than a single finger grip (i.e. single finger across all four vibration actuators).

Nevertheless, this prior study was carried out in a static configuration – that is, participants sat on a chair and the cane tip contacted a wooden surface statically, without any motion. This experiment has now been extended in our work, by examining the effectiveness of the proposed tactile rendering methods in a more practical setting. Seven legally blind white cane users (three female, four male) took part in this study. As suggested in [28], the four finger grip configuration was utilized. Thus, participants placed four fingers on the four individual vibrating rubber rings, as depicted above in Fig. 2.3. In the tactile rendering method, the actuators were displaying vibration patterns consecutively. Note that the direction of tactile flow showed no influence in identification accuracy in our prior work; therefore, we employed an outward direction of the tactile flow. In contrast to the previously examined static situation, participants now stood up, walked in place, and swept the Eyecane again across tactile surface samples. In this study, three different ground surfaces were employed: wooden floor, cobblestone paving, and directional TGSI. The surface samples were selected in a pseudo-random order and presented to each subject, with 240 trials in total. As soon as a subject verbally reported a distance level, as displayed by the tactile pattern, the experimenter recorded the answer and then initiated the next stimulus trial. Subjects were allowed to take a break in between, as desired.

The mean correct identification rate for all signals on all surface samples was 99.12%. For the wooden floor, the cobblestone paving, and the directional TGSI, it was found to be 99.1%, 99.1% and 99.17%, respectively. For only 14 out of 1680 trials, incorrect responses were given. These were found only for the distance levels 2 and 3, for which only the middle and the ring finger were stimulated.

3.4 Detection Accuracy of Distances Encoded by Vibration

In a final study, we examined how well depth ranges, displayed as vibrations to users, could be identified. For this, a ToF sensor with a resolution of 176×144 pixels (SR4000, MESA Imaging) was employed. The depth camera was held by the experimenter and depth images of obstacle scenes were captured in a static situation. The modulation frequency was set to 15 MHz; the maximum distance at which obstacles could be detected was 10 m. Initially, the information about depth in the environment is given with respect to the location of the camera. Based on the angle of the camera, the horizontal distance of any visible obstacle in the depth image can be calculated. After resampling, a pixelized depth image of obstacles ahead is obtained, in which correct horizontal and vertical Cartesian distances are stored.

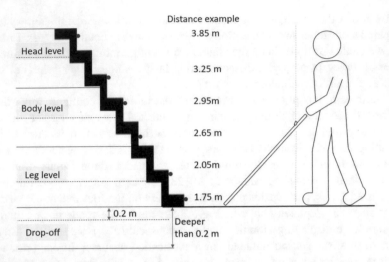

Fig. 2.9 Four/eight height levels of obstacle depth information. Per level, the horizontal distance between the closest point on the obstacle and the frontal plane of the user is specified

For the display in the following study, the overall covered vertical height in the depth image is subdivided equidistantly into separate levels – using a total of either four or eight levels. In the former case, from top to bottom (according to human body sections), we consider the *head level*, the *body level*, the *leg level*, and a *drop-off*. In the latter case of eight sections, each of these previous four levels is subdivided further into two subparts – an upper and a lower section. The general idea of the vertical subdivision of height into levels is illustrated in Fig. 2.9. Note that the bottom drop-off level is subdivided into a *shallow* and a *deep* drop-off; the former denoting a vertical distance of less than 20 cm, and the latter one equal or greater to 20 cm, respectively. The former could be a curb or a step of a descending stair, while the latter may be a deeper hole in the ground or descending stairs. Regarding horizontal distance, the depth pixel with the smallest distance within a respective height level interval is employed. Note that in order to reduce processing time, as well as the cognitive workload of a user who would be receiving the corresponding signals, only a small 10-pixels-wide column around the depth image vertical centerline is considered for this. As an example, in Fig. 2.9 the obtained horizontal distances to the steps of a stair are given in meters.

For our user study, depth images of fifteen different environments with obstacles were obtained. Figure 2.10 depicts three example images of acquired depth data of such scenes. The depth is color-coded; also note two vertical lines in the center indicating the source region for the depth values. Finally, normal photographs of the surroundings are also provided. Any vibrotactile distance signals would be based on the visible depth information. Assuming four levels, in the first example (sequential) signals would be generated for the head, body, and leg levels, respectively. Since the distances per level are getting smaller, the signal intensities would increase.

Fig. 2.10 Examples pictures of obstacles and depth profiles. Top: depth information of time-of-flight camera; Bottom: color images of charge-coupled device camera

There would be no signal for the drop-off level since no vertical downwards step is present. The second and third example are typical for scenes where the obstacles at the body or head level may not be detectable by a conventional white cane. Vibratory feedback in the second example would be at the body level only. In the third example, a stronger intensity would be experienced at the head level, and lighter feedback at the leg level.

A study was carried out with the described system, examining how well users could perceive depth details of obstacles encoded via the vibrotactile display of the Eyecane. Five visually impaired subjects (one female, four male) participated in the study. Depth was covered in a range from 0.8 to 5 m. The latter was subdivided equally into 8 depth intervals – each measuring 0.525 m. Eight different vibration intensities were employed to encode these eight depth intervals. The previously described ERMs were activated with pulse width modulation, using linearly decreasing duty cycles, ranging from 100% for the shortest distance to 20% for the longest distance. Note that distance values larger than 5 m were not displayed.

Participants were asked to hold the handle of the Eyecane with the four fingers grip. In the case of using four height levels, each vibration motor mapped exactly to one level; that is, from the index finger mapping to the head level, to the little finger mapping to the drop-off. In the case of eight levels, the first four sublevels (head and body) are displayed to the four fingers first, followed by a second set of signals, mapping the remaining four levels (leg and drop-off) to the fingers. That is, in this case, each finger receives two consecutive vibratory signals.

Each ERM is activated for 250 ms, with an inter-stimulus interval of 150 ms. Thus, the total duration of each tactile depth rendering is 1450 ms for the four and 3050 ms for the eight level case, respectively. Moreover, if no obstacle is detected, no signal is displayed during the respective time interval. In each trial, such sets of cues indicating the depth in each pseudo-randomly selected scene (out of fifteen) were presented to participants three times consecutively. An auditory beep initiated the start of each trial. Participants were then asked to verbally state the perceived distance ranges – either four or eight, depending on the condition. In total, 30

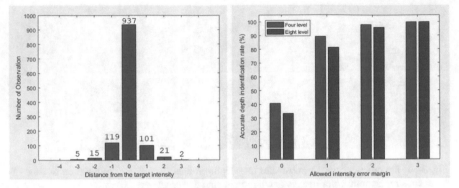

Fig. 2.11 Results of depth interval study; subjects' deviations from correct intensity levels (left); accurate identification rates, for different intensity error margins (right)

such trials for the four-level and 15 trials for the eight-level configuration were presented to each subject, respectively. The verbal responses were recorded by the experimenter.

The correct identification rate of all distance range renderings, considered in separate, is about 78%. However, it has to be stated that 53.75% of all presented cases where *no signal* ones, i.e. no vibration was displayed; it can be argued that correct identification in this case is trivial. As an alternative, measuring correct identification rate per depth image (i.e. all distance ranges over all height levels), the accuracy was found to be lower. As shown in Fig. 2.11 (right), a complete set of distance values was only correctly identified in 40.7% and 33.3% of the cases, in the four and eight level condition, respectively. Nevertheless, further analysis of the type of errors (left) indicates that in the majority of cases false responses were only off by one interval step. Larger estimation errors, of two or three intervals, were less common. The correct identification rate, when allowing an error margin in the depth interval estimates, would show an improvement, as can be expected (see again Fig. 2.11). Moreover, note that the absence of a signal was never misclassified. Finally, due to the lack of an absolute intensity reference, participants had to rely on the initial training phase for calibration. This sometimes led to cases where random depth signals, that coincidentally were consecutive, were all perceived as being shifted (e.g. consecutive depth signals for interval 6, 7, 8 were reported as 5, 6, 7).

4 Haptic Devices for Exploration of Visual Data

In the past decades, touchscreen displays manifested themselves as an integral and ubiquitous part of everyday life. The scope of applications ranges from general personal use to highly specialized tasks. Although the advances in mobile and tablet technology are vast, the integration of complex haptic feedback, except for vibratory cues, is generally not considered for these devices, leaving a gap in terms of closed-loop physical interaction. Despite the fact that an increasing number of visually-impaired persons own a smartphone, they mainly rely on auditory feedback with these. However, the latter is limited regarding the presentation of some visual content, such as images or shapes.

In order to address the mentioned shortcomings, we have developed a set of shape displays combined with surface haptics rendering systems based on mobile touchscreen tablets. In the following sections, we will first overview the devices. Thereafter, we will discuss rendering methods developed for the interfaces, and finally, we will outline corresponding user experiments and initial results. More details on these systems can be found in [45] and [46].

4.1 Hardware Overview

4.1.1 Early Prototype

Our first developed prototype can be seen as an extension of the TPaD device of the Tablet Project (see Fig. 2.12 (right)). It incorporates two modes of haptic display – texture rendering via piezoelectric surface haptics and shape display via a motion platform. As a tablet, we employ the Asus Nexus^{TM} 7, which is connected to an IOIO board via Bluetooth, and to the TPaD circuit board via micro-USB cable. The microcontroller and the amplifier are integrated into a single circuit board. The latter generates the necessary output to control the actuators of the haptic display components. All rendering computations, as well as the peripheral device control, are carried out on the tablet.

The TPaD surface haptic display comprises a piezoelectric bending element that creates the necessary actuation to achieve a squeeze film effect. It comprises five 8 mm wide piezoceramic layers, glued onto a passive support layer made of glass. A crucial component is the amplifier that generates the required high voltage signals

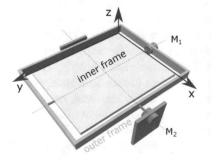
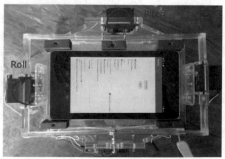

Fig. 2.12 Mechanical design of initial prototype (left); top-down view of final assembled prototype (right). (Obtained from [45] (courtesy of L. Chhong Shing))

of up to 100 Vpp to drive the TPaD. Its operation is frequency dependent, wherefore the amplifier must be tuned to operate at the resonance frequency of the glass. For tactile rendering, the tablet determines the desired amplitude, which is transmitted to the microcontroller. The latter then outputs a pulse-width-modulation (PWM) signal through the amplifier to the TPaD.

The shape display has two degrees of freedom. The general mechanical design is illustrated in Fig. 2.12 (left). It consists of two frames – the outer being rotated by the first motor and the inner by the second. The inner frame is seated inside the outer one; motor activation results either in a direct rotation about the X-axis or the Y-axis, respectively. The employed actuators are standard low-cost servomotors available from Parallax. Finally, the surface haptics tablet is placed on top of the inner frame, which is dimensioned such as to hold it in place. The shape display device component is mechanically limited to rotations of approximately $\pm 15°$ around the axes.

Due to the fixed center of rotation about each axis, the device is best suited to displaying changing contact normals at the center of the tablet. Note that at locations different from the center, artifacts perpendicular to the interaction plane are introduced – a user's finger is either pushed upwards or loses contact with the tablet when it moves. This can be avoided by adding additional degrees of freedom to the motion platform, which led to the development of the designs outlined in the next sections. Nevertheless, for contact locations close to the center of rotation usable haptic feedback can be generated, as addressed below in a user study.

4.1.2   SurfTics 1.0

In order to overcome the limitations of the initial prototype, a new device – named *SurfTics* 1.0 – has been developed. It combines the previously mentioned TPaD tactile surface display with a 3DoF Revolute-Revolute-Spherical (RRS) motion platform (similar to [47]), thus supporting rotation as well as translation of the shape display. This setup allows for the rendering of shapes in 3DoF, at non-centered positions on the touchscreen (see Fig. 2.13).

The mechanical setup comprises of a base plate, three stepper motors with encoders, three RRS manipulator arms, and a top plate with a NanoSuction pad as a universal tablet mount. The electrical setup consists of an ATmega1280 8-bit microcontroller (AM), three drive units (DU), and a customized board (IOIO). Data transfer is realized via UART connections between the boards. The main processing unit is the AM, which calculates the inverse kinematic algorithm of the finger position data and coordinates the three drive units. The latter each consist of an ATmega328 (AN), a stepper motor driver (Texas Instruments DRV8825), a quadrature encoder (1'000 increments), and a stepper motor (NEMA17, $0.8°$/step). The AN is employed for running a position control loop and handling the UART communication. Quadrature encoders are used for precise position feedback, as stepper motors are prone to slippage, especially when running at high load. More details on the device design can be found in [48].

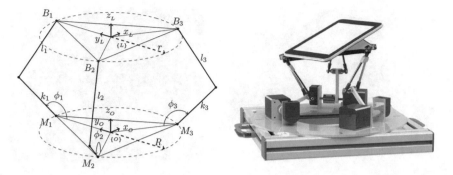

Fig. 2.13 Mechanical design of the SurfTics 1.0 device (left); CAD-rendering of device (right). (©2017 IEEE. Reprinted, with permission, from [48])

The workspace allows for a 20 mm translation along the Z-axis and $\pm 30°$ rotation about the X- and the Y-axis, respectively. Based on the stepper motor resolution (400 steps per revolution times 32 – achieved by employing the microstepping mode of the driver circuit) the minimum and maximum velocity can be calculated as $v_{max} = 24.4$ rpm and $v_{min} = 0.995$ rpm, respectively. The 1'000 CPR quadrature encoder delivers a resolution of $0.27°$ for the rotations.

The improved device design permits more diverse user interaction and haptic feedback. User experiments showed that the new design and the combined feedback benefits identification tasks, as will be overviewed below.

4.1.3   SurfTics++

The previously described SurfTics version already exhibited a reasonably improved performance; however, it still included a number of drawbacks. Firstly, vibrations were generated when the shape display was moving. These were undesired and interfered with the vibrations created by the tactile display. The vibrations were presumably caused by the employed stepper motor. Secondly, quite audible acoustic noise originating from the drive system was present. This was likely coming from the PWM signals for driving the stepper motors; these being in the audible range. Thirdly, the upper speed limit of 30 rpm of the drive units is a limitation for displaying sharp edges. Therefore, a new version of the SurfTics hardware was developed (dubbed *SurfTics++*), which on the one hand improves on the maximum bearable user force, dynamic range, and motion smoothness, and on the other hand reduces the audible noise in the system. The amended design will be described in detail next (also see Fig. 2.14).

For the further enhanced version, the three actuated revolute joints, with attached lever arms, are distributed equally at 120° positions, at a radius of 100 mm from the center point of the base plate. The three lower lever arms (from the motors to

Fig. 2.14 CAD-rendering of SurfTics++ device (left); final realization of hardware (right). (Obtained from [46] (courtesy of F. Enneking))

the revolute joints) measure 60 mm in length; the three upper lever arms (from the revolute joints to the spherical joints of the top plate) are 90 mm in length. The three spherical joints attached to the top plate are also distributed equally in 120° steps, also at a radius of 100 mm from the center point of the top plate.

The most crucial part of the setup is the drive system, each unit consisting of a motor, an encoder for tracking the motor's position, and a controller for providing the needed high power control signals. The required maximum motor torque can be calculated from an assumed maximum finger force requirement of 25 N. In addition to that, the motors also have to bear the weight of the Nexus 7 tablet and the top plate weighing 0.62 kg. The resulting overall maximum force of 31.08 N is distributed over the three motors, depending on the finger position during interaction. It is assumed that in the worst case, the portion of the overall force born by a single motor can reach 66% of the overall force. Hence, considering the 0.060 m lever arms, the maximum torque requirement for the selected motors is 1.24 Nm. In addition to this requirement, the motors should also be backdrivable, with as little resistance as possible.

In contrast to the stepper motors used in SurfTics 1.0, brushed DC motors can provide a variable torque and are in general backdrivable. This is also true for brushless DC motors, but their control is more difficult and drivers are significantly more expensive. Thus, we employ brushed DC motors in the SurfTics++, specifically the Maxon RE 35 118778. This motor is highly backdriveable, showing negligible resistance or cogging torque. Its lower speed limit is 13 rpm at a 4% PWM duty cycle. In order to tweak the speed range to a desired lower speed limit of 1 rpm, a 13:1 gear reduction via capstan drives is employed. In addition, the motors come with AVAGO HEDS 5540 A11 quadrature encoders mounted on the extended motor shaft on the back, with a resolution of 500 CPR. According to this, in total an encoder resolution of 26'000 countable steps per revolution is achieved. As a motor driver, the Pololu 20 A, 5.5–50 V single motor controller, was selected. It comprises a current sense pin as well as a drive-coast operation mode. This hardware allows for driving PWM signals at frequencies of up to 40 kHz.

The workspace of the device is given by 20 mm translation along the Z-axis, and $\pm 30°$ rotation about the X- and the Y-axes. The dynamic range of the drives is 1–24 rpm. For force rendering, the usable stiffness range is 1.37–6.00 Nmm^{-1}. The positioning accuracy of the drive system was found to be 0.48°.

4.2 Rendering

The kinesthetic shape display component of the various described SurfTics versions is able to present 2.5D images on the tablet. For this depth map images can be employed that are similar to a topographic map; with the color/gray level scale representing the height at each pixel/location. Depending on the employed bit-depth for encoding height, different resolutions of height levels are available. In addition to the height, local orientation can also be, for instance, obtained as height field gradients and displayed. Note that due to gradient discontinuities at edges, directional Gaussian smoothing usually has to be included for this.

Regarding the desired Z-elevation of the tablet, note that its inclination also has an effect on the perceived height at the contact location of the fingertip. To counteract this effect, one can employ a simple compensation function:

$$z_{comp} = x \tan(\phi_Y) + y \tan(\phi_X), \tag{2.1}$$

with the ϕ_i being the rotations around the tablet axes, and x, y giving the contact position in tablet space. By subtracting this value from the current Z-elevation, the inclination angle effects are decoupled from the rendering.

Similar to the rendering of height, texture information can also be encoded in grayscale images. Pixel intensity can be mapped directly to the amount of friction. Preliminary tests with the combined setup indicated that it is more difficult for a user to discriminate between continuous changes of friction than to perceive a discontinuous transition between levels. Therefore, in the experiments described below, we rendered friction only at some edges as hard transitions. These are given by dark and bright regions in the grayscale images and mapped to low and high friction.

4.3 User Experiments

In the following, we will describe two experiments that were carried out with the described hardware. In the first, we examine how well users can discriminate between concave and convex shapes, using the initial prototype as well as SurfTics 1.0. In the second study, we explore how well subjects can differentiate between various 2.5D geometries, using the SurfTics 1.0.

4.3.1 Curvature Detection Experiment, Comparing Prototype to SurfTics 1.0

The goal of this experiment has been to evaluate the basic rendering performance of two previously described display prototypes – in particular, we have compared the initial prototype to the SurfTics 1.0. Note that the former one is limited in the display due to a fixed center of rotation.

Four different shapes, being either convex or concave, were presented to study participants; the former were given by quadratic functions, extruded on the surface along one axis. The functions were set such as to result in a base width of either 3 cm or 4 cm, and a height of 1 cm. The virtual objects were placed at the center of the workspace, with the ridge/trough extending along the Y-axis. For the early prototype, the object was also randomly shifted along the X-axis, to allow for variation between different trials. The maximum rotation was limited to $\pm 15°$ for both devices.

The concave and convex shapes were presented in four different conditions. In the first, the early prototype was employed, displaying height and curvature; but with a fixed center of rotation. In the remaining three, the SurfTics 1.0 device was employed; the rendering with that one was either height-only, orientation-only, or a combination of both cues (see also the sketch in Fig. 2.15).

Ten sighted volunteers – 2 females, 8 males, average age 27 – participated in the study. All participants were naive to the design and the concept of the devices. All were sighted, with no sensory deficits, and happened to be right-handed. Two participants reported being familiar with the employed haptic technology.

Participants were requested to sign a consent form prior to the experiment. Subsequently, they were blindfolded and asked to wear a noise-canceling headphone, to mask any confounding cues. Before the actual experiment, participants were allowed to familiarize themselves with the devices as well as the task. After the training phase, one stimulus was randomly selected and displayed, using the device specific to the condition. Participants were permitted to explore the virtual object without any time constraints. After each trial, participants had to indicate which type of curvature (concave/convex) and which size they perceived. Also, the trial time was recorded.

Firstly, we examine the shape identification results. A Shapiro-Wilk test revealed that the data were not normally distributed per group. Therefore, a Friedman

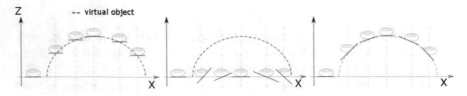

Fig. 2.15 Rendering conditions with the SurfTics 1.0, with fingertip moving along convex shape example – height-only (left), orientation-only (middle), combination (right). (Obtained from [45] (courtesy of L. Chhong Shing))

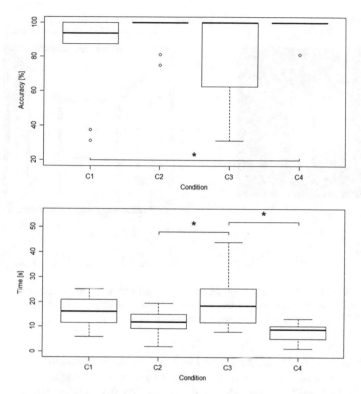

Fig. 2.16 Box plots summarizing the results of the shape identification tasks (top) and the task performance time (bottom), for all four conditions (C1–C4) C1 and C4 present curvature and height by employing the early prototype and the SurfTics 1.0 respectively. C2 and C3 provide height information only and curvature only through the SurfTics 1.0 device. (Obtained from [45] (courtesy of L. Chhong Shing))

non-parametric test was carried out, yielding a significant effect of the different conditions ($\chi^2(3) = 9.766$, p = 0.02). A Wilcoxon post-hoc test indicated a statistically significant difference between the early prototype rendering and the combined shape display (W = 26, p = 0.037). Moreover, the task completion time was normally distributed (Shapiro-Wilk). Repeated measures ANOVA indicated significant differences between the conditions (F(3,36) = 5.574, p = 0.003). A post-hoc Tukey test showed differences of the height-only rendering with the other two conditions employing the SurfTics 1.0 (p < 0.05). No significant difference was found in the size discrimination. The main results are visualized in Fig. 2.16.

4.3.2   Width Identification Experiment with the SurfTics 1.0

A further study has been carried out to explore how well users could identify the width of 2.5D geometries via the SurfTics 1.0 device. When users moved a

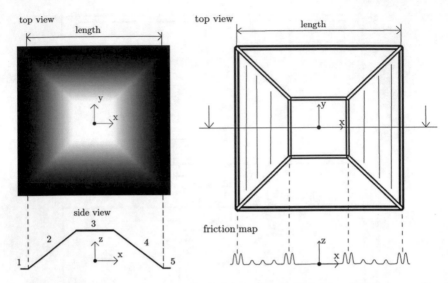

Fig. 2.17 Sketch of shape rendering, for display of a pyramid frustum (left); note numbered regions with varying orientation. Sketch of surface friction rendering of edges/transitions (right). (©2017 IEEE. Reprinted, with permission, from [48])

finger across the tablet, height as well as local normal orientation of the contacted object surface were rendered via the shape display device component. In addition, surface haptics could be included when contacting the surface depending on the experimental condition, as will be outlined below. An overview of these rendering processes is indicated in Fig. 2.17.

Nine sighted participants – three female, six male, average age 27 – took part in the experiment. The haptic device was employed to render pyramid frustums, of equal height in Z-direction, but with varying width in the base (see Fig. 2.17). The geometries – denoted as S1 to S4 – had a width (i.e. side length) of 40, 50, 60, and 70 mm, thus resulting in a matching decrease in slope on the sides.

Participants initially familiarized themselves with the setup and the task. Thereafter, in a training phase, a blindfold and a headphone playing white noise were donned, blocking out visual and auditory cues. Subjects were then asked to explore the four shapes and were informed verbally by the experimenter on the size. This process lasted until the participants felt comfortable with the setup and confident about correctly carrying out the task.

Subsequently, the actual trials took place. The different samples were presented, each one eight times, in pseudo-random order. The index finger of a blindfolded participant's dominant hand was guided to the center of the rendered shape by the experimenter. In each trial, a participant was allowed to explore a shape for a maximum of 15 s, before being asked to report on the perceived width of the shape. The responses of the participants were recorded by the experimenter. All trials were carried out in direct succession, without any feedback on correctness. Furthermore, note that four participants started the first half of the trials with surface haptics

Table 2.2 Confusion matrices (response vs. stimulus) showing results for – condition of shape rendering without surface haptics (left) and with surface haptics (right). The latter exhibits better performance

Without condition					With condition				
Stim. \ Resp.	S1	S2	S3	S4	Stim. \ Resp.	S1	S2	S3	S4
S1	66.7%	26.4%	5.6%	1.3%	S1	83.3%	15.3%	1.4%	0.0%
S2	22.2%	50.0%	27.8%	0.0%	S2	12.5%	65.3%	16.7%	5.5%
S3	8.3%	25.0%	54.2%	12.5%	S3	0.0%	25.0%	59.7%	15.3%
S4	1.4%	8.3%	29.2%	61.1%	S4	0.0%	0.0%	22.2%	77.8%

enabled (*With* condition), while the remaining ones started without this feedback (*Without* condition). After completing the first half, the mode switched for the groups.

Overall, in the experiment we found an average identification rate of 58%, for the condition without surface friction display, and of 71%, including the friction rendering. The recorded data were successfully tested for normality, and further analysis via a paired t-test indicated statistical significance of the results ($t(8) = 2.879$; $p < 0.05$). This hints at the beneficial effect of including friction rendering for the identification task. The overall data are illustrated in confusion matrices for both experimental conditions in Table 2.2. It is interesting to note that when no surface haptic feedback was included, incorrect replies were even found for maximum differences, i.e. S1 was identified as S4, and/or vice versa.

5   Discussion and Conclusion

In this chapter, we have outlined various aspects related to haptically-assisted interfaces for persons with visual impairments. We focused on two different threads – the use of ETAs as mobility assistance systems, as well as setups for the haptic display of visual information on mobile devices.

In the former domain, haptic feedback to support both a specific sweeping and a specific scanning mode in ETAs was investigated, including experiments with actual white cane users. The first two studies illustrated that the tactile rendering methods presented in [28] result in high identification rates of detecting distance levels, not only in the static but also in the dynamic condition. Moreover, the results were also confirmed in the presence of different ground surface textures.

Studying distance identification during motor activity is an important extension since any haptic signals that encode distance information should be recognizable by users during dynamic use of the cane (such as sweeping). According to Post et al. [49], vibration perception may degrade during dynamic movements. In this context, the proposed tactile rendering method was still found to be capable of delivering corresponding distance levels during sweeping. In addition, having properly isolated vibration stimuli per fingertip also proved to be a critical factor in the ETA design, to avoid stimulating the whole handle.

In the additional study focusing on the scanning mode, participants remained stationary only. However, this was following our idea of switching between modes; more detailed information about obstacles would be obtained in the scanning mode after a user was to detect the presence of an obstacle in the sweeping mode. A user would then stop and point the ETA at such obstacles to obtain additional information. Results of the scanning study revealed that the accuracy of depth perception of obstacles was initially not very high. We conjecture that this is due to the large number of intensity levels provided; eight distance regions were employed, without intensive user training. However, it should be noted that reducing the number of distance levels may not provide sufficient information about obstacles, e.g. it might not be possible to accurately detect individual steps of a staircase. Nevertheless, using a larger range of distance levels will require a substantial amount of user training. This will be examined further in future work; it may be a possibility to adaptively adjust the number of distance levels. The option to personalize distance detection settings and haptic signals according to white cane user preferences should also be incorporated.

In the latter domain, we introduce three prototypes for displaying visual data on mobile tablets. For haptic rendering, a surface haptic display is combined with a shape display mechanism. Preliminary user studies indicate that such devices could be used for mediating visual information via the haptic sensory channel. However, considerable additional work is required; especially, also user studies with visually impaired persons. Only simple shapes were employed in the user studies, as in similar related research. A direct extension will be to explore how more complex shapes, as well as mixed contents, could be perceived on such augmented mobile tablets by users.

Acknowledgements We would like to express our gratitude to our students Chhong Shing Lee, Thomas Hausberger, Michael Terzer, and Florian Enneking who developed the described SurfTics versions. We also thank Prof. Roger Gassert for guiding and supporting the research of Dr. Kim at ETH Zurich. We also thank the Swiss foundation Access for All, especially, Mathias Deschler and Stefan Schneller for supporting the studies. A part of the work outlined in this chapter was also supported by the Swiss Commission for Technology and Innovation (Project 11393.1 PFNM-NM).

References

1. Gallois: Anoculoscope. Bulletin Le Valentin Hauy (Octobre 1883)
2. Gault, R.H.: Hearing through the sense organs of touch and vibration. Journal of the Franklin Institute 204(3), 329–358 (1927)
3. Linvill, J., Bliss, J.: A direct translation reading aid for the blind. Proceedings of the IEEE 54(1), 40–51 (02 1966)
4. Bach-y Rita, P., Collins, C., Saunders, F., White, B., SCADDEN, L.: Vision substitution by tactile image projection. Nature 221(5184), 963–964 (04 1969)

5. Kim, Y., Moncada-Torres, A., Furrer, J., Riesch, M., Gassert, R.: Quantification of long cane usage characteristics with the constant contact technique. Applied ergonomics 55, 216–225 (2016)
6. Manduchi, R., Kurniawan, S.: Mobility-related accidents experienced by people with visual impairment. AER Journal: Research and Practice in Visual Impairment and Blindness 4(2), 44–54 (2011)
7. Kay, L.: An ultrasonic sensing probe as a mobility aid for the blind. Ultrasonics 2(2), 53–59 (1964)
8. Pyun, R., Kim, Y., Wespe, P., Gassert, R.: Characterization of proximity sensors for the design of electronic travel aids. In: Association for the Advancement of Assistive Technology in Europe (AAATE) Conference. pp. 714–719 (2013)
9. Borenstein, J., Koren, Y.: High-speed obstacle avoidance for mobile robots. In: Proceedings IEEE International Symposium on Intelligent Control 1988. pp. 382–384 (1988)
10. Pyun, R., Kim, Y., Wespe, P., Gassert, R., Schneller, S.: Advanced augmented white cane with obstacle height and distance feedback. In: 2013 IEEE 13th International Conference on Rehabilitation Robotics (ICORR). pp. 1–6. IEEE (2013)
11. Cassinelli, A., Sampaio, E., Joffily, S., Lima, H., Gusmão, B.: Do blind people move more confidently with the tactile radar? Technology and Disability 26(2, 3), 161–170 (2014)
12. Gallo, S., Chapuis, D., Santos-Carreras, L., Kim, Y., Retornaz, P., Bleuler, H., Gassert, R.: Augmented white cane with multimodal haptic feedback. In: 2010 3rd IEEE RAS & EMBS International Conference on Biomedical Robotics and Biomechatronics. pp. 149–155. IEEE (2010)
13. Meers, S., Ward, K.: A vision system for providing the blind with 3D colour perception of the environment. In: Proceedings of the Asia-Pacific Workshop on Visual Information Processing. No. 1438 (2005)
14. Hesch, J.A., Mirzaei, F.M., Mariottini, G.L., Roumeliotis, S.I.: A laser-aided inertial navigation system (l-ins) for human localization in unknown indoor environments. In: 2010 IEEE International Conference on Robotics and Automation. pp. 5376–5382. IEEE (2010)
15. Hicks, S.L., Wilson, I., Muhammed, L., Worsfold, J., Downes, S.M., Kennard, C.: A depth-based head-mounted visual display to aid navigation in partially sighted individuals. PloS one 8(7), e67695 (2013)
16. Obermoser, S., Klammer, D., Sigmund, G., Sianov, A., Kim, Y.: A pin display delivering distance information in electronic travel aids. In: IEEE International Conference on Biomedical Robotics and Biomechatronics (Biorob). pp. 236–241. IEEE (2018)
17. Wei, X., Phung, S.L., Bouzerdoum, A.: Pedestrian sensing using time-of-flight range camera. In: CVPR 2011 WORKSHOPS. pp. 43–48. IEEE (2011)
18. Gassert, R., Kim, Y., Oggier, T., Riesch, M., Deschler, M., Prott, C., Schneller, S., Hayward, V.: White cane with integrated electronic travel aid using 3D TOF sensor (Mar 2012), https://patents.google.com/patent/US8922759B2/en, uS 8922759B2
19. Farcy, R., Bellik, Y.: Locomotion assistance for the blind. In: Keates, S., Langdon, P., Clarkson, P.J., Robinson, P. (eds.) Universal Access and Assistive Technology. pp. 277–284. Springer London, London (2002)
20. Ward, J., Meijer, P.: Visual experiences in the blind induced by an auditory sensory substitution device. Consciousness and cognition 19(1), 492–500 (2010)
21. Kajimoto, H., Kanno, Y., Tachi, S.: Forehead electro-tactile display for vision substitution. In: EuroHaptics (2006)
22. Wang, Y., Kuchenbecker, K.J.: Halo: Haptic alerts for low-hanging obstacles in white cane navigation. In: 2012 IEEE Haptics Symposium (HAPTICS). pp. 527–532. IEEE (2012)
23. Spiers, A.J., Dollar, A.M.: Design and evaluation of shape-changing haptic interfaces for pedestrian navigation assistance. IEEE Transactions on Haptics 10(1), 17–28 (Jan 2017)
24. Spiers, A.J., van der Linden, J., Wiseman, S., Oshodi, M.: Testing a shape-changing haptic navigation device with vision-impaired and sighted audiences in an immersive theater setting. IEEE Transactions on Human-Machine Systems 48(6), 614–625 (Dec 2018)

25. Spiers, A., Harwin, W.: http://www.personal.reading.ac.uk/~shshawin/haptic_torch/
26. Sarakoglou, I., Garcia-Hernandez, N., Tsagarakis, N.G., Caldwell, D.G.: Integration of a tactile display in teleoperation of a soft robotic finger using model based tactile feedback. In: 2012 IEEE/RSJ International Conference on Intelligent Robots and Systems. pp. 46–51. IEEE (2012)
27. Kim, S.C., Kim, C.H., Yang, T.H., Yang, G.H., Kang, S.C., Kwon, D.S.: Salt: Small and lightweight tactile display using ultrasonic actuators. In: RO-MAN 2008-The 17th IEEE International Symposium on Robot and Human Interactive Communication. pp. 430–435. IEEE (2008)
28. Kim, Y., Harders, M., Gassert, R.: Identification of vibrotactile patterns encoding obstacle distance information. IEEE Transactions on Haptics 8(3), 298–305 (July 2015)
29. Moustakas, K., Nikolakis, G., Kostopoulos, K., Tzovaras, D., Strintzis, M.G.: Haptic rendering of visual data for the visually impaired. IEEE MultiMedia 14(1), 62–72 (Jan 2007)
30. Sjöström, C., Jönsson, B.: The Phantasticon: To use the sense of touch to control a computer and the world around you. In: 4th European Conference for the Advancement of Assistive Technology. vol. 3, pp. 273–277 (1997)
31. Fritz, J.P., Barner, K.E.: Design of a haptic data visualization system for people with visual impairments. IEEE Transactions on Rehabilitation Engineering 7(3), 372–384 (Sep 1999)
32. Raza, M.U., Israr, A., Zhao, S., MacDonald, B.: A tactile palette to translate graphics for the visually impaired. Draft – Disney Research & National Braille Press (2017)
33. Memeo, M., de Jesus Oliveira, V.A., Nedel, L.P., Maciel, A., Brayda, L.G.: Tactile treasure map: Integrating allocentric and egocentric information for tactile guidance. In: AsiaHaptics. pp. 369–374 (2016)
34. Memeo, M., Brayda, L.G.: How geometrical descriptors help to build cognitive maps of solid geometry with a 3DOF tactile mouse. In: EuroHaptics. pp. 75–85 (2016)
35. Sinclair, M., Pahud, M., Benko, H.: TouchMover 2.0 – 3D touchscreen with force feedback and haptic texture. In: 2014 IEEE Haptics Symposium (HAPTICS). pp. 1–6 (Feb 2014)
36. Winfield, L., Glassmire, J., Colgate, J.E., Peshkin, M.: T-PaD: Tactile pattern display through variable friction reduction. In: WorldHaptics Conference (WHC'07). pp. 421–426 (March 2007)
37. Mullenbach, J., Shultz, C.D., Piper, A.M., Peshkin, M.A., Colgate, J.E.: TPad Fire: Surface haptic tablet. In: Haptic and Audio Interaction Design. pp. 1–3 (2013)
38. Salbu, E.: Compressible squeeze films and squeeze bearings. Journal of Mechanical Design 86(2), 355–364 (1964)
39. Wiesendanger, M.: Squeeze Film Air Bearings Using Piezoelectric Bending Elements. Ph.D. thesis, Ecole Polytechnique Federale de Lausanne (01 2001)
40. Leonard, R.M., et al.: Statistics on vision impairment: A resource manual. Lighthouse International (2001)
41. Leuenberger, K., Gassert, R.: Low-power sensor module for long-term activity monitoring. In: 2011 annual international conference of the IEEE engineering in medicine and biology society. pp. 2237–2241. IEEE (2011)
42. Assistive products for blind and vision-impaired persons – Tactile walking surface indicators. Standard, International Organization for Standardization (Jan 2019), https://www.iso.org/standard/76106.html, iSO 23599:2019
43. Schenkman, B.N.: Identification of ground materials with the aid of tapping sounds and vibrations of long canes for the blind. Ergonomics 29(8), 985–998 (1986)
44. Cutter, J.: Independent Movement and Travel in Blind Children: A Promotion Model. Critical Concerns in Blindness, Information Age Publishing, Incorporated (2007), https://books.google.at/books?id=-f0nDwAAQBAJ
45. Chhong Shing, L.: Haptic exploration of images for the visually impaired (2017), Bachelor thesis, University of Innsbruck
46. Enneking, F.: Development and evaluation of SurfTics, a haptic 3DoF shape display with tactile texture feedback. Master's thesis, Management Center Innsbruck (2017)

47. Di Gregorio, R.: The 3RRS Wrist: A new, simple and non-overconstrained spherical parallel manipulator. Journal of Mechanical Design 126, 850–855 (09 2004)
48. Hausberger, T., Terzer, M., Enneking, F., Jonas, Z., Kim, Y.: SurfTics – kinesthetic and tactile feedback on a touchscreen device. In: 2017 IEEE World Haptics Conference (WHC). pp. 472–477 (June 2017)
49. Post, L., Zompa, I., Chapman, C.: Perception of vibrotactile stimuli during motor activity in human subjects. Experimental brain research 100(1), 107–120 (1994)

Chapter 3
Maps as Ability Amplifiers: Using Graphical Tactile Displays to Enhance Spatial Skills in People Who Are Visually Impaired

Fabrizio Leo, Elena Cocchi, Elisabetta Ferrari, and Luca Brayda

Abstract This chapter reviews several findings that provide strong evidence for the effectiveness of pin-array tactile displays in enhancing spatial skills in people who are visually impaired in educational and rehabilitative contexts. Two main scenarios of use of these displays will be described: geometry and spatial memory training, and orientation and mobility training. The advantages of using pin-array tactile displays over standard rehabilitative methods in terms of increased flexibility and versatility will be discussed.

1 Introduction

According to recent estimates, 1.3 billion people have some kind of visual impairment worldwide. 36 million are blind and 216 million have moderate to severe visual impairment [1]. Although it has been suggested that the majority of visual impairment is avoidable [2], other factors such as the ageing of the population might considerably increase the number of blind and moderately or severely vision impaired up to 115 and 588 million people, respectively, by 2050 [1].

This growing population has to tackle two main issues. One of these is the fruition of all the graphical content sighted people are accustomed to. For instance, students who are blind cannot access visual information through standard textbooks or see the drawings of a teacher on a white board. Similarly, adults who are blind cannot see graphs in a newspaper or the trendline of their bank account. The second criticism is related to navigation in indoor and outdoor environments. Sighted

F. Leo (✉) · E. Ferrari · L. Brayda
Robotics, Brain and Cognitive Sciences Department, Fondazione Istituto Italiano di Tecnologia, Genoa, Italy
e-mail: fabrizio.leo@iit.it; luca.brayda@iit.it

E. Cocchi
Istituto David Chiossone, Genoa, Italy
e-mail: cocchi@chiossone.it

© Springer Nature Switzerland AG 2020
T. McDaniel, S. Panchanathan (eds.), *Haptic Interfaces for Accessibility, Health, and Enhanced Quality of Life*, https://doi.org/10.1007/978-3-030-34230-2_3

persons are very much helped by prior graphical information of unknown spaces. They can easily access Google® maps, birds-eye maps in shopping centers and museums. Unfortunately, individuals who are blind cannot use this information with the same ease. On the one hand, it has been generally proposed that persons who are blind can effectively navigate using their spared senses (e.g. Ref. [3]). This seems true especially when the environment is familiar or the missing visual information is completely replaced by tactile or auditory cues, at least for the most relevant landmark information. This situation, unfortunately, rarely occurs [4, 5]. On the other hand, there are evidences showing that people who are blind, particularly those who are congenitally blind, might have spatial cognition difficulties (see for a review, Ref. [6]). For instance, they tend to develop egocentric frames of reference [7, 8], which is a less flexible modality to orient themselves than the allocentric one, particularly when unpredictable obstacles make necessary a change in the path [9–11].

The combination of the issues described above can unfortunately lead to psychological and social problems such as isolation, loneliness, anxiety, and depression [12, 13], and seems to increase unemployment [14] as well as the mortality rate compared to the sighted population [15].

This imposes the urgency to find technological and rehabilitative solutions to tackle the needs of persons who are visually impaired.

In this chapter, we present an overview of traditional rehabilitative solutions for persons who are blind, focusing on the use of the sense of touch as a sensory replacement for vision. Then, we will present findings showing how new technological innovations, such as those represented by graphical pin-array tactile displays, can be effectively used to deliver graphical information to people who are visually impaired of various age and spatial skills.

The first part of the chapter is dedicated to spatial cognition. We show how a graphical tactile display can be used to implement *serious games* to teach geometry and to train spatial working memory skills in young persons who are visually impaired. The ability to correctly recognize geometric shapes, and, more generally, the ability to memorize spatial representations, are not only important prerequisites to keep pace with sighted fellows at school, but play a crucial role in the development of spatial orientation [16, 17]. Prior findings have also shown that spatial working memory, needed for learning geometry, is a necessary component of human navigation (e.g. Ref. [18]). Brain imaging studies have shown overlapping activations in spatial working memory and in navigation tasks in areas such as lateral parietal regions and right hippocampus (e.g. Refs. [19, 20]).

While the first part deals with subjects and practices comfortably developed at a desk, the second part is, instead, dedicated to motion. We will show how a graphical tactile display can be used as a tool in orientation and mobility. We will derive what kind of tactile cues might be more useful when exploring a map precedes the actual navigation in the real environment for a target reaching task. Then, we will provide evidences on how pin-array displays can help people who are visually impaired to form spatial cognitive maps and reduce navigation errors in real indoor and outdoor environments. Finally, we will discuss some possible applications of these findings in rehabilitation and real-life scenarios.

2 How Graphical Tactile Display Can Improve Geometry and Spatial Memory Abilities in Persons Who Are Visually Impaired

Mathematicians have proposed that geometry is the main theory of space [21]. Certainly, geometrical shapes are parts of our lives as they appear almost everywhere. They are extensively used in science, art, marketing and visual social conventions (e.g. road signs). Learning geometry is considered a difficult challenge for many students and it may be the biggest challenge at school for pupils who are visually impaired, mainly because their known lack of understanding of many spatial concepts [22]. Whether such lack of understanding is due to lack of vision or, more intriguingly, to lack of accessible instruments substituting vision, is still a matter of debate.

In effect, in some European schools (e.g. in Norway), students who are blind can get exemptions for geometry classes because teachers consider teaching the subject matter to this population nearly impossible [23]. Geometry is indeed heavily reliant on visual content, in turn triggering spatial imagination, which students who are blind cannot access. Similarly, spatial concepts such as "four walls that meet a ceiling" cannot easily be experimented with using the spared senses [22]. On the one hand, the difficulties students who are blind have with geometry are due to the fact that the school system is tailored for sighted persons. Youngsters who are blind may present specificities related to their lack of vision and subsequent intensive training in haptics [24, 25]. For instance, they do not show the prototype effect, that is, the facilitation sighted persons have when recognizing geometrical shapes close to a prototype, e.g. a canonical triangle which stands on its base [24] although this might be due to the fact that a haptic prototype might be different from a visual prototype [26]. Furthermore, persons who are blind might have issues with other skills involved in spatial learning such as haptic orientation discrimination [27], spatial imagination [28] and mental rotation [29].

In addition, research has investigated another domain which is related to spatial content processing, that is, spatial memory, particularly spatial working memory. Conflicting evidences have been found concerning the ability of those who are visually impaired compared to those who are sighted. Vecchi et al. [30] showed no influence of vision loss when the task requires passive storage in working memory, that is, when information is stored and recalled as presented without mental modification. Other evidences showed instead that the absence of vision might affect performance in certain complex tasks requiring transformation and integration of spatial information [31], memorization of a large amount of data [32], or multiple patterns of information, such as recalling the location of tactile targets in two different configurations [33].

Despite the difficulties, development of spatial skills in youngsters who are blind can be favored by providing teachers with effective support tools [34–36]. Most of the current standard tools to deliver graphical information to persons who are visually impaired take advantage of the sense of touch. Swell or embossed paper, thermoformed surfaces or textured materials are often used to create two-dimensional tactile pictures [37]. While visual images represent the information as variations in color or contrast, a tactile image uses the elevation of the image, compared to the background, as a cue.

Standard Approaches in Tactile Graphics

Textured material – This method is the simplest and cheapest as it does not require any special equipment. A raised image is created by attaching different kinds of materials such as clothes, strings, cardboards, etc., to a substratum [38]. In this way, different levels of elevation and textures can be used as identifiers of object properties. As drawbacks, the process to produce a single image is time-consuming and it is very difficult to replicate a single image in multiple copies.

Swell paper – The most common type of tactile image, also known as microcapsule or Minolta paper. Swell paper contains microcapsules which are heat sensitive elements that, when passed under a fuser, swell creating a raised image. Creating an image using this technique is quite fast, but images tend to degrade with repeated use, and the elevation of the shapes cannot be controlled, as it is quite uniform at about 0.5 mm.

Thermoform – Another largely used type of tactile image. Thermoform paper (although actually made of plastic) is created through a heat and vacuum process producing a plastic copy from a master of various materials. The plastic paper on which copies are produced, named *Brailon*®, is available in many sizes and in three thicknesses. Thermoform paper is very robust but, unfortunately, the production process is time-consuming and rather expensive.

Braille embossing – Tactile images are produced using the printing technology of modified Braille embossers, which use hammers to punch out individual dots of Braille cells. The production process is quick but the elevation of the dots is rather small, continuous lines cannot be drawn, and, as in the case of microcapsule, the paper is not resistant over time.

Ink-jet – This recent technology uses special UV-cured ink which is printed in a multilayer fashion using a special ink-jet printer. The advantages are the possibility of printing using different substrates (plastic, swell paper, paper, Braillon, aluminium) at different elevations, and the high resolution that allows printing of fine details.

Other Limitations of the Standard Approaches

Although the standard methods of creating raised images are still largely used nowadays, they unfortunately present several limitations. First of all, they allow neither the capability to present information dynamically nor to easily adapt it to the needs of individual users. Secondly, the presentation of raised line patterns requires, most of the times, the presence of a person assisting the person who is visually impaired. Lastly, producing tactile images using most of the techniques described above is expensive and time consuming.

Technological Innovations to Create and Display Tactile Images
Recently, some effort has been done to overcome the issues related to the traditional ways of producing tactile graphics. Technological tools have appeared over time with the goal of presenting graphical information to those who are visually impaired. Some of them used tactile displays [39–47], haptic interaction technologies [48–50] or haptic vibrational feedback coupled with mobile devices [51, 52].

A particularly promising technology seems to be the tactile display formed by pin-arrays. A pin-array display is composed of an array of individually actuated pins (also called *dots* or *taxels*) that can be raised above a flat substrate under computer or smartphone control [53]. One of the advantages of this technology is that the pins apply both normal and tangential force to the fingertip as the finger actively moves across the surface of the display, preserving both the force and geometry cues of real surfaces [54]. Furthermore, pin-array displays also allow a global, holistic perception as the whole hand can be placed on the image [54].

Most studies using pin-array displays have investigated the ability of individuals who are visually impaired to understand shapes [43, 45, 55–59]. Nevertheless, little work has been done to evaluate the effectiveness of pin-array displays as learning tools in geometry and spatial memory training.

In our lab, we have designed *serious games* specifically for youngsters who are visually impaired. In general, when the training of some skill is needed in children and adolescents, using a game-like approach has been shown to be an effective method [60, 61].

We then field-tested these *serious games* in a training paradigm for youngsters who are visually impaired: one of these was based on the recognition and localization of simple geometrical shapes in noise; the second involved the memorization of single or double spatial configurations. Both training paradigms used a Hyperbraille® as graphical tactile display. It is a multi-line Braille display provided by Metec AG, composed by an array of 30 by 32 pins. The pins are spaced at intervals of 2.5 mm, and they can raise above the surface by 0.7 mm. The device can be connected via USB cable to a standard laptop.

2.1   Geometrical Shapes Recognition and Localization in Noise

Leo et al. [62] conducted a four-sessions geometry training using a pin-array display. Visually impaired children and adolescents had to recognize and locate a target shape (triangle, rectangle or square) presented with three distractors, that is, shapes which differed from a canonical geometrical figure because of some lowered pins (see Fig. 3.1). Importantly, the level of difficulty was adapted to the ability of each participant at the beginning of each training session. In particular, the degree of similarity between canonical and distractors shapes could differ. This choice avoided using a task too simple (which could trigger boredom as a first strong

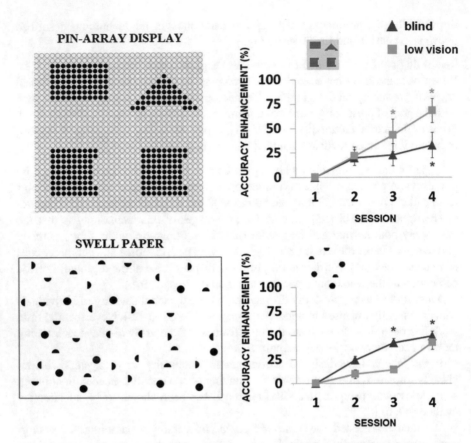

Fig. 3.1 Upper left column: 'Shapes recognition' test using the pin-array display with a rectangle (top left) as target and three other shapes as distractors. The level of difficulty, coded as the degree of similarity between target and distractors, varied depending on the ability of the child. Lower left column: 'Tactile symbol recognition and enumeration in noise' test using swell paper. Upper right column: Normalized accuracy enhancement across sessions in the 'shapes recognition' test using the pin-array display. Lower right column: Normalized accuracy enhancement across sessions in the 'tactile symbol recognition and enumeration in noise' test using swell paper. Asterisks indicate a significantly larger accuracy enhancement relative to the baseline. *$p < .05$

experimental bias) or too difficult (which could trigger frustration as a second bias), and therefore, allowed observation of learning effects across sessions. The training with the pin-array display was compared to training using swell paper in which the youngsters had to count as quickly and accurately as possible the number of circles surrounded by half-circle distractors in a limited time (Fig. 3.1).

Results showed that youngsters who are visually impaired, even the congenitally blind, could understand the shapes presented with the pin-array display. More importantly, they improved during the training (Fig. 3.1). The improvement using the pin-array was similar to the improvement they obtained with the training using swell paper. This suggests that, although with a lower resolution, a graphical display can be effective in implementing simple geometry training. Furthermore, using the pin-array display allowed automation of the procedure to tailor the level of challenge of the training to the needs of a single user, finally increasing the autonomy of those with visual impairments.

2.2 Improving Spatial Working Memory Using Graphical Tactile Displays

Leo et al. [63] followed a similar methodology, but this time three groups of participants (blind, very low vision, and sighted youngsters) performed a spatial working memory task. They had to remember the locations of targets presented inside one (single-matrix task) or two matrices (double-matrix task, see Fig. 3.2).

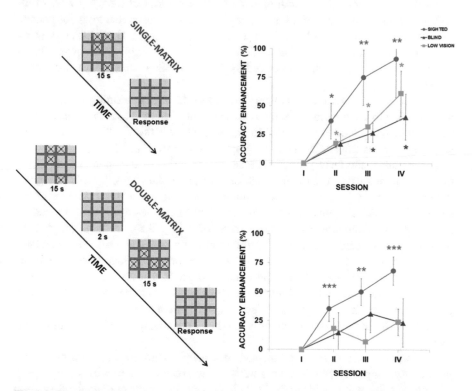

Fig. 3.2 Upper left column: Timeline of the single-matrix task with a 4 by 4 grid and four targets. As soon as the targets disappeared, the participants had to touch the empty supposed locations of the grid where they thought the targets were. Lower left column: Timeline of the double-matrix task. In this case, the participants had to touch the empty supposed locations of the targets of the first matrix followed by the supposed locations of the second matrix. Upper right column: Normalized accuracy enhancement across sessions in the single-matrix task. Lower right column: Normalized accuracy enhancement across sessions in the double-matrix task. Asterisks indicate a significantly larger accuracy enhancement relative to the baseline. *$p < .05$; **$p < .01$; ***$p < .001$

Fig. 3.3 Example of
proprioceptive exploration

A similar task, although not administered with technological tools, on adults only
and with no training involved, has already been shown as challenging for persons
who are visually impaired [33]. As in Leo et al. [62], the level of difficulty at the
beginning of the training was tailored on the ability of a single participant. Unlike
Leo et al. [62], the difficulty was coded by the size of the matrix and the number of
targets.

Results showed that the youngsters improved during the training in the single-
matrix task. On the contrary, when the task was more complex, as in the case of the
two matrices, the sighted continued to improve whereas the visually impaired only
showed a non-significant tendency for improvement (see Fig. 3.2).

Interestingly, the three groups differed also in terms of the spontaneous exploration strategies used to scan the matrices. In particular, the blindfolded sighted
youngsters, at the end of the exploration, tended to place their fingers at the
supposed target locations and to 'freeze' their hand position, as if to code a
proprioceptive simultaneous memory trace (see Fig. 3.3). In fact, the authors
named it *proprioceptive strategy*. Furthermore, this strategy positively correlated
with the performance in the task. Importantly, the proprioceptive strategy was not
exclusively used by the sighted. The blind youngsters that used it improved more
than their counterparts who did not use it. However, it has to be noted that the blind
participants used this strategy much less frequently than their sighted peers.

Results also showed that using only one hand negatively correlated with performance.

To summarize, this study confirmed that people who are blind do not easily
process two separate spatial configurations, as shown in a previous study [33].
Importantly, this was not due to difficulty in understanding the graphical information
displayed with the pin-array matrix. When the presented matrix was only one, the
performance of the blind participants was indeed not significantly different than the
performance of their sighted peers.

However, the fact that certain exploration strategies were associated with better
performance, and that the blind participants were able to use them, suggests the

exciting possibility that a rehabilitative intervention aimed at increasing their use might boost the spatial working memory of those who are visually impaired.

3   How Graphical Tactile Displays Can Improve Orientation and Mobility Abilities in People Who Are Visually Impaired

In the previous section, we have seen that geometrical ability and spatial working memory can be trained in the visually impaired using pin-array displays. This is a relevant result because these skills constitute the basis for more complex abilities such as navigation. Navigating an unknown environment requires understanding spatial relationships between one's self, close and far objects [64]. The mental representation including the aforementioned information is known as a *spatial cognitive map* [65, 66].

Despite the traditional viewpoint that the development of a spatial cognitive map requires a visual input [67, 68], the observation that the congenitally blind can represent spatial information, as well as understand the spatial relations between objects in a traveled route [5, 69, 70], supports the idea that other senses, such as touch and audition, allow development of an effective cognitive map. This is also well in line with Bryant's [71] suggestion of a single spatial representation system that provides a common representational format for the different input modalities of vision, audition, touch, and language. Furthermore, a recent study showed functional evidences of an equivalence between spatial images from touch and vision [72, 73].

However, other findings showed that congenitally blind persons might have poorer spatial abilities than their sighted counterparts (see for a review, Ref. [6]). For instance, their distance estimation in the locomotor space might be impaired [74]; they are less accurate in inferring spatial relationships in large-scale environments [75]; and they prefer *route-like* than *survey* representations [8, 76]. A route representation describes the space from an egocentric perspective by using an intrinsic frame of reference (e.g. "in front of you", "to your right", etc.). A survey description represents the space in an allocentric fashion because it is based on a *bird's-eye* view, independent from the observer [77]. Therefore, a survey representation takes advantage of coordinates such as 'West', 'East', 'South', 'North'. Both strategies permit efficient navigation, but a survey representation of the space usually allows greater flexibility than a route representation (e.g. Ref. [78]) and is therefore associated with better spatial performance not only in the sighted (e.g. Ref. [9]) but also in the visually impaired (e.g. Ref. [79]). For instance, a survey representation makes it easier for a person to take shortcuts or to deviate from the usual path if an unattended obstacle appears along the way [9, 11]. Findings highlighted some possible explanations about the preference for route knowledge in the visually impaired. For instance, Loomis and co-authors have shown that, when forced to acquire the information serially, the recognition of objects through sight was equivalent to recognition through touch [80]. In other words, people who are blind prefer route representations because they mostly acquire serial information while navigating thanks to the sense of touch, audition and kinesthetic information (e.g. Ref. [81]). Nevertheless, it is important to note that individuals who are blind can also develop and use survey representations (e.g. Ref. [82]).

The spatial information about a place can be acquired with the direct exploration of the environment, and also, beforehand, through the consultation of a map.

Visual and Tactile Maps

Maps are diagrammatic representations of an area showing physical features in an easily accessible format, usually as a visual bi-dimensional image. Maps are scaled, smaller representations of the real environment ranging from the map of a room to larger maps of outdoor spaces, continents, or even the observable universe.

As such, maps usually contain only the essential spatial information that is displayed to represent faithfully the spatial relationships of features within the area. The selection of the relevant spatial information is especially important in the case of tactile maps. Tactile maps are raised images that are designed to convey spatial information to people who are visually impaired. Haptic exploration of tactile maps is indeed serial whereas visual exploration is more global and holistic. Serial exploration takes more time and imposes a larger cognitive and memory load. Hence, tactile maps need to represent only essential information.

As in visual maps, the physical features in tactile maps are represented by point, line, area symbols, and frequently, texture. The meaning of these symbols depends on the scale of the map. For instance, a point symbol could represent a bus stop on a small scale map or even a city on a large scale map. A line symbol can represent for instance a road, a wall, or a border, and an area symbol could represent squares, parks, or even countries.

Types of Tactile Maps

In all types of tactile maps, the information is represented as elevation or texture. However, they change for material and process to create a raised image. The standard production techniques are the same we have described in a previous section of this chapter, that is, swell paper, Thermoform, Braille embossing and ink-jet. The limitations of these techniques are the same we have seen for the representation of general graphics.

Briefly, the design process of the map is complex and time consuming. Secondly, the information is static and cannot change dynamically over time. Lastly, the material (printers, special paper, ink, etc.) is quite expensive. Electronic maps represent a tentative to overcome these issues, and today, maps are almost entirely digital, with layers of information being combined and customized [83]. Clearly, the development of digital, three-dimensional maps is enlarging the gap between maps accessed with vision and those accessed by touch. There is no reason why tactile maps should not be digital as well, but the way to render digital tactile maps with technological means that stimulate the sense of touch is a debated issue [84].

Use of Tactile Maps

Several guidelines for the design and preparation of tactile graphics and maps have been developed during the years, some of them being specific for the type of production process (e.g. [85]). Studies also investigated to what extent traditional swell paper or Thermoform tactile maps can be useful to users who are visually impaired. Two pioneering works [86, 87] showed that visually impaired children and adults successfully used a tactile map to navigate an urban area and a college campus, respectively. More recently, a series of studies from Ungar have shown that those who are visually impaired can understand and use tactile maps [88], which can be proficiently employed to estimate directions [89], distances [90] and self-location [91]. Tactile and audio-tactile maps have also been proven to be more effective than the mere direct experience with the environment [92] or other methods such as verbal description [93, 94], although the visually impaired may need some training to use them effectively because of their difficulty in modeling scale factors, i.e. translating distances from a map to the environment it represents [90].

Less is known about the usability of tactile maps presented with graphical pin-array tactile displays. These kinds of display are indeed quite recent as the first pioneering devices were built in the eighties of the last century [54, 95]. Secondly, there are several competing technologies of pin-matrix displays which can be based on electromechanical actuators, voice-coil motors, shape memory alloys, or on the reshaping of polymer materials [96]. These different types of pin-array displays may lead to different resolutions and also to varying tactile sensations for the end users. Finally, high resolution pin-matrix displays (e.g. Braille displays) are still quite expensive to-date. The cost for a single line 16 character unit is over $2000, and over $55,000 for a pin-matrix display composed of 7200 dots [95].

Most of the previous studies investigating the effectiveness of pin-array displays either did not test participants who are blind [97] or only recorded qualitative self-reports from blind computer users [98–101].

Zeng, Miao and Weber [102] assessed the ability of visually impaired in understanding maps of real environments presented with a pin-array matrix and their subsequent ability to create pre-journey routes based on those maps. They found that their accuracy and speediness in learning the pin-array map was similar to that one achieved with swell paper maps as well as their ability to make pre-journey routes. However, they did not investigate the actual navigation performance that followed the learning of the map.

Here, we present three studies designed to assess the effectiveness of tactile maps presented beforehand with graphical pin-array tactile displays followed by the actual navigation in the real environment without technologic aids. The first study was performed with a Hyperbraille. The second and third studies were performed with BlindPAD.

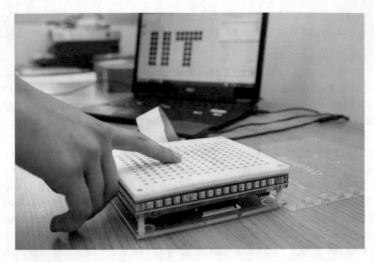

Fig. 3.4 The BlindPAD

3.1   BlindPAD

BlindPAD is a multi-line tactile display composed of 192 pins on an 8 mm pitch [59, 103, 104]. Each pin can be independently raised or lowered in about 20 ms and the whole display can be refreshed in under 2 s, under the control of the 12 by 16 array of electromagnetic actuators. The device can be connected via wireless to a standard laptop (see Fig. 3.4).

3.2   On Identifying Useful Cues in Programmable Tactile Maps

In this study, we aimed at identifying which cues are more useful in tactile maps presented with pin-array displays. We chose reaching tasks because they are goal-directed and require planning an exploration strategy of the real environment. Therefore, they are ideal to test whether or not a map was understood and which cues might have led to poor understanding.

Two groups of visually impaired explored tactile maps of a room, displayed on a pin-array display. The maps included three different cues (see Fig. 3.5): a triangle representing the starting position of the participant inside the room, a square representing the target position to be reached and a semicircle, drawn around the start position, representing the distance between starting position and target. The semicircle represented a standard distance known to the participant: either the arm length or the typical length of a white cane. Both target distance and starting position varied across trials. After the exploration of the maps, the participants performed the actual navigation task inside the room (see Fig. 3.5).

Tactile map exploration   **Navigation task**

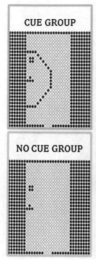

Fig. 3.5 Task procedure: in a first phase the participants explored the tactile map and in the second phase they performed the navigation task

Results showed that the visually impaired who received the cue about distance were more accurate in localizing the target position in the navigation task compared to the group who did not receive this cue. In particular, their error was significantly lower in the horizontal dimension whereas the performance of the two groups was similar along the vertical dimension (see Fig. 3.6). This means that the two groups were equally accurate in estimating the distance of the targets whereas, relative to the angle of direction, the participants who received the additional cue were more precise.

The usefulness of the cues and the overall map readability were also confirmed by the results of the questionnaires about the map exploration. In general, the participants were able to estimate the distance between starting position and target based on the map information using a five-point scale (see Fig. 3.7). The participants who received the additional cue were also able to discriminate between levels of distance between targets placed at arm and targets placed at cane distance. On the contrary, the control group was not able to discriminate between these two conditions.

All participants scored the usefulness of the cues that were shown on the maps. Their main impression was that the cue indicating the start position was the most useful as it allowed them to choose in which direction they had to move. The participants that explored the tactile maps with the additional cue also reported that this cue might be most useful in an outside environment. Therefore, tactile cues different than real objects do make tactile maps more understandable, much like meridians and parallels clarify the mutual position of countries. However, we chose

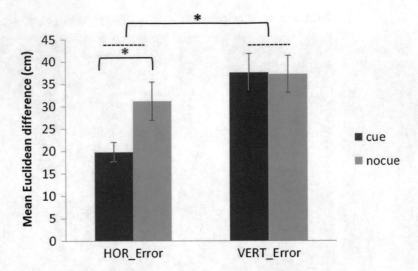

Fig. 3.6 Mean and Standard Error of Euclidean difference calculated between coordinates of position reached by participants and coordinates of target positions. Black asterisk indicates a significantly smaller horizontal than vertical error. Red asterisk indicates a significantly smaller horizontal error in the cue group. *$p < .05$

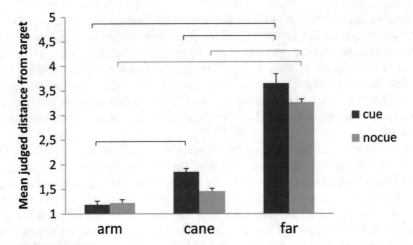

Fig. 3.7 Mean and Standard Error scores attributed by participants after the exploration of tactile map concerning the distance between start position and target position on a five-point scale (1 = very close, 5 = very far). Square brackets indicate significant comparisons, $p < .001$

cues for which people who are blind already have internal models, and we have shown that they are effective. Which standard tactile cues about scale factors might be more useful is still a matter of debate.

3.3 Tactile Feedback with a Pin-Array Display Improves Self-Location Precision in People Who Are Blind

Brayda et al. [103] followed the same approach of assessing how individuals who are blind can use pin arrays to learn offline a tactile map and, then, do a navigation task without technological aids.

In particular, they presented a map on a pin-array display showing an indoor environment and a position to be reached by two groups of visually impaired participants. Both groups repeated three times the exploration of the map, its externalization (i.e. the description of a map that has been previously memorized) using two methods (LEGO® buildings and verbal questionnaire) and a navigation task. While one group received tactile feedback showing the position reached in the previous trial and the true target position, the other group could only re-explore the same map (Fig. 3.8) with only the true target position. In this way, the authors could investigate whether using an updated map is associated with a lower navigation error when locating a target in a real environment, rather than the simple reviewing of the same unchanged map (which represents the limited capabilities of static tactile maps used by people who are blind so far).

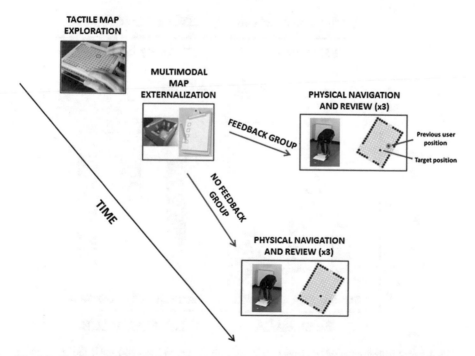

Fig. 3.8 Timeline of the experiment: both groups explored the map of the room on the pin-array display, then externalized the map with LEGO and with a questionnaire. Finally, they navigated in the real room three times. While the feedback group received feedback about the previous reached position (in green) and the target position, the no feedback group only reviewed the initial map with only the target position

Results showed that those who are visually impaired were able to form an accurate spatial cognitive map as measured by the map externalization methods. Furthermore, the group that received the additional information about the location reached in the previous trial was more accurate and fast in the following trials compared to the group that could only review the original map (see Fig. 3.9). These

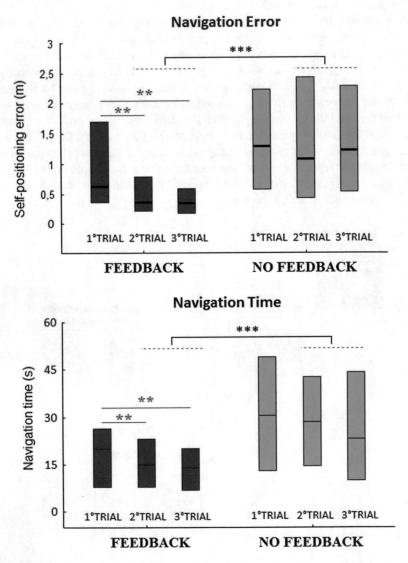

Fig. 3.9 Navigation error (m) for each trial in the feedback group (red bars) and in the no-feedback group (green bars). Red asterisks indicate a significant reduction of the self-positioning error in the second and third trial in the feedback group. Grey asterisks indicate a significantly larger self-positioning error in the second and third trial in the no-feedback compared to the feedback group. **$p < .01$; ***$p < .001$

results show that visually impaired individuals can take advantage of allocentric information, as suggested by previous studies [82, 105]. In fact, in half of the trials, the map was shown with the entrance door on top which forced participants to rely only on an allocentric representation. Despite of this added difficulty, blind participants were able to improve their navigation performance.

Results also indicated that people who are blind are able to use technological solutions for navigation *before* navigating and not only during navigation (or after navigation as a mere rehearsal mean). In other words, they can form effective spatial cognitive maps of simple areas, they can orient themselves in these places and they can take advantage of allocentric information to modify their path whenever needed. This finding evidences also the possibility of using pin-array displays not only as navigation but also as learning and rehabilitative aids. The absence of a reliable navigation improvement in the no feedback group suggests that the classical 'learning by repetition' strategy might be less effective than a strategy based on the role of updated feedback provided with a technological tool.

Interestingly, accuracy and completion time of the map externalization methods, in particular, the LEGO reconstruction, correlated with the navigation performance suggesting that navigation ability can be predicted by the degree of accuracy in forming a mental representation of the environment.

3.4   People Who Are Blind Improve Their Sense of Orientation and Mobility in Large Outdoor Spaces by Means of a Tactile Pin-Array Matrix

Since Brayda et al. [103] tested the effect of an updated sensory feedback only in a small room, Leo et al. [106] investigated whether this effect generalized to large outdoor spaces (Fig. 3.10).

Results showed again that only the visually impaired who received the updated sensory feedback significantly improved across trials (Fig. 3.11). This finding reinforces the idea that tactile feedback about self-location can act as a novel reference point of the map from which the visually impaired can make further spatial inferences. Primarily, people who are blind can estimate the self-location error, which is the Euclidean distance between the supposed previous location of the target and the true target location. This allows him/her to improve his/her performance in the following attempts. Furthermore, the results showed that the effect of the feedback is evident not only in small, well-controlled, indoor settings but also in more noisy and ecological outdoor environments. This is an important finding because it indicates that updated tactile feedback could represent an effective tool for navigation, both in rehabilitation and real-life situations, improving the spatial abilities of individuals who are blind and favoring their autonomy.

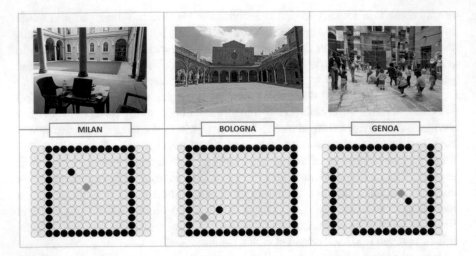

Fig. 3.10 Upper panels: the three locations where the experiments were performed. Lower panels: example of maps used in those locations. The black taxel inside the spaces represents the virtual target location; the green taxel represents the tactile feedback provided to the feedback group. (From Leo et al. [106])

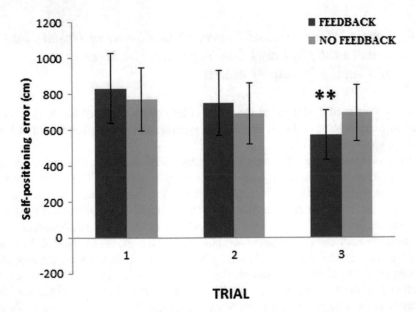

Fig. 3.11 Self-positioning error (in cm) during the three trials. Asterisks indicate a reduction in the self-positioning error in trial 3 compared to trial 1 in the feedback group. **$p < .01$. (From Leo et al. [106])

4 Conclusions

The findings described in this chapter show that pin-array tactile displays are promising educational and aid tools. They can be used regardless of age and degree of visual disability. They seem effective in showing both generic spatial content and maps for navigation. The challenge is now to test these displays using more complex tasks and develop applications aimed at solving practical issues in accessing graphical information.

Acknowledgments We thank Silvia Dini, Tania Violin, Claudia Vigini, Marisa Lococciolo, Anna Gettani, Elisabetta Capris and Claudio Cassinelli of the Istituto Chiossone for their organizational, logistic and testing support in Genoa and for their organizational activity in Bologna and Milan. We thank Damiano Storelli, Giovanni Cellucci and Fernando Torrente of the Istituto Cavazza in Bologna for the logistic and testing support in Bologna. We thank Laura Muro and Franco Lisi of the Istituto dei Ciechi in Milan for the logistic and testing support in Milan. For the Istituto Italiano di Tecnologia, we thank Alberto Inuggi for his technical support, Angelo Raspagliesi and Sara Nataletti for their help in testing in Genoa, Giulio Sandini for his strategy and financial support. Finally, we wish to thank all the participants and their families.

References

1. Bourne, R.R.A., Flaxman, S.R., Braithwaite, T., Cicinelli, M.V., Das, A, Jonas, J.B., et al.: Magnitude, temporal trends, and projections of the global prevalence of blindness and distance and near vision impairment: a systematic review and meta-analysis. Lancet Glob Health 5(9), e888–e897 (2017)
2. Pizzarello, L., Abiose, A., Ffytche, T., Duerksen, R., Thulasiraj, R., Taylor, H., Faal, H., Rao, G., Kocur, I., Resnikoff, S.: VISION 2020: The Right to Sight: A Global Initiative to Eliminate Avoidable Blindness. Archives of Ophthalmology 122(4), 615–620 (2004) doi:https://doi.org/10.1001/archopht.122.4.615.
3. Klatzky, R.L., Golledge, R.G., Loomis, J.M., Cicinelli, J.G., Pellegrino, J.W.: Performance of Blind and Sighted Persons on Spatial Tasks. J Vis Impair Blind 89(1), 70–82 (1995)
4. Chandler, E., Worsfold, J.: Understanding the requirements of geographical data for blind and partially sighted people to make journeys more independently. Appl Ergon 44(6), 919–928 (2013)
5. Loomis, J.M., Klatzky, R.L., Golledge, R.G., Cicinelli, J.G., Pellegrino, J.W., Phyllis, F.A.: Non visual navigation by blind and sighted: Assessment of path integration ability. J Exp Psychol Gen 122(1), 73–91 (1993)
6. Pasqualotto, A., Proulx, M.J.: The role of visual experience for the neural basis of spatial cognition. Neurosci Biobehav 36(4), 1179–1187 (2012)
7. Iachini, T., Ruggiero, G., Ruotolo, F.: Does blindness affect egocentric and allocentric frames of reference in small and large scale spaces? Behav Brain Res 273, 73–81 (2014)
8. Noordzij, M.L., Zuidhoek, S., Postma, A.: The influence of visual experience on the ability to form spatial mental models based on route and survey descriptions. Cognition 100(2), 321–342 (2006)
9. Lawton, C.A.: Gender differences in way-finding strategies: Relationship to spatial ability and spatial anxiety. Sex Roles 30(11–12), 765–779 (1994)
10. Millar, S.: Self-Referent and Movement Cues in Coding Spatial Location by Blind and Sighted Children. Perception 10(3), 255–264 (1981)

11. Millar, S.: Models of Sensory Deprivation: The Nature/Nurture Dichotomy and Spatial Representation in the Blind. Int J Behav Dev 11(1), 69–87 (1988)
12. Ribeiro, M.V.M.R., Hasten-Reiter Júnior, H.N., Ribeiro, E.A.N., Jucá, M.J., Barbosa, F.T., Sousa-Rodrigues, C.F. de.: Association between visual impairment and depression in the elderly: a systematic review. Arq Bras Oftalmol 78(3), 197–201 (2015)
13. Senra, H., Barbosa, F., Ferreira, P., Vieira, C.R., Perrin, P.B., Rogers, H., et al.: Psychologic Adjustment to Irreversible Vision Loss in Adults. Ophthalmology 122(4), 851–861 (2015)
14. Bell, E.C., Mino, N.M.: Employment Outcomes for Blind and Visually Impaired Adults. J Blind Innov Res 5(2) (2015)
15. Crewe, J.M., Morlet, N., Morgan, W.H., Spilsbury, K., Mukhtar, A.S., Clark, A., et al.: Mortality and hospital morbidity of working-age blind. Br J Ophthalmol 97(12), 1579–1585 (2013)
16. Lee, S.A., Spelke, E.S.: Children's use of geometry for reorientation. Dev Sci 11(5), 743–749 (2008)
17. Shusterman, A., Ah Lee, S., Spelke, E.S.: Young children's spontaneous use of geometry in maps. Dev Sci 11(2), F1–F7 (2008)
18. Garden, S., Cornoldi, C., Logie, R.H.: Visuo-spatial working memory in navigation. Appl Cogn Psychol 16(1), 35–50 (2002)
19. Salmon, E., Van der Linden, M., Collette, F., Delfiore, G., Maquet, P., Degueldre, C., et al.: Regional brain activity during working memory tasks. Brain 119(5), 1617–1625 (1996)
20. Grön, G., Wunderlich, A.P., Spitzer, M., Tomczak, R., Riepe, M.W.: Brain activation during human navigation: gender-different neural networks as substrate of performance. Nat Neurosci 3(4), 404–408 (2000)
21. Berthelot, R., Salin, M.H.: The role of pupils' spatial knowledge in the elementary teaching of geometry. In: Mammana, C., Villani, V., Eds. Perspectives on the Teaching of Geometry for The 21st Century, pp. 71–77. Kluwer, Dordrecht, The Netherlands (1998)
22. Dick, T., Kubiak, E.: Issues and Aids for Teaching Mathematics to the Blind. In: Vol. 90, The Mathematics Teacher, pp. 344–349. National Council of Teachers of Mathematics (1997)
23. Klingenberg, O.G.: Geometry: Educational implications for children with visual impairment. Philos Math Educ J. 20(15) (2007)
24. Theurel, A., Frileux, S., Hatwell, Y., Gentaz, E.: The Haptic Recognition of Geometrical Shapes in Congenitally Blind and Blindfolded Adolescents: Is There a Haptic Prototype Effect? PLoS One 7(6), e40251 (2012)
25. Heller, M.A., Gentaz, E.: Psychology of Touch and Blindness. New York and London: Psychology Press, UK (2014)
26. Woods, A.T., Moore, A., Newell, F.N.: Canonical Views in Haptic Object Perception. Perception 37(12), 1867–1878 (2008)
27. Postma, A., Zuidhoek, S., Noordzij, M.L., Kappers, A.M.L: Haptic orientation perception benefits from visual experience: Evidence from early-blind, late-blind, and sighted people. Percept Psychophys 70(7), 1197–1206 (2008)
28. Noordzij, M.L., Zuidhoek, S., Postma, A.: The influence of visual experience on visual and spatial imagery. Perception 36(1), 101–112 (2007)
29. Ungar, S., Blades, M., Spencer, C.: Mental rotation of a tactile layout by young visually impaired children. Perception 24(8), 891–900 (1995)
30. Vecchi, T., Monticellai, M.L., Cornoldi, C.: Visuo-spatial working memory: Structures and variables affecting a capacity measure. Neuropsychologia 33(11), 1549–1564 (1995)
31. Cornoldi, C., Vecchi, T.: Mental imagery in blind people: The role of passive and active visuo-spatial processes. Touch, Represent Blind 143–181 (2000)
32. Vecchi, T.: Visuo-spatial Imagery in Congenitally Totally Blind People. Memory 6(1), 91–102 (1998)
33. Vecchi, T., Tinti, C., Cornoldi, C.: Spatial memory and integration processes in congenital blindness. Neuroreport 15(18), 2787–2790 (2004)
34. Dell, A.G., Newton, D., Petroff, J.: Assistive Technology in the Classroom: Enhancing the School Experiences of Students with Disabilities. Pearson (2016)

35. Barraga, N.C., Erin, J.N.: Visual impairments and learning. Proed, Austin, Texas (2001)
36. Klingenberg, O.G.: Conceptual Understanding of Shape and Space by Braille-Reading Norwegian Students in Elementary School. J Vis Impair Blind 106(8), 453–465 (2012)
37. Theurel, A., Witt, A., Claudet, P., Hatwell, Y., Gentaz, E.: Tactile picture recognition by early blind children: the effect of illustration technique. J Exp Psychol Appl 19(3), 233–240 (2013)
38. Edman, P.K.: Tactile graphics. American Foundation for the Blind, New York, New York, USA (1992)
39. Landau, S., Gourgey, K.: A new approach to interactive audio/tactile computing: The Talking Tactile Tablet. In: Proceedings of Technology and Persons with Disabilities Conference. California State University, Northridge (2003)
40. Borges, A., Jansen, L.R.: Blind people and the computer: an interaction that explores drawing potentials. In: Proceedings of SEMENGE'99-Seminario de Engenharia. Universidade Federal Fluminense (1999)
41. Vidal-Verdú, F., Hafez, M.: Graphical tactile displays for visually-impaired people. IEEE Trans Neural Syst Rehabil Eng. 15(1), 119–130 (2007)
42. Xu, C., Israr, A., Poupyrev, I., Bau, O., Harrison, C.: Tactile display for the visually impaired using TeslaTouch. In: Proceedings CHI EA '11, pp. 317–322 (2011)
43. Motto Ros, P., Dante, V., Mesin, L., Petetti, E., Del Giudice, P., Pasero, E.: A new dynamic tactile display for reconfigurable braille: implementation and tests. Front Neuroeng 7, 6 (2014)
44. Chouvardas, V.G., Miliou, A.N., Hatalis, M.K., et al.: Tactile displays: a short overview and recent developments. In: Proc 5th Int Conf Technol Autom, pp. 246–251 (2005)
45. Pietrzak, T., Crossan, A., Brewster, S.A., Martin, B., Pecci, I.: Exploring Geometric Shapes with Touch. In: Gross, T., Gulliksen, J., Kotzé, P., Oestreicher, L., Palanque, P., Prates, R.O., and Winckler, M. (eds.) Human-Computer Interaction – INTERACT 2009, pp. 145–148. Springer Berlin Heidelberg (2009)
46. Rastogi, R., Pawluk, D.T.V.: Dynamic tactile diagram simplification on refreshable displays. Assist Technol 25(1), 31–38 (2013)
47. Rastogi, R., Pawluk, T.V.D., Ketchum, J.: Intuitive Tactile Zooming for Graphics Accessed by Individuals Who are Blind and Visually Impaired. IEEE Trans Neural Syst Rehabil Eng 21(4), 655–663 (2013)
48. McLaughlin, M.L., Sukhatme, G., Hespanha, J.: Touch in Virtual Environments: Haptics and the Design of Interactive Systems. Prentice-Hall (2001)
49. Rassmus-Gröhn, K.: User-Centered Design of Non-Visual Audio-Haptics. PhD Thesis, Lund University (2008)
50. Brayda, L., Campus, C., Memeo, M., Lucagrossi, L.: The importance of visual experience, gender and emotion in the assessment of an assistive tactile mouse. IEEE Trans Haptics 8(3), 279–286 (2015)
51. Buzzi, M.C., Buzzi, M., Leporini, B., Senette, C.: Playing with Geometry: A Multimodal Android App for Blind Children. In: Proc 11th Biannu Conf Ital SIGCHI Chapter, pp. 134–137 (2015)
52. Giudice, N.A., Palani, H.P., Brenner, E., Kramer, K.M.: Learning non-visual graphical information using a touch-based vibro-audio interface. In: Proceedings of the 14th international ACM SIGACCESS conference on Computers and accessibility – ASSETS '12, p. 103. ACM Press, New York, New York, USA (2012)
53. Wall, S.A., Brewster, S.: Sensory substitution using tactile pin arrays: Human factors, technology and applications. Signal Processing 86(12), 3674–3695 (2019)
54. O'Modhrain, S., Giudice, N.A., Gardner, J.A., Legge, G.E.: Designing media for visually-impaired users of refreshable touch displays: Possibilities and pitfalls. IEEE Trans Haptics 8(3), 248–257 (2015).
55. Leo, F., Baccelliere, C., Waszkielewicz, A., Cocchi, E., Brayda, L.: Tactile Symbol Discrimination on a Small Pin-array Display. In: Proc 2018 Work Multimed Access Hum Comput Interface – MAHCI'18, pp. 9–15 (2018)

56. Fritz, J., Way, T., Barner, K.: Haptic representation of scientific data for visually impaired or blind persons. In: Proc CSUN Conf Technol Disabil (1996)
57. Shimojo, M., Shinohara, M., Fukui, Y.: Shape identification performance and pin-matrix density in a 3 dimensional tactile display. In: Proceedings of IEEE 1997 Annual International Symposium on Virtual Reality, pp. 180–187. IEEE Comput. Soc. Press (1997)
58. Brewster, S., Brown, L.M.: Tactons: structured tactile messages for non-visual information display. In: 5th Australas User Interface Conf 2004 (v28), pp. 15–23 (2004)
59. Besse, N., Rosset, S., Zarate, J.J., Ferrari, E., Brayda, L., Shea, H.: Understanding graphics on a scalable latching assistive haptic display using a shape memory polymer membrane. IEEE Trans Haptics 11(1), 30–38 (2017)
60. Ritterfeld, U., Cody, M., Vorderer, P.: Serious games: Mechanisms and effects. Serious Games: Mechanisms and Effects, pp. 1–530. Routledge, New York/London: Routledge (2009)
61. Breuer, J., Bente, G.: Why so serious? On the relation of serious games and learning. J Comput Game Cult 4(1), 7–24 (2010)
62. Leo, F., Cocchi, E., Brayda, L.: The Effect of Programmable Tactile Displays on Spatial Learning Skills in Children and Adolescents of Different Visual Disability. IEEE Trans Neural Syst Rehabil Eng 25(7), 861–872 (2017)
63. Leo, F., Tinti, C., Chiesa, S., Cavaglià, R., Schmidt, S., Cocchi, E., et al.: Improving spatial working memory in blind and sighted youngsters using programmable tactile displays. SAGE Open Med 6, 205031211882002 (2018). http://journals.sagepub.com/doi/10.1177/2050312118820028
64. Siegel, A.W., White, S.H.: The Development of Spatial Representations of Large-Scale Environments. Adv Child Dev Behav 10, 9–55 (1975)
65. Tolman, E.C.: Cognitive maps in rats and men. Psychol Rev 55(4), 189–208 (1948)
66. Strelow, E.R.: What is needed for a theory of mobility: Direct perceptions and cognitive maps—lessons from the blind. Psychol Rev 92(2), 226–248 (1985)
67. Gibson, J.J.: Visually controlled locomotion and visual orientation in animals. Br J Psychol 49(3), 182–194 (1958)
68. Gibson, J.J.: The Ecological Approach to Visual Perception. Psychology Press (2014)
69. Thinus-Blanc, C., Gaunet, F.: Representation of space in blind persons: vision as a spatial sense? Psychol Bull. 121(1), 20–42 (1997)
70. Rieser, J.J., Guth, D.A., Hill, E.W.: Sensitivity to Perspective Structure While Walking without Vision. Perception 15(2), 173–188 (1986)
71. Bryant, D.J.: Representing Space in Language and Perception. Mind Lang 12(3–4), 239–264 (2007)
72. Giudice, N.A., Betty, M.R., Loomis, J.M.: Functional equivalence of spatial images from touch and vision: evidence from spatial updating in blind and sighted individuals. J Exp Psychol Learn Mem Cogn 37(3), 621–634 (2011)
73. Campus, C., Brayda, L., De Carli, F., Chellali, R., Famà, F., Bruzzo, C., et al.: Tactile exploration of virtual objects for blind and sighted people: the role of beta 1 EEG band in sensory substitution and supramodal mental mapping. J Neurophysiol 107(10), 2713–2729 (2012)
74. Gaunet, F., Thinus-Blanc, C.: Early-Blind Subjects' Spatial Abilities in the Locomotor Space: Exploratory Strategies and Reaction-to-Change Performance. Perception 25(8), 967–981 (1996)
75. Herman, J.F., Chatman, S.P., Roth, S.F.: Cognitive mapping in blind people: Acquisition of spatial relationships in a large-scale environment. J Vis Impair Blind 77(4), 161–166 (1983)
76. Millar, S.: Understanding and Representing Space: Theory and Evidence From Studies with Blind and Sighted Children. Oxford University Press, UK (1994)
77. Taylor, H.A., Tversky, B.: Spatial mental models derived from survey and route descriptions. J Mem Lang 31(2), 261–292 (1992)
78. O'Keefe, J., Nadel, L.: The hippocampus as a cognitive map. Oxford: Clarendon Press (1978)

79. Schmidt, S., Tinti, C., Fantino, M., Mammarella, I.C., Cornoldi, C.: Spatial representations in blind people: The role of strategies and mobility skills. Acta Psychol (Amst) 142(1), 43–50 (2013)
80. Loomis, J.M., Klatzky, R.L., Lederman, S.J.: Similarity of Tactual and Visual Picture Recognition with Limited Field of View. Perception 20(2), 167–77 (1991)
81. Postma, A., Zuidhoek, S., Noordzij, M.L., Kappers, A.M.L.: Differences between early blind, late blind and blindfolded sighted people in haptic spatial configuration learning and resulting memory traces. In: Vol. 36, Perception, pp. 1253–1265 (2007)
82. Tinti, C., Adenzato, M., Tamietto, M., Cornoldi, C.: Visual experience is not necessary for efficient survey spatial cognition: evidence from blindness. Q J Exp Psychol. 59(7), 1306–1328 (2006)
83. Haklay, M., Weber, P.: OpenStreetMap: User-Generated Street Maps. IEEE Pervasive Comput 7(4), 12–18 (2008)
84. Levesque, V., Petit, G., Dufresne, A., Hayward, V.: Adaptive level of detail in dynamic, refreshable tactile graphics. In: 2012 IEEE Haptics Symposium (HAPTICS), pp. 1–5. IEEE (2012)
85. Picard, D.: VISUO-TACTILE ATLAS. Organisation mondiale de la propriété intellectuelle, France (2012)
86. Leonard, J.A., Newman, R.C.: Spatial Orientation in the Blind. Nature 215(5108), 1413–1414 (1967)
87. Bentzen, B.L.: Production and Testing of an Orientation and Travel Map for Visually Handicapped Persons. New Outlook Blind (1972). https://eric.ed.gov/?id=EJ066516
88. Ungar, S., Blades, M., Spencer, C.: The role of tactile maps in mobility training. Br J Vis Impair 11(2), 59–61 (1993)
89. Ungar, S., Blades, M., Spencer, C., Morsley, K.: Can visually impaired children use tactile maps to estimate directions? J Vis Impair Blind 88(3), 221–233 (1994)
90. Ungar, S., Blades, M., Spencer, C.: Teaching visually impaired children to make distance judgments from a tactile map. J Vis Impair Blind 91(2):163–174 (1997)
91. Ungar, S., Blades, M., Spencer, C.: The construction of cognitive maps by children with visual impairment. Constr Cogn maps 32(1951), 247–273 (1996)
92. Cattaneo, Z., Vecchi, T.: Blind vision : the neuroscience of visual impairment. MIT Press (2011)
93. Brambring, M., Weber, C.: Tactual, verbal and exploratory information for geographic orientation of the blind. Z Exp Angew Psychol 28(1):23–37 (1981)
94. Papadopoulos, K., Koustriava, E., Koukourikos, P.: Orientation and mobility aids for individuals with blindness: Verbal description vs. audio-tactile map. Assist Technol 30(4), 191–200 (2018)
95. Hyperbraille: The project. http://hyperbraille.de/project/ (2019)
96. Zeng, L., Weber, G.: Exploration of Location-Aware You-Are-Here Maps on a Pin-Matrix Display. IEEE Trans Human-Machine Syst 46(1), 88–100 (2016)
97. Velazquez, R., Fontaine, E., Pissaloux, E.: Coding the Environment in Tactile Maps for Real-Time Guidance of the Visually Impaired. In: 2006 IEEE International Symposium on MicroNanoMechanical and Human Science, pp. 1–6. IEEE (2006)
98. Ivanchev, M., Zinke, F., Lucke, U.: Pre-journey Visualization of Travel Routes for the Blind on Refreshable Interactive Tactile Displays. In: Miesenberger, K., Fels, D., Archambault, D., Peňáz, P., and Zagler, W. (eds.) Computers Helping People with Special Needs, pp. 81–88. Springer International Publishing (2014)
99. Zeng, L., Weber, G.: Audio-Haptic Browser for a Geographical Information System. In: Miesenberger, K., Klaus, J., Zagler, W., and Karshmer, A. (eds.) Computers Helping People with Special Needs, pp. 466–473. Springer Berlin Heidelberg (2010)
100. Zeng, L., Weber, G.: ATMap: Annotated Tactile Maps for the Visually Impaired. In: Esposito, A., Esposito, A.M., Vinciarelli, A., Hoffmann, R., and Müller, V.C. (eds.) Cognitive Behavioural Systems, pp. 290–298. Springer Berlin Heidelberg (2012)

101. Schmitz, B., Ertl, T.: Interactively Displaying Maps on a Tactile Graphics Display. In: SKALID 2012-Spatial Knowledge Acquisition with Limited Information Displays: 13 (2012)
102. Zeng, L., Miao, M., Weber, G.: Interactive Audio-haptic Map Explorer on a Tactile Display. Interact Comput 27(4), 413–429 (2014)
103. Brayda, L., Leo, F., Baccelliere, C., Ferrari, E., Vigini, C.: Updated Tactile Feedback with a Pin Array Matrix Helps Blind People to Reduce Self-Location Errors. Micromachines 9(7), 351 (2018)
104. Zarate, J.J., Shea, H.: Using Pot-Magnets to Enable Stable and Scalable Electromagnetic Tactile Displays. IEEE Trans Haptics 10(1), 106–112 (2017)
105. Morrongiello, B.A., Timney, B., Humphrey, G.K., Anderson, S., Skory, C.: Spatial Knowledge in Blind and Sighted Children. J Exp Child Psychol 59(2), 211–233 (1995)
106. Leo, F., Violin, T., Inuggi, A., Raspagliesi, A., Capris, E., Cocchi, E., et al.: Blind Persons Get Improved Sense of Orientation and Mobility in Large Outdoor Spaces by Means of a Tactile Pin-Array Matrix. In: CHI'19 Workshop on Hacking Blind Navigation. Glasgow, Scotland (2019)

Chapter 4
Haptics for Sensory Substitution

Bijan Fakhri and Sethuraman Panchanathan

Abstract Using one sensory modality to compensate for a modality that is unavailable is called Sensory Substitution and it is useful and often necessary for conveying some types of information effectively to people with disabilities. Using haptics to substitute for other modalities provides unique benefits as the tactile modality is incredibly flexible and underutilized. This chapter explores general purpose, media-focused, and interactive applications of Haptic Sensory Substitution as well as the future of Haptic Sensory Substitution and its implications for assistive technology.

1 Introduction to Sensory Substitution

Sensory Substitution (SS) is the process of delivering a signal from the domain of one sensory modality to an alternative sensory modality, for example circumventing the auditory modality with the haptic modality. The purpose of SS is often to circumvent an impaired modality via an alternative one so that a person can experience stimuli from the impaired modality. Formally, we will refer to the modality being replaced as the *source modality* and the modality that the signal is being delivered to the *target modality*. In other words, Sensory Substitution is a method by which people who are blind can see by hearing, or people who are deaf can hear via touch. This revolutionary idea, that people can learn to experience sensations grounded in one modality via another, was pioneered by the late Dr. Bach-y-Rita in the 1960s. The notion that the signals eminating from the receptors of one modality can be interpreted in the brain as stimuli from another domain was novel and spurred the development of methods and systems in the last four

B. Fakhri (✉)
Center for Cognitive Ubiquitous Computing, Arizona State University, Tempe, AZ, USA
e-mail: bfakhri@asu.edu

S. Panchanathan
Arizona State University, Tempe, AZ, USA
e-mail: panch@asu.edu

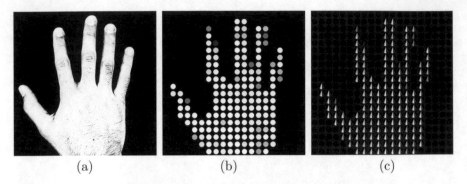

(a)   (b)   (c)

Fig. 4.1 (**a**) Original image of hand (**b**) Original image converted into electrical activations based on the brightness of that portion of the image (**c**) activations are converted into solenoid positions to stimulate the skin: solenoids stimulate the skin in proportion to their activation

decades harnessing this phenomenon to treat disability, enhance education, and enrich people's lives.

The objective of a Sensory Substitution Device (SSD) is to transform a signal from the source domain into a form that can be perceived by the target modality. In the famous case of Dr. Bach-y-Rita's TVSS system, the source domain was visual and the target domain haptics [4]. Dr. Bach-y-Rita showed that people who are blind could, with training, learn to interpret visual stimuli projected on their back as tactile stimuli using the Tactile Vision Sensory Substitution device (TVSS). The device consisted of a dental chair, retro-fitted with 400 solenoid actuators that would press upon the user's back when seated. The solenoids were controlled by a camera system that converted images to electrical signals: a bright portion of an image would result in solenoids pressing against the back of the user in the corresponding location. This is illustrated in Fig. 4.1a–c.

After training, the TVSS allowed user who are blind to recognize household objects without touching them. These results had massive implications for the field of neuroscience and that of assistive technology – technology to improve the lives of people with disabilities, demonstrating that through clever uses of technology sensory impairments can be circumvented. Researchers later went on to develop Sensory Substitution Devices (SSDs) to substitute vision with hearing [46], vestibular with tactile [65], and hearing for tactile [59] with impressive initial results. While fantastic medical advances have resulted in sensory prostheses such as the Cochlear Implant (CI) [48] and retinal prosthesis [9]; SSDs provide a great alternative for circumventing the loss of a sensory modality as surgical procedures are often prohibitively expensive and always invasive. This chapter will focus on SSDs with a target modality of haptics, or Haptic SSDs.

2   Advantages and Limitations of Haptics for Sensory Substitution

The skin is the largest organ on the body, making touch one of the most versatile modalities to design SSDs for. Designers have a wide range of options with respect to where to place devices: some SSDs have even been designed for the tongue. Because of the plentiful real-estate, haptic SSDs can be designed to impart minimal obstruction to other crucial functions of the senses. For example, haptic actuators can be placed on places such as the back, upper arms, or waistline, locations that are not often used in day-to-day activity.

Touch also happens to be underutilized as a communication medium for technology. Designing touch-based SSDs has the added benefit of likely not interfering with other communication mediums to cause sensory overload. A person who is blind for example is unlikely to accept obstructing their hearing with a vision-to-auditory SSD, but is more likely to if the target modality is one that is not already being highly utilized, such as touch. Using haptics not only avoids interfering with a modality already in use, but may allow for a higher effective cognitive bandwidth due to the multi-channel nature of adding haptics.

Current models of the human memory and attention system portray the different modalities as semi-independent channels to one's attention. The Baddeley multi-channel model for example allocates different sensory inputs unique and semi-independent subsystems of working memory and independent processing systems for such each modality [74]. Consequently, sensory signals of different modalities can more effectively make use of the human cognitive bandwidth than the same information presented to a single modality; this phenomena is called the "modality effect". Sensory overload occurs when the attention system is overwhelmed and because touch is often underutilized in daily tasks, taking advantage of it can augment attentional bandwidth while successfully averting sensory overload. For this reason, the haptic modality has received substantial interest in military (high cognitive load) settings and haptic-vestibular SSDs have been developed for pilots flying in low-visibility settings [78].

The sense of touch though does exhibit inherent limitations. One such limitation is the limited information capacity of haptics. It is estimated that the visual system has a capacity of 4.3 Mbits/s [28] and the auditory system 8 Kbits/s [26]. In comparison, the haptic modality is estimated to have a mere 600–925 bits/s of capacity [58]. This implies an upper bound on the amount of information an SSD can convey through the sense of touch, and consequently an upper bound on the fidelity of information one can access from a higher-bandwidth modality through haptics. It was not so minuscule though that users could not use it to substitute vision and perform basic vision tasks [4].

While touch allows for a wide variety of locations, sophisticated interaction often requires multiple tactile actuators to convey complex information. The skin imposes a minimum spacing requirement between factors to maintain discernability and this spacing is a function of the location of the body the tactors are placed as well as the kind of stimulation (pressure, vibration, temperature, etc) that will be applied. For example, on the human back the minimum discriminable separation of vibration stimuli is about 11 mm [18]. Consequently, actuators must often be adequately spaced out on the body, taking up more space than a device relying on a more concentrated modality like vision or hearing. The design of the device and the signal processing is crucial for effective use and adoption as an SSD and haptic SSDs can largely be categorized into 3 categories: general purpose (Sect. 3), media readers (Sect. 4), and interactive devices (Sect. 5) which are explored in the following sections.

3   General Purpose Sensory Substitution

General purpose Sensory Substitution is intended to circumvent a source modality via the target modality outright, making it a complete substitute for the source modality. This is in contrast to application specific SS, where a device or technique transforms a signal from the source modality into the domain of the target modality in such a way that is tailored to the application. Oftentimes there exists a tradeoff between efficiency and generality: the more general an SS method, the more training is required, while more application-specific methods are often learned more quickly.

The first and likely most famous implementation of general purpose vision sensory substitution occurred in 1969, when Dr. Paul Bach-y-Rita and his team developed the Tactile Vision Sensory Substitution (TVSS), demonstrating that with a somewhat long training period (up to 150 h), users of the device could recognize common objects as well as motion, gradients, and shadows at a distance [4, 83]. While impressive, the work had a long way to go towards vision-to-tactile SS that could truly replace vision, let alone be a practical solution for daily activity. The device was incredibly bulky, having been constructed from a dental chair, hundreds of solenoids, camera equipment and electrical amplifiers. The device's resolution was also too low to discern fine detail and long training times were required for proficiency. Furthermore, the system lacked color detection and sported a field of view was that was narrow and fixed. All of these problems made it impractical for real-world use such as navigation, reading, etc.

Some of these issues were addressed in later devices. For example the "Rabbit Display", developed by the MIT Media Lab, made use of a tactile illusion called "saltation" in order to increase the effective resolution of a low resolution tactile display [75]. Saltation (also known as the "cutaneous rabbit") is an illusory sensation of touch felt in between the location of where the stimuli was actually applied to the skin [23] and can be achieved by timing the stimuli in a specific manner. The authors emphasized that the display would be useful in conveying direction information to users such as pilots (such as a vestibular SSD) or to help people with navigation. Because of the low resolution nature of the display (3×3), it can be inferred that it can be made relatively small and lightweight, making it a viable option for mobile applications and more socially acceptable. Saltation can even evoke sensation away from the body [54], and may be used in the future to "extend" displays off of the body. The low-resolution nature of the display though limits the detail that can be conveyed, even if saltation is employed to increase perceived resolution. Generalizing this technique to a larger, finer display is not trivial though, as inducing saltation requires haptic stimuli to be presented to the skin in specific timings and patterns, limiting the representable patterns of the display and thus the informational content.

Further improving on acuity and portability, researchers in 2001 developed a Tongue Display Unit (TDU) for vision-to-electrotactile Sensory Substitution applications. The device converts images from a digital camera into electrical signals that are applied to the tongue in a similar manner to how Bach-y-Rita's TVSS converted image information into tactile stimulation (illustrated in Fig. 4.1c). While unconventional, the tongue was chosen as the site for the TDU for both its sensitivity to electric current and density of receptors, making it better suited for discerning fine details than a user's back. Researchers showed that users of the Tongue Display Unit were able to achieve a visual acuity of 20/860 on a standard "Tumbling E" visual acuity test and 20/430 after 9 h of training, generalizing much better than the original TVSS [57, 68].

The same group went on to use the TDU as a rehabilitation device for people with vestibular conditions affecting their balance, renaming the TDU the BrainPort. Researchers used the BrainPort to convey balance information to people who had lost their sense of balance, substituting it with electrotactile stimulation and saw marked improvements in balance, some users being able to stop using the device entirely while retaining their newfound balance [65]. This group demonstrated that haptic SSDs can not only be used as sensory substitutes but also as rehabilitation devices.

With all of these advances since the original TVSS, there are some limitations that remain untouched such as color distinction and stereo vision. Vision to tactile SSDs also are still cumbersome for practical daily use as the state-of-the-art implementations (BrainPort) require the display to be in the mouth limiting social interactions and possibly exacerbating stigma towards users. There has been more success in general vision substitution with the auditory system as the target modality. Blind users have even been able to navigate with SSDs such as the "vOICe", which stands for "oh I see" [46, 82] and experience color with EyeMusic [1], a system that abstracts images into tones and sounds of instruments hence the name. The discrepancy in performance between vision-to-haptic and vision-to-auditory SSDs is likely to do the information capacity discrepancy being smaller between vision and auditory versus vision and haptics. Auditory-to-Haptic Sensory Substitution though enjoys a similar advantage over vision-to-haptic.

Some of the earliest attempts at general auditory-to-haptic SS were made by the Audiological Engineering Corp in the 1980s. The group designed what are now known as the Tactaid devices. The devices partition audio data into a varying number of bands based on the model of Tactaid device; for example, Tactaid VII uses seven bands and conveys activity in the bands to the user via seven unique vibrotactile actuators. Researchers evaluated the devices with users who had hearing impairments and found that users were able to discern syllables and showed "enhanced monosyllabic word recognition" but users did not report significant subjective improvements in recognition of speech [30]. A more recent and more successful method for auditory-to-haptic SS was developed in 2014 by researchers at Rice University. Instead of using just seven tactors, researchers developed a suit called the VEST containing 26 eccentric rotating mass (ERM) motors, developing patterns involving groups of 9 vibrotactile motors in a square array that conveyed directional "sweeps". They found that the spatiotemporal sweep patterns were more distinguishable than just spatial or static patterns alone. Combining the VEST with speech processing methods (compressing and converting the speech into haptic patterns), users were able to discern speech much more clearly than ever before, distinguishing words at much higher accuracies than with the Tactaid devices [16, 59]. General purpose Sensory Substitution devices explore the limits of perception but are rarely ever widely adopted as assistive technology. Instead, application specific SSDs tend to have more success as practical aids for daily use.

4 Media

While the written word enabled mass communication, standard media formats are not accessible to the entirety of society. People with visual impairments often have difficulty accessing communication mediums due to their design being reliant on vision. Haptic SSDs for reading media are designed to convey the information in media that is visual or text-based to the sense of touch. Examples of this include devices for reading text, exploring images, and understanding maps, which are all important for and individual's education and independence.

4.1 Language and Communication

The most famous, and arguably most successful Sensory Substitution technique for reading is Braille, the tactile, two column, three row cells of raised dots were invented by Louis Braille and published in 1829. Alphanumeric characters are converted into tactile representations, where each letter or number is assigned a Braille code occupying one Braille cell. These cells can be read and written and are the standard reading and writing system for individuals who are blind in many countries. It has been shown that after extensive practice Braille users can achieve a reading rate of 90 wpm [76]. Visual reading rates are about 200 words per minute for comparison. Written text though had to be translated into Braille before it was accessible and was often bulkier than the original material. Almost a decade and a half after the invention of Braille, refreshable Braille displays emerged as a solution to the size and heft of translated works. Refreshable Braille displays typically consist of a row of refreshable Braille cells where the dots are controlled by a piezoelectric bimorph cantilever that is activated by an electric potential [72]. Modern refreshable Braille cells are 2×4 in contrast to the original 2×3 cells, with the additional 2 dots used for cursor position and other indicators, according to the American Foundation for the Blind (AFB) [73]. Modern refreshable displays sport between 18 and 84 Braille cells and can interface with computers via bluetooth, while also utilizing input controls for typing and navigating [8, 21, 69]. In response to the era of touchscreens, methods for typing in Braille have been evolved to be compatible with the flat featureless surfaces of touchscreens [42]. Many touchscreen consumer devices today allow for Braille typing using solely the display, eliminating the requirement for additional hardware.

Other less successful methods for tactile communication were developed such as Vibratese, a tactile language based on vibrations on the body that varied in amplitude and duration to communicate alphanumeric symbols. Invented by F. A. Geldard in 1957, test subjects trained in Vibratese were able to achieve a reading rate of up to 60 words per minute using the system [61]. The language never saw widespread adoption likely due to Braille having already been the standard.

Another alternative to the refreshable Braille cell called $STRESS$ emerged in 2005. The device uses vertical stacks of piezoelectric plates that deform with an electric current. The user places their fingers on top of the stack so that their fingers are perpendicular to the individual piezoelectric plates and the plates bend in response to applied electric current to create different sensations at the finger tips. Researchers saw promising preliminary results in creating "virtual Braille" with the $STRESS$ device, a 1-dimensional version of Braille [37]. Researchers then went on to explore more complicated game based use-cases with the technology [81] detailed in Sect. 5.4.

Unfortunately, while technological improvements continue to advance the defacto tactile communication method Braille, literacy is in decline. The National Federation of the Blind (NFB) in a 2009 report stated that the Braille literacy rate has dwindled to less than 10% of individuals who are blind in the United States [56]. The NFB report states that Braille education is critical to literacy and employment among individuals who are blind, and while screen-readers have facilitated computer access their existence likely inhibits Braille adaption. The NFB is calling for Braille adoption to be elevated in priority for those who teach individuals who are blind.

One of the issues with Braille though is that non-digital text must be transcribed before becoming accessible, and in response several technologies emerged to read written characters beginning in the 1960s. The Optohapt for example used photosensitive sensors to detect characters on paper (on a retrofitted typewriter). The characters were passed through the sensor at a rate of 70 characters per minute creating electrical signals that were sent to vibrating actuators located at 9 spatially dispersed bodily sites [22]. During the same period, a competing device proposed by Linvill and Bliss called the Optacon (OPtical-to-TActile CONversion), was developed. The device consisted of a capture module (a wand-like device) fitted with an 8×12 array of photosensitive cells can be placed on a page to be read with a user's dominant hand. The user then places a finger from their other hand on the

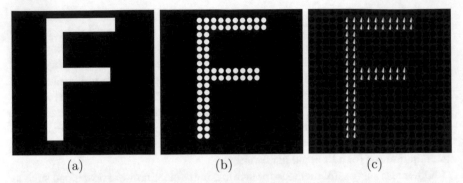

Fig. 4.2 (a) Original image representation of letter **F** (b) Letter converted into electrical activations (c) activations converted into solenoid positions to stimulate the skin

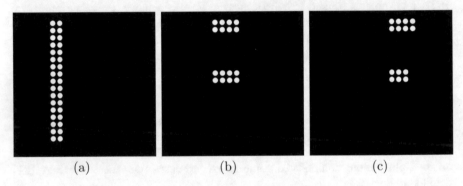

Fig. 4.3 Sliding window presentation of the letter **F**. A user would be exposed to the sliding window stimuli over about 1 s

actuator. The actuator is an array of 24 × 6 pins that the finger rests on that move up and down in response to signals from the capture module [7]. The authors claimed that a reading rate of 50 wpm could be achieved with 160 h of training.

The researchers behind the TVSS also explored different ways to display letters using the device instead of visual information, comparing static haptic patterns and dynamic ones for each letter. Letters were converted to tactile stimuli (illustrated in Fig. 4.2a–c). They found that a sliding window approach was most successful for accurate letter discrimination among participants in a user study, achieving an accuracy of 51% correct letter discrimination [40]. This conclusion (that spatiotemporal patterns are more discriminable than static ones) has been supported by later work by the developers of the VEST [59] and LRHI [19]. The sliding window approach only exposed a user to a portion of the letter at any one moment, but the whole letter would be presented over a duration of 1 s, illustrated in Fig. 4.3, imposing a maximum reading rate of 60 characters per minute (60 cpm).

The TVSS implementation of a character reader seemed like overkill (400 actuators), and with a limit of 60 cpm it did not show much promise as a media reading device. More recently lower resolution displays have been explored for communicating written characters. Researchers developed a low resolution tactor array of 9 vibration actuators placed 3×3 on the back rest of a chair. Representations of letters were "traced" over the tactors as if the letters were being dynamically "drawn" on the user's back. Patterns varried over space and time and participants were able to achieve an accuracy of 87% for letter and number recognition [85]. This was vastly higher than the accuracies achieved with the TVSS (51%) with far fewer actuators. This leads us to believe that for abstract information representation a more "coded" scheme may be more useful than attempting to reproduce the characteristics of the visual content faithfully. Although the hardware requirements are vastly reduced and accuracies improved, the dynamic patterns may still be too slow for use in real time, implying that a different coding scheme similar to Braille may be more practically useful.

Braille-like devices are still superior it seems when it comes to reading and writing using haptics and while Braille may be currently in decline, emerging technology in the space of refreshable Braille displays has appeared as recently as 2017 in the form of non-mechanical, air actuated displays in contrast to piezoelectric designs. This new technology uses fluids to make bubbles in the display as the dots, and it is being integrated with a traditional touchscreen tablet. This technology appeared in 2017 in the form of the Blitab (a play on words combining "blind" and "tablet") and is purported to have 14 rows of 23 6-dot Braille cells [50]. This 2-dimensional display paves the way for richer human-computer-interaction and possibly a reemergence of Braille literacy.

4.2 Visual Content Readers

Apart from language systems, there has been growing interest in the development of haptic devices for understanding traditionally visual information such as images, graphs, and maps. Students with visual impairments are often at a disadvantage in academic settings because the content is in an inaccessible format. Even when text is transcribed or conveyed via a media reading device, images continue to present a challenge to students and teachers. An intuitive method for representing two-dimensional information using haptics are "raised paper diagrams". These diagrams are often made from "swell paper", which expands in an oven-like device where it has been printed on creating a tactile surface [53]. An example of such a diagram is illustrated in Fig. 4.4a, b. A similar method for creating 2D tactile visualizations that allows an end-user to reconfigure a diagram is in the form of moldable wax-based rods called "Wikki Stix", shown in Fig. 4.5. Users can scan them with their fingers to feel the features of the visualization. While useful, Wikki Stix and raised paper diagrams still requires a translation from an original image for instructional purposes. It is also often difficult to incorporate sufficient information density due to the physical limitations of the media. Descriptions are often added by a teacher or caption to aid in comprehension of the visualizations, but a more elegant solution has been developed in the form of the Talking Tactile Tablet (T3). The T3 consists

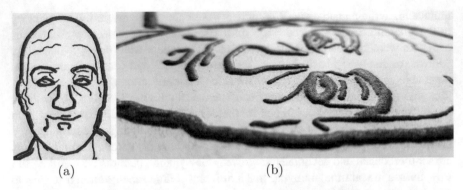

(a)   (b)

Fig. 4.4 (**a**) Raised paper diagram of a man's head on white paper (**b**) The same diagram viewed close and at an angle. Note that several different heights and thicknesses are possible on such diagrams

Fig. 4.5 Wikki Stix used for conveying visual-spatial information via haptics. They are flexible and waxy, making them easily configurable and stationary on surfaces

of a tactile diagram that can be felt overlaying a touch sensitive screen. When a user presses the tactile map they are presented with auxiliary audible information to complement the tactile map [36]. An even more fleshed out version uses a smartphone and 3D printed overlays to perform a similarly multimodal experience to the T3, is called TacTILE. Authors of the TacTILE developed a complete toolchain for the rapid development of such devices [25].

More elaborate attempts to make visual information accessible began appearing in the late 1990s. Japanese researchers Ikei et al. attempted to convey an image's textures via haptics by constructing a 5×10 pin finger display driven by piezo-electric actuators (similar to refreshable Braille displays). The pins though were not static like their Braille counterparts, but vibrated at 250 Hz at varying amplitudes to mimic tactile textures. Researchers converted close-up images of textured surfaces such as a bamboo woven basket, thatch basket, painted wall, and a rug to haptic textures by converting the images to pin intensities on their finger display. A user study revealed that using their technique sighted users were able to correctly identify the image belonging to the texture being displayed on the finger pad more than 90% of the time [27]. The high recognition accuracies and straight-forward method for converting images to tactile representations was promising, as generalizing to other domains would be relatively simple, although no study was performed with individuals who are blind and thus had no visual reference for the textures they were experiencing. Ikei's method worked for any arbitrary texture but had no sense of "space" that is required to accurately convey most visualizations.

Researchers Wall and Brewster sought to solve this problem in 2006 when they developed a graphical diagram reading system by integrating the VTPlayer mouse with a digital drawing tablet and used the stylus to interact with the graph. The VTPlayer mouse is a computer mouse that is augmented with two 4×4-pin Braille cells. The user would point on the tablet with the stylus and receive textured information of what they were pointing at with the VTPlayer on their non-dominant hand. Complementary audio feedback would also be available if the user pressed the buttons on the VTPlayer [79]. Earlier, in 2005, Wall and Brewster performed a psychophysical study comparing the TVPlayer mouse, the WingMan Force Feedback mouse, and classic raised paper for use in image understanding. They used a simple line gradient discrimination task: a line was displayed and participants were asked to discriminate the gradient of the line using the three devices. While the force feedback mouse outperformed the VTPlayer, the raised paper was superior. Interestingly, the authors surmised that this is likely due to the combination of proprioceptive and tactile cues that neither the VTPlayer or WingMan mouse provide at the same time [80], which likely led them to develop the 2006 graph reading system using a stylus as well as the tactile feedback from the VTPlayer mouse.

Emerging techniques in Computer Vision have given way to much more comprehensive automatic image understanding systems. Facebook's automatic image captioning [49] generates captions automatically from images. Google "Lookout" is a mobile assistive technology app that allows users to point their phone camera at objects they would like information about [11]. As of this writing, Google Lookout describes objects in the scene by giving audio descriptions such as "Trash can, 12-o'clock", but allows the user very little freedom to explore a visual scene in an interactive way. Microsoft's Seeing AI is slightly more sophisticated, augmented with the ability to read text, documents, people, scenes, money, and give illumination descriptions (color, brightness) [52]. All of these 1-shot methods though do not allow users to interact with images in the same way sighted individuals do, by exploring the image over time. Combining these very powerful image understanding techniques with a proprioceptive and tactile interface would likely lead to a more effective and meaningful image understanding tool.

5 Interactive Applications

Sensory Substitution devices for interactive applications are designed to function in environments that are, interactive. An interactive environment responds or changes with respect to the user's behavior. For example, a video game is interactive while a textbook is not. Thus SSDs for interactive applications must contend with the demands of interactive environments, that is latency sensitivity, sensory overload, and diverse and dynamic situations. This section will explore SSDs designed for interactive applications of mobility and travel, interactive instructional systems, social interactions, and virtual interactive environments.

5.1   Instructional Systems

Instructional SSDs are those that are intended to be used for learning; more specifically they are intended to be used for learning in dynamic environments that react to user input, in contrast to media reading SSDs that are intended to be used to convey information about static sources such as books and illustrations.

5.1.1   Mobility Learning

Sighted people can look up images of a location and quickly acquaint themselves with the flow of the environment. Unfortunately, those for who images are inaccessible do not have such a luxury and can not benefit from the vast amounts of visual data that is available online. Furthermore, familiarity with an environment is often more important for people with visual impairments than sighted individuals. To address this issue, virtual environments that model locations that are of interest and allow people with visual impairments to interact with those environments may benefit people with visual impairments by allowing them to familiarize themselves with the novel location before visiting in person. These systems are referred to as "Mobility Training" systems.

On such system developed at the University of Colorado at Colorado Springs is called MoVE: Mobiltiy Training in Haptic Virtual Environment [70]. Its purpose is to enable people who are blind to explore a model of new environments haptically. The system is iterative, a user explores the virtual model, then explores the physical location and repeats this process to fine-tune their understanding of the space, intuitively learning the relationship between the rendered world and the real world. MoVE uses SensAble Inc's PHANToM force feedback device, allowing users to interact with the virtual environment by poking around with the PHANToM (shown in Fig. 4.6), receiving force feedback when they contact objects. In a preliminary study, researchers found that user who are blind were quickly able to discriminate simple virtual objects such as spheres versus planes. While this approach is promising, the iterative nature has yet to be tested for individuals with visual impairments.

Sharkey et al. devised a more comprehensive approach using a force feedback joystick, audio feedback, and a "guiding computer agent" to create and explore virtual environments before exploring their real counterparts they were modelled after. The force feedback encoded information about texture, objects via force-fields, and structural boundaries while the audio component added descriptions of the scene as well as of the user's orientation in space to aid in navigation. They

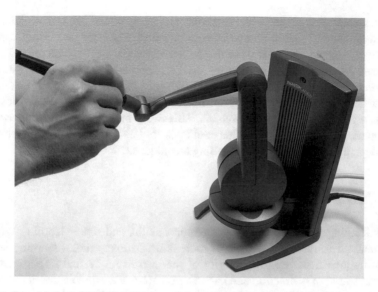

Fig. 4.6 Sensable Inc's PHANToM Desktop, a force feedback device for haptic applications

found that users were able to accurately and quickly learn to navigate in the virtual environment and when presented with the physical version quickly generalized what they had learned to the real environment [71]. Later came Omero, combining haptic and acoustic feedback with user preferences to learn the layout of new locations similar to the Skarkey system. Researchers tested the system with people with visual impairments and received positive subjective feedback; those with really low vision were not as successful as the system made extended use of visualizations on a monitor [13].

Lahav et al. developed a similar system for cognitive mapping via a multimodal approach and compared the performance of users who are blind in real-world navigation tasks versus other users who did not have access to the technology, expressing that users who had access to the technology developed more complete and accurate cognitive maps of the environment [35]. Researchers used a multisensory virtual environment (MVE) that individuals who are blind could explore before exploring a physical environment (laid out in the same way). The MVE provided haptic force feedback and audio feedback of obstacles in the environment. Researchers found that individuals who are blind and were allowed to use the MVE developed more complete and accurate cognitive maps of the environment than those who were not given access to the MVE.

A more realistic approach was designed by Tzovaras et al. in 2009: a mixed reality system for training/educating people who are blind using a virtual white cane via the CyberGrasp device. Using a virtual white cane, trainees were able to traverse a life-sized virtual replica of an environment. Researchers enhanced the experience by providing realistic haptic feedback of cane collisions with virtual objects and realistic audio feedback [77]. This method provided the most realistic approach as users employed skills to navigate the real environment almost identically to the virtual one but may not have been the most effective for generating complete cognitive maps of the environment. A direct comparison of this mixed reality real-scale method and the non-virtual reality methods above would be a welcome addition to the literature to unviel specific advantages and disadvantages of the two approaches. Furthermore, all of these Mobility Training systems require designers to model the environments beforehand, effectively reducing the pool of available environments to a small batch. This could possibly be rectified with crowdsourcing and integration with 2D to 3D modelling techniques.

5.1.2 Motor Learning

Motor learning is the development of motor skills, and motor learning tools are tools that aid in the development of such skills. In many motor learning settings demonstrations make up the majority of the instruction. Visual impairments can hinder this kind of instruction and haptic SSDs provide a valuable avenue to replace visual instruction. Motor learning systems may also provide feedback with respect to a user's movement in real-time, something that an instructor may not be able to give. Furthermore, some users may not be receptive to touch-based feedback from an instructor and may feel more comfortable with a device's feedback to correct motor movements. In the absence of an in-person instructor, or when an instructor does not have time to devote to a single student, an SSD that conveys motor skill information would be also be useful to most users.

In 2002, Yang et al. designed a suit for VR-based motor learning covering the torso with a vibrotactile display called POS.T. Wear. Employing a technique called "Just Follow Me" (JFM), the researchers used the POS.T. Wear to convey movement information of nearby objects to the wearer. The JFM metaphor consists of a "ghostly master" (illustrated in Fig. 4.7) that is overlayed onto the trainee's body in the virtual environment. The master will then guide the trainee by performing the correct movements to be learned by the trainee. Yang et al. used JFM and the POS.T. Wear to study a user's obstacle awareness in virtual worlds and later as a motor learning tool [86].

A more intuitive haptic motor learning approach called Mapping of Vibrations to Movement (MOVeMENT) was developed by McDaniel et al. Instead of the ghostly master avatar approach in JFM, MOVeMENT seeks to map haptic stimulation to basic movements of the human body in an intuitive fashion. MOVeMENT is novel in that it is not application specific and can generalize to almost any motor learning activity. By targeting basic movements, MOVeMENT is capable of generalizing to almost any complex movement. Basic movements were developed by dividing the body via three planes that span three-dimensional space (sagittal, frontal, and horizontal planes). The planes ground the fundamental movements: extension or flexion is movement that increases or decreases respectively a joint angle in the sagittal plane, abduction or adduction refers to movement occurring in the frontal plane towards or away from the sagittal plane (respectively), and pronation or

Fig. 4.7 A visualization of
the ghostly master metaphor.
A trainee (solid) feels the
ghost (transparent) as it
moves through the trainee's
body while performing an
instructional movement.
(Original image from [84])

supination is rotation of a joint angle towards or away from the body from within
the horizontal plane. McDaniel et al. designed haptic patterns to code for these
five fundamental movements and used them as building blocks to describe more
complex movements to a user using a push-pull metaphor to illicit movement in a
certain plane. Participants in a preliminary study found the patterns intuitive and
were able to discriminate them with high accuracy [44].

5.2 Social

Social interaction is crucial to the well-being of individuals and this of course
applies to people with disabilities. Unfortunately, many disabilities preclude individuals from equitable inclusion in all aspects of social activity. This can be due
to practical issues or even socially constructed expectations of social interaction.
Towards enriching the lives of people with disabilities by enabling a more equitable
social experience, many researchers have sought to develop systems to rectify some
of these inadequacies.

Researchers at Arizona State University for example have developed several SSD
technologies for use in social situations. The "Haptic Belt" (shown in Fig. 4.8)
paired with a face detection system conveys the direction and distance of other
people during a social interaction [43]. Tactile Rhythm was also explored in order to
convey interpersonal distances to individuals who are blind [45]. These are coarse
details of social interactions that are less accessible to people who are blind, but
there are also very important fine details of social interaction that people who are
blind miss out on too. An example of this would be facial expressions. At the same

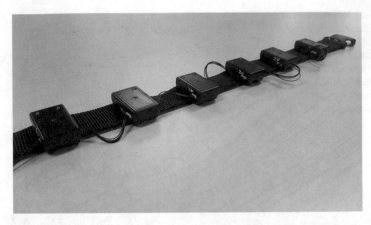

Fig. 4.8 Haptic Belt developed at the CUbiC Lab at Arizona State University [67]. The belt was designed to be modular and can be extended to fit more or fewer tactors connected in series. The location of the tactors can also be modified by simply sliding them along the belt

lab, researchers developed the "VibroGlove" a glove to convey facial expressions to people who are blind [33]. A chair-based approach was also explored, showing promise of conveying facial expression information via "Facial Action Units", a system for describing facial expressions by their structural parts [5]. This culminated in a project called the Social Interaction Assistant (SIA), a person-centered SS system that combines active learning computer vision system with haptic tactors that convey information to users they might otherwise miss [60]. A user would wear a camera similar to the shown in Fig. 4.9a and receive haptic feedback from the camera using devices such as the VibroGlove and Haptic Belt (Fig. 4.8).

5.3 Electronic Travel Aids (ETAs)

Mobility is a crucial component of independence, agency, and wellness. Vision disabilities account for a large portion of these mobility issues, and it is of no surprise because navigation itself is a complicated processes requiring visual integration over time and space and a strong dependence on memory. Researchers have determined that efficiently storing and recalling the relationship of landmarks in space is essential to spacial cognition, and thus navigation [55], and because vision provides a method for establishing landmarks in 3D space it can be inferred that it is heavily reliant on for navigation [17]. For this reason, a large number of SSDs have been developed to aid those with issues navigating. The most popular Sensory Substitution device for mobility is the "white cane", shown in Fig. 4.10a, b. This device is used to transform information that would traditionally be acquired via vision to the haptic, priopreceptive, and auditory modalities. With the white cane, users scan the ground in front of them with the cane in sweeping motions in order to

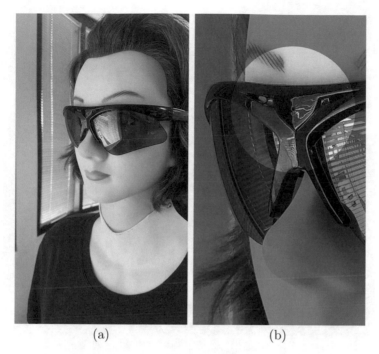

(a)   (b)

Fig. 4.9 (a) Mannequin wearing sunglasses mounted with a pinhole digital camera (b) close-up of pinhole camera

detect obstacles in their path by colliding with them. Users can often infer not just the existence of an obstacle but also some of the obstacle's properties via the tactile effects felt on contact as well as the sound emenating from the collision.

There are though drawbacks to the traditional "white cane" such as the limited range at which users can detect obstacles. White canes typically have a range of 1.5 m in front of the user. A user must also collide with an object in order to detect it, which can be troublesome if the object is a person, dog, or something fragile. Users can also miss obstacles with the cane due to gaps in their sweeping pattern. White canes also can only detect obstacles at or below waist level, leaving the user vulnerable to obstacles like overhanging tree branches [66]. Researchers have instinctually sought to improve upon the white cane to remedy some of these issues.

One of the first attempts to augment the white cane was in 1945 with the "Laser Cane". This device augmented a traditional cane with three gallium arsenide infrared laser rangefinders to detect obstacles and dropoffs at different distances. It was capable of detecting obstacles at several different angles, including an angle pointing upwards from the handle of the cane, so that users could detect obstacles above their waist and avoid tree branches. Haptic and (optionally) audio feedback was delivered to the user based on the level and distance of a detected obstacle. The device was developed with continuous feedback from travelers who are blind and was finished in 1974 [6]. While such a cane was novel, both laser and battery

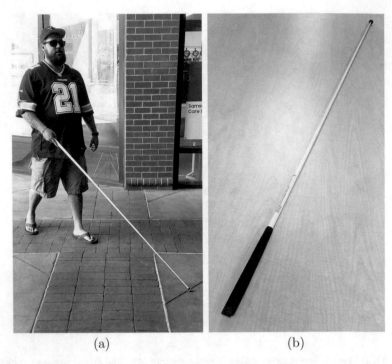

(a)   (b)

Fig. 4.10 (**a**) PhD student Bryan Duarte navigating with white cane (**b**) close-up of white cane

technology of the period restricted usage to a mere three hours per charge. The "Laser Cane" was one of the first attempts to give users information about obstacles before a collision, but it did so in a very coarse way, giving little information in the way of bearing (angle with respect to travel).

A method for detecting the bearing of objects was developed in 2002 by Dr. Roman Kuc. The device used two sonar range finders that together are used to infer the bearing of detected obstacles. Wrist-worn vibration motors vibrate with respect to the bearing of the obstacle, giving the user distance and direction information [34]. Several other "smart" canes were developed. Researchers at the Indian Institute of Technology performed a study, and found that their ultrasonic "Smart Cane" increased obstacle awareness, decreased collision prevelance, and increased mean detection distance as compared to traditional white canes in a navigation task [20]. Similar attempts at building smart canes are prevalent [2, 47] and are commonly variants of each other but [24] takes the most elaborate approach whereby the cane is equipped with wheels and "drives" a user around. The device introduces modes such as "goal finding", where the device navigates for the user, providing turn by turn directions. This device though has not been verified by a study.

A more nuanced approach is the caneless ETA, removing altogether the need for a white cane. One such configuration is called the "Haptic Radar", a self-contained headworn headband augmented with sensors that detects obstacles and intuitively conveys them to the wearer via haptics. The array of sensors each convey obstacle distance information for a path emanating from the sensor (there are several circling the head) [64]. Researchers found that participants tasked with a navigation task navigated more confidently with a Haptic Radar than without [10]. Caneless systems may be advantages as they may reduce stigma induced by the iconic white cane. With GPS becoming ubiquitous, turn-by-turn directions have become life-changing for those needing directional and situational assistance. Most turn-by-turn directions are conveyed using the device's screen and are often accompanied by audio, but haptic solutions may offer a better alternative to convey this information.

5.4 Virtual

Virtual worlds are a rich part of the modern experience. Whether it be games, simulations, or educational environments, virtual worlds are becoming commonplace with the advent of consumer VR and widespread gaming hardware. One of the issues is that most virtual environments are developed with vision being the primary interaction modality, effectively excluding many individuals from participation. While some non-visual video games exist, they are few and far between and almost always rely solely on audio feedback. A few examples of modern video games accessible without vision are FEER, an "Endless Runner" game [51, 63], Timecrest: The Door, a story-based game with multiple endings and dynamic storylines [3, 14] and A Blind Legend, a first person fighting game for both PC and Android [15]. While a handful of games can be played with audio only, the majority of video games and virtual environments remain inaccessible to individuals who are blind. Haptic implementations may provide solutions to this problem.

Developers in the Haptics Laboratory of McGill University in 2006 developed a game of "Memory" using the $STRESS^2$ tactile display [81], a more ergonomic version of the original $STRESS$ 1D haptic display [37]. Instead of images or text to memorize, the "cards" consisted of unique haptic patterns, making for an interesting spin on the classic game of Memory. Likewise, researchers at Arizona State University designed a 2D spatial game based around the Low Resolution Haptic Display (LRHD), a chair affixed with a 4×4 array of vibrotactile motors. The point of the game was to find the goal 2D top-down environment. The user's position was displayed on the haptic chair as well as the goal using unique vibration patterns and the user could move in the environment using a computer mouse peripheral to find the goal. A study using the game found that users were able to learn how to play the game quickly and their performance increased markedly as they played [19]. An image of the Low Resolution Haptic Display is shown in Fig. 4.11. These games are in contrast to audio-only games as they are haptic-only games.

Several devices and systems have been developed as SSDs for virtual environments. Some of these SSDs substitute vision for touch, while others substitute virtual touch for physical touch. For example, in 1998 researchers employed a

Fig. 4.11 The Low
Resolution Haptic Display, a
4 × 4 array of vibration
motors mounted vertically on
acoustic foam for compliance
and damping [19]

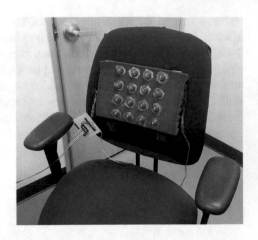

force feedback joystick called the Impulse Engine 3000 as an interface to virtual textures and objects. Researchers demonstrated a statistically significant relationship between the virtual texture's perceived roughness with the physical analogue and found that participants who were blind were more discriminating than sighted ones using their system [12]. More complex interaction such as discriminating the angle and identity of objects proved more difficult to discern with the system. Researchers found similar results in 1999 using the PHANToM force feedback device (pictured in Fig. 4.6) [29]. Again, simple textures were rendered convincingly but the technology was not convincing for object recognition. The primary limitation with these implementations is that only a single point of contact with the "virtual world" is possible, making the interactions akin to poking around with your finger in virtual space.

In response to these problems, researchers proposed non-realistic haptic rendering (NRHR). They argued realistic rendering can be too complicated to parse haptically and non-realistic haptic rendering can make things simpler, giving researchers the chance to eliminate distracting details while emphasizing the important information [31]. To do this, they mapped 3D models onto 2D planes which they argued were easier to navigate. The researchers also propose a different method for guided navigation in virtual environments: a haptic guide. Guiding forces are given to the user as force vectors placed on the PHANToM's stylus [32]. Similarly, in 2012 researchers using the VTPlayer Mouse developed and tested directional cues via the Braille-like cells. Participants found the cues intuitive and easy to learn [62]. This body of research implies that directional guides are useful in navigating virtual environments haptically.

Towards navigating virtual environments "naturally", in 2013 researchers developed the Virtual EyeCane. The virtual cane gives users an auditory signal with respect to the closest object the cane is pointed at in the virtual world [41], making this system a Virtual Electronic Travel Aid (VETA), similar to the first Laser Cane but unhindered by the limitations of rangefinding in the physical world. A more comprehensive approach was taken by Zhao et al. in 2018 in development of the "Canetroller", which is a virtual cane that gives realistic auditory and haptic feedback in the virtual world so that people who are blind can translate their cane skills to VR. The Canetroller realistically simulates cane forces, impact vibrations, and impact sounds [87]. Besides the EyeCane and Canetroller, there have not been any significant attempts to make accessible to people with visual impairments virtual worlds on equal footing, in essence to take a visual world and present it using an SSD such that they can interact in much the same way as their sighted counterparts. Virtual worlds by their very nature provide mechanisms for making them accessible as object detection and semantic segmentation are less complicated in those environments.

6 Conclusion: Future Trends and Trajectory

Some standout implementations of Haptic Sensory Substitution are Bach-y-Rita's TVSS, the BrainPort, and the Eagleman and Novich's VEST, showing the true raw representational power of the modality, but they also reveal some limitations. For the TVSS, long training hours, a chair-based design with many actuators, and lack of fine details hinder its use in real-world applications. While the BrainPort tackles the portability and details issues somewhat, it still suffers from the practical concerns of requiring the display to be placed on a user's tongue. For auditory substitution, the VEST is impressive in its ability to convey speech, but other more subtle aspects of hearing are still missing, such as localization via stereo hearing. Further strides in the realm of Haptic Sensory Substitution are more likely to arise with clever integrations with emerging signal processing tools and clever delivery techniques.

In the realm of vision-to-haptic SS, strides in Computer Vision show promise for enabling more effective Sensory Substitution. For example, object detection has made great strides, as well as depth estimation from monocular images. Having access to both depth and object identities from monocular images could drastically improve ETAs by allowing ones that rely on depth information to use only a camera instead of lasers, sonar, infrared, or stereo cameras. Figure 4.12b illustrates the impressive performance of emerging depth estimation models (MegaDepth). The methods underlying the image understanding applications from Sect. 4.2 utilizing neural networks also show great promise in augmenting haptic SSD technology. Combining these powerful models with a proprioceptive and tactile interface would likely lead to a more effective and meaningful image understanding tool that can be used both in the physical world but even more so in virtual environments, possibly making all visual virtual worlds sufficiently accessible to people with visual impairments.

Automated or semi-automated methods of architecture modeling [39] also show promise for alleviating the manual design requirements of mobility training systems. Combined with systems such as the CaneTroller and crowdsourcing, could make familization with novel environments in the safety of one's home practical and accessible to people who are blind, having applications for education in the form of virtual field trips.

(a) (b)

Fig. 4.12 (a) Original image of an office (b) depth image from model trained on the MegaDepth depth dataset [38]

Acknowledgements The authors thank Arizona State University and the National Science Foundation for their funding support. The preparation of this chapter is supported by the National Science Foundation under Grant No. 1828010 and 1069125.

References

1. Abboud, S., Hanassy, S., Levy-Tzedek, S., Maidenbaum, S., Amedi, A.: EyeMusic: Introducing a 'visual' colorful experience for the blind using auditory sensory substitution. Restorative Neurology and Neuroscience 32(2), 247–257 (2014)
2. Ando, B., Baglio, S., Marletta, V., Valastro, A.: A Haptic Solution to Assist Visually Impaired in Mobility Tasks. IEEE Transactions on Human-Machine Systems 45(5), 641–646 (2015)
3. Apple: Timecrest: The Door. https://apps.apple.com/za/app/timecrest-the-door/id1027546326 (2015), https://apps.apple.com/za/app/timecrest-the-door/id1027546326
4. Bach-Y-Rita, P., Collins, C.C., Saunders, F.A., White, B., Scadden, L.: Vision substitution by tactile image projection. Nature 221(5184), 963–964 (1969)
5. Bala, S., McDaniel, T., Panchanathan, S.: Visual-to-tactile mapping of facial movements for enriched social interactions. 2014 IEEE International Symposium on Haptic, Audio and Visual Environments and Games, HAVE 2014 – Proceedings pp. 82–87 (2014)
6. Benjamin, J.M.: The laser cane. Bulletin of prosthetics research pp. 443–50 (1974), http://www.ncbi.nlm.nih.gov/pubmed/4462934
7. Bliss, J.C., Katcher, M.H., Rogers, C.H., Shepard, R.P.: Optical-to-Tactile Image Conversion for the Blind. IEEE Transactions on Man-Machine Systems 11(1), 58–65 (1970)

8. Bucchieri, V.: Apparatus and method for presenting and controllably scrolling Braille text (2013), https://patents.google.com/patent/US8382480B2/en

9. Caspi, A., Dorn, J.D., McClure, K.H., Humayun, M.S., Greenberg, R.J., McMahon, M.J.: Feasibility study of a retinal prosthesis:Spatial vision with a 16-electrode implant. Archives of Ophthalmology 127(4), 398–401 (2009)

10. Cassinelli, A., Sampaio, E., Joffily, S.B., Lima, H.R., Gusmão, B.P.: Do blind people move more confidently with the Tactile Radar? Technology and Disability 26(2-3), 161–170 (2014)

11. Clary, P.: Lookout: an app to help blind and visually impaired people learn about their surroundings. https://www.blog.google/outreach-initiatives/accessibility/lookout-app-help-blind-and-visually-impaired-people-learn-about-their-surroundings/ (2018), https://www.blog.google/outreach-initiatives/accessibility/lookout-app-help-blind-and-visually-impaired-people-learn-about-their-surroundings/

12. Colwell, C., Petrie, H., Kornbrot, D., Hardwick, A., Furner, S.: Haptic virtual reality for blind computer users. In: Assets '98 Proceedings of the third international ACM conference on Assistive technologies. pp. 92–99. Marina del Rey, California, USA (1998), https://dl.acm.org/citation.cfm?id=274515

13. De Felice, F., Renna, F., Attolico, G., Distante, A.: A haptic/acoustic application to allow blind the access to spatial information. Proceedings – Second Joint EuroHaptics Conference and Symposium on Haptic Interfaces for Virtual Environment and Teleoperator Systems, World Haptics 2007 pp. 310–315 (2007)

14. DMNagel: Timecrest: The Door. https://www.applevis.com/apps/ios/games/ timecrest-door (2017), https://www.applevis.com/apps/ios/games/timecrest-door

15. Dowino: A Blind Legend. https://play.google.com/store/apps/details?id=com.dowino.ABlindLegend (2019), https://play.google.com/store/apps/details?id=com.dowino.ABlindLegend&hl=en

16. Eagleman, D.: Plenary talks: A vibrotactile sensory substitution device for the deaf and profoundly hearing impaired. In: 2014 IEEE Haptics Symposium (HAPTICS). pp. xvii–xvii (2014), http://ieeexplore.ieee.org/document/6775419/

17. Ekstrom1, A.D.: Why vision is important to how we navigate. Hippocampus 73(4), 389–400 (2015)

18. Eskildsen, P., Morris, A., Collins, C.C., Bach-y Rita, P.: Simultaneous and successive cutaneous two-point thresholds for vibration. Psychonomic Science 14(4), 146–147 (1969)

19. Fakhri, B., Sharma, S., Soni, B., Chowdhury, A.: A Low Resolution Haptic Interface for Interactive Applications. HCI International pp. 1–6 (2019)

20. Fallis, A.: 'Smart' Cane for the Visually Impaired: Design and Controlled Field Testing of an Affordable Obstacle Detection System. 12th International Conference on Mobility and Transport for Elderly and Disabled Persons 53(9), 1689–1699 (2010)

21. Freedom Scientific Inc.: Freedom Scientific Braille Displays and Keyboards. http://www.freedomscientific.com/ (2018), http://www.freedomscientific.com/

22. Geldard, F.A.: Cutaneous coding of optical signals: The optohapt. Perception & Psychophysics 1(11), 377–381 (1966)

23. Geldard, F.A., Sherrick, C.E.: The cutaneous "rabbit": A perceptual illusion. Science 178(4057), 178–179 (1972)

24. GHARIEB, W., NAGIB, G.: Smart Cane for Blinds. Proc. 9th Int. Conf. on AI Applications (August), 253–262 (2015), c:%5CUsers%5Cjessica%5CBIBLIOTECA%5Cdesign%5CMetodologiaexperimental-GuiBonsiepe.pdf%5Cnhttp://www.researchgate.net/profile/Gihan_Nagib/publication/255615346_Smart_Cane_for_Blinds/links/542020190cf241a65a1b065a.pdf%5Cnhttp://www.researchgate.net/publi

25. He, L., Wan, Z., Findlater, L., Froehlich, J.E.: TacTILE: A Preliminary Toolchain for Creating Accessible Graphics with 3D-Printed Overlays and Auditory Annotations. Proc. 19th Int. ACM SIGACCESS Conf. Comput. Access. pp. 397–398 (2017), https://doi.org/10.1145/3132525.3134818

26. Homer Jacobson: The Informational Capacity of the Human Ear. Science 112(2901), 143–144 (1950), http://science.sciencemag.org/content/112/2901/143

27. Ikei, Y., Wakamatsu, K., Fukuda, S.: Vibratory tactile display of image-based textures. IEEE Computer Graphics and Applications 17(6), 53–61 (1997)
28. Jacobson, H.: The informational capacity of the human eye. Science 113(2933), 292–293 (1951)
29. Jansson, G., Petrie, H., Colwell, C., Kornbrot, D.: Haptic virtual environments for blind people: Exploratory experiments with two devices. The International Journal of Virtual Reality 3(4), 8–17 (1999), https://pdfs.semanticscholar.org/348e/45107167a0325051e60c883c153572a127e4.pdf%0Ahttp://www.ijvr.org/issues/pre/4-1/2.pdf
30. Karyn, G., Gina, M., Alessandra, M., Robert, C., Peter, B., Graeme, C.: A comparison of Tactaid II+ and Tactaid 7 use by adults with a profound hearing impairment. Ear and Hearing 20(6), 471–482 (1999), http://www.scopus.com/inward/record.url?eid=2-s2.0-0033436316&partnerID=40&md5=1058dec9323d2a378c6fcb685db9acc5
31. König, H., Schneider, J., Strothotte, T.: Haptic Exploration of Virtual Buildings Using Non-Realistic Haptic Rendering. Society pp. 377–384 (2000)
32. König, H., Schneider, J., Strothotte, T.: Orientation and Navigation in Virtual Haptic-Only Environments. In: Paelke, V., Volbracht, S. (eds.) Proceedings User Guidance in Virtual Environments. pp. 123–136. Shaker Verlag, Aachen, Germany (2001)
33. Krishna, S., Bala, S., McDaniel, T., McGuire, S., Panchanathan, S.: VibroGlove: an assistive technology aid for conveying facial expressions. In: CHI '10 Extended Abstracts on Human Factors in Computing Systems. pp. 3637–3642 (2010), https://doi.org/10.1145/1753846.1754031
34. Kuc, R.: Binaural sonar electronic travel aid provides vibrotactile cues for landmark, reflector motion and surface texture classification. IEEE Transactions on Biomedical Engineering 49(10), 1173–1180 (2002)
35. Lahav, O., Mioduser, D.: Construction of cognitive maps of unknown spaces using a multi-sensory virtual environment for people who are blind. Computers in Human Behavior 24(3), 1139–1155 (2008)
36. Landau, S., Wells, L.: Merging Tactile Sensory Input and Audio Data by Means of The Talking Tactile Tablet. Proc. Eurographics'03 2(60), 414–418 (2003)
37. Lévesque, V., Pasquero, J., Hayward, V., Legault, M.: Display of virtual braille dots by lateral skin deformation: feasibility study. ACM Transactions on Applied Perception 2(2), 132–149 (2005)
38. Li, Z., Snavely, N.: MegaDepth: Learning Single-View Depth Prediction from Internet Photos. Proceedings of the IEEE Computer Society Conference on Computer Vision and Pattern Recognition pp. 2041–2050 (2018)
39. Liu, C., Wu, J., Furukawa, Y.: FloorNet: A unified framework for floorplan reconstruction from 3D scans. In: European Conference on Computer Vision. vol. 11210 LNCS, pp. 203–219 (2018)
40. Loomis, J.M.: Tactile letter recognition under different modes of stimulus presentation. Perception & Psychophysics 16(2), 401–408 (1974)
41. Maidenbaum, S., Levy-Tzedek, S., Chebat, D.R., Amedi, A.: Increasing accessibility to the blind of virtual environments, using a virtual mobility aid based on the "EyeCane": Feasibility study. PLoS ONE 8(8) (2013)
42. Mascetti, S., Bernareggi, C., Belotti, M.: TypeInBraille: A Braille-based Typing Application for Touchscreen Devices. In: The proceedings of the 13th international ACM SIGACCESS conference on Computers and accessibility. pp. 295–296. Dundee, Scotland, UK (2011)
43. McDaniel, T., Krishna, S., Balasubramanian, V., Colbry, D., Panchanathan, S.: Using a haptic belt to convey non-verbal communication cues during social interactions to individuals who are blind. HAVE 2008 – IEEE International Workshop on Haptic Audio Visual Environments and Games Proceedings (October), 13–18 (2008)
44. McDaniel, T., Villanueva, D., Krishna, S., Panchanathan, S.: MOVeMENT: A framework for systematically mapping vibrotactile stimulations to fundamental body movements. HAVE 2010 – 2010 IEEE International Symposium on Haptic Audio-Visual Environments and Games, Proceedings pp. 13–18 (2010)

45. McDaniel, T.L., Krishna, S., Colbry, D., Panchanathan, S.: Using tactile rhythm to convey interpersonal distances to individuals who are blind. CHI Extended Abstracts pp. 4669–4674 (2009), https://dl.acm.org/citation.cfm?id=1520718
46. Meijer, P.B.: An Experimental System for Auditory Image Representations. IEEE Transactions on Biomedical Engineering 39(2), 112–121 (1992)
47. Menikdiwela, M.P., Dharmasena, K.M., Abeykoon, A.M.S.: Haptic based walking stick for visually impaired people. 2013 International Conference on Circuits, Controls and Communications, CCUBE 2013 pp. 1–6 (2013)
48. Merzenich, M.M., Michelson, R.P., Pettit, C.R., Schindler, R.A., Reid, M.: Neural Encoding of Sound Sensation Evoked by Electrical Stimulation of the Acoustic Nerve. Annals of Otology, Rhinology & Laryngology 82(4), 486–503 (1973)
49. Metz, C.: Facebook's AI Is Now Automatically Writing Photo Captions. https://www.wired.com/2016/04/facebook-using-ai-write-photo-captions-blind-users/ (2016), https://www.wired.com/2016/04/facebook-using-ai-write-photo-captions-blind-users/
50. Metz, R.: BLITAB. https://www.technologyreview.com/s/603336/this-500-tablet-brings-words-to-blind-users-fingertips/ (2017), https://www.technologyreview.com/s/603336/this-500-tablet-brings-words-to-blind-users-fingertips/
51. Meyer, I., Mikesch, H.: FEER the Game of Running Blind. http://www.feer.at/index.php/en/home/ (2018), http://www.feer.at/index.php/en/home/
52. Microsoft: Seeing AI. https://www.microsoft.com/en-us/ai/seeing-ai (2018), https://www.microsoft.com/en-us/ai/seeing-ai
53. Miller, I., Pather, A., Milbury, J., Hathy, L., O'Day, A., Spence, D.: Guidelines and Standards for Tactile Graphics, 2010. http://www.brailleauthority.org/tg/web-manual/index.html (2011), http://www.brailleauthority.org/tg/web-manual/index.html
54. Miyazaki, M., Hirashima, M., Nozaki, D.: The "Cutaneous Rabbit" Hopping out of the Body. Journal of Neuroscience 30(5), 1856–1860 (2010), http://www.jneurosci.org/cgi/doi/10.1523/JNEUROSCI.3887-09.2010
55. Monacelli, A.M., Cushman, L.A., Kavcic, V., Duffy, C.J.: Spatial disorientation in Alzheimer's disease: The remembrance of things passed. Neurology 61(11), 1491–1497 (2003), https://pdfs.semanticscholar.org/e957/14321d7fb821b421f2897496ccd1d10fed60.pdf
56. National Federation of the Blind Jernigan Institute: The Braille Literacy Crisis in America, Facing the Truth, Reversing the Trend, Empowering the Blind. Tech. rep., National Federation of the Blind, Baltimore Maryland (2009)
57. Nau, A., Bach, M., Fisher, C.: Clinical Tests of Ultra-Low Vision Used to Evaluate Rudimentary Visual Perceptions Enabled by the BrainPort Vision Device. Translational Vision Science & Technology 2(3), 1 (2013), http://tvst.arvojournals.org/Article.aspx?doi=10.1167/tvst.2.3.1
58. Novich, S.D., Eagleman, D.M.: Using space and time to encode vibrotactile information: toward an estimate of the skin's achievable throughput. Experimental Brain Research 233(10), 2777–2788 (2015)
59. Novich, S.D.: Sound-to-Touch Sensory Substitution and Beyond (2015), https://scholarship.rice.edu/handle/1911/88379
60. Panchanathan, S., Chakraborty, S., McDaniel, T.: Social Interaction Assistant: A Person-Centered Approach to Enrich Social Interactions for Individuals with Visual Impairments. IEEE Journal on Selected Topics in Signal Processing 10(5), 942–951 (2016)
61. Pasquero, J.: Survey on communication through touch. McGill Centre for Intelligent Machines 6(August), 1–28 (2006), http://scholar.google.com/scholar?hl=en&btnG=Search&q=intitle:Survey+on+Communication+through+Touch#0
62. Pietrzak, T., Pecci, I., Martin, B.: Static and dynamic tactile directional cues experiments with VTPlayer mouse. In: Proceedings of the Eurohaptics conference. pp. 63–68. Paris, France (2006)
63. Régo, N.: The Game of Running Blind in FEAR. https://coolblindtech.com/the-game-of-running-blind-in-fear/ (2018), https://coolblindtech.com/the-game-of-running-blind-in-fear/

64. Riener, a., Hartl, H.: Personal Radar: a self-governed support system to enhance environmental perception. Proceedings of BCS-HCI 2012 pp. 147–156 (1974), http://dl.acm.org/citation.cfm? id=2377933%5Cnpapers://c80d98e4-9a96-4487-8d06-8e1acc780d86/Paper/p15116

65. Bach-y Rita, P., Danilov, Y., Tyler, M., Grimm, R.J.: Late human brain plasticity: vestibular substitution with a tongue BrainPort human-machine interface. Plasticidad y Restauracion Neurologica 4(1-2), 31–34 (2005), http://www.medigraphic.com/pdfs/plasticidad/prn-2005/prn051_2f.pdf%5Cnhttp://www.ncbi.nlm.nih.gov/pubmed/15194608%0Ahttp://doi.wiley.com/10.1196/annals.1305.006

66. Rosen, S.: Long Cane Techniques Study Guide Step-By-Step A Guide to Mobility Techniques. https://tech.aph.org/sbs/04_sbs_lc_study.html#4, https://tech.aph.org/sbs/04_sbs_lc_study.html#4

67. Rosenthal, J., Edwards, N., Villanueva, D., Krishna, S., McDaniel, T., Panchanathan, S.: Design, implementation, and case study of a pragmatic vibrotactile belt. In: IEEE Transactions on Instrumentation and Measurement. vol. 60, pp. 114–125 (2011)

68. Sampaio, E., Maris, S., Bach-y Rita, P.: Brain plasticity: 'Visual' acuity of blind persons via the tongue. Brain Research 908(2), 204–207 (2001)

69. Schmidt, R.N., Lisy, F.J., Prince, T.S., Shaw, G.S.: Refreshable braille display system (1998), https://patents.google.com/patent/US6354839B1/en

70. Semwal, S.: MoVE: Mobiltiy training in haptic virtual environment. Tech. rep., University of Colorado at Colorado Springs, Colorado Springs (2001), https://pdfs.semanticscholar.org/243e/b3d64990f34d0b126b36132acc17e4f50737.pdf%0Ahttp://citeseerx.ist.psu.edu/viewdoc/download?doi=10.1.1.24.6646&rep=rep1&type=pdf

71. Sharkey, P., Sik Lanyi, C., Standen, P., University of Reading. ICDVRAT, D.o.C.: Multisensory virtual environment for supporting blind persons' acquisition of spatial cognitive mapping, orientation, and mobility skills (1993), 279 (2002)

72. Smithmaitrie, P., Kanjantoe, J., Tandayya, P.: Touching force response of the piezoelectric Braille cell. Disability and Rehabilitation: Assistive Technology 3(6), 360–365 (2008)

73. Stageberg, S.: The Device That Refreshes: How to Buy a Braille Display. https://www.afb.org/aw/5/6/14669 (2004), https://www.afb.org/aw/5/6/14669

74. Sweller, J., Ayres, P., Kalyuga, S.: Amassing information: The information store principle. In: Cognitive Load Theory (2011)

75. Tan, H.Z., Pentland, A.: Tactual. Disptays for Wearabte Computing. Personal Technologies pp. 225–230 (1997)

76. Troxel, D.: Experiments in Tactile and Visual Reading. IEEE Transactions on Human Factors in Electronics 8(4), 261–263 (1967), https://ieeexplore.ieee.org/stamp/stamp.jsp?tp=&arnumber=1698280

77. Tzovaras, D., Moustakas, K., Nikolakis, G., Strintzis, M.G.: Interactive mixed reality white cane simulation for the training of the blind and the visually impaired. Personal and Ubiquitous Computing 13(1), 51–58 (2009)

78. Van Erp, J., Self, B.: Tactile Displays for Orientation, Navigation and Communication in Air, Sea and Land Environments, vol. 323 (2008)

79. Wall, S., Brewster, S.: Feeling What You Hear: Tactile Feedback for Navigation of Audio Graphs. In: CHI 2006 Proceedings, Disabilities. pp. 1123–1132 (2006)

80. Wall, S.A., Brewster, S.: Sensory substitution using tactile pin arrays: Human factors, technology and applications. Signal Processing 86(12), 3674–3695 (2006)

81. Wang, Q., Levesque, V., Pasquero, J., Hayward, V.: A haptic memory game using the STRESS 2 tactile display. In: Proceedings of CHI 2006. p. 271 (2006)

82. Ward, J., Meijer, P.: Visual experiences in the blind induced by an auditory sensory substitution device. Consciousness and Cognition 19(1), 492–500 (2010)

83. White, B.W., Saunders, F.A., Scadden, L., Bach-Y-Rita, P., Collins, C.C.: Seeing with the skin. Perception & Psychophysics 7(1), 23–27 (1970)

84. WikiHow: How to Swing a Golf Club. https://www.wikihow.com/Swing-a-Golf-Club (2019), https://www.wikihow.com/Swing-a-Golf-Club

85. Yanagida, Y., Kakita, M., Lindeman, R.W., Kume, Y., Tetsutani, N.: Vibrotactile letter reading using a low-resolution tactor array. Proceedings – 12th International Symposium on Haptic Interfaces for Virtual Environment and Teleoperator Systems, HAPTICS pp. 400–406 (2004)
86. Yang, U., Jang, Y., Kim, G.J.: Designing a VibroTactile Wear for Close Range Interaction for VRbased Motion Training. Icat 2002 pp. 4–9 (2002), http://citeseerx.ist.psu.edu/viewdoc/summary?doi=10.1.1.103.5793
87. Zhao, Y., Bennett, C.L., Benko, H., Cutrell, E., Holz, C., Morris, M.R., Sinclair, M.: Enabling People with Visual Impairments to Navigate Virtual Reality with a Haptic and Auditory Cane Simulation. Proceedings of the 2018 CHI Conference on Human Factors in Computing Systems – CHI '18 pp. 1–14 (2018), http://dl.acm.org/citation.cfm?doid=3173574.3173690

Part II
Haptics for Health and Wellbeing

Chapter 5
Haptics in Rehabilitation, Exergames and Health

Mohamad Hoda, Abdulmotaleb El Saddik, Philippe Phan, and Eugene Wai

Abstract It is well known that home exercise is as good as the rehabilitation center. Unfortunately, passive devices such as dumbbells, elastic bands, stress balls, and tubing, which have been widely used for home-based upper-body rehabilitation, do not provide therapists with the information needed to monitor patients' progress, identify impairments, and suggest treatments. Moreover, the lack of interactivity of these devices turns rehabilitation exercises into boring, unpleasant tasks. In this chapter, we introduce a family of exergame rehabilitation systems aimed at solving the aforementioned problems. The systems combine recent rehabilitation approaches with efficient, yet affordable, skeleton tracking input technologies, and multimodal interactive computer environment. In addition, the systems provide real-time feedback to stroke patients, summarize feedback after each session, and predict the overall recovery progress. Moreover, these systems show a new style of rehabilitation that motivates patients by engaging family and friends in the rehabilitation process and allowing therapists to remotely assess the progress of patients and adjust the training strategy accordingly. The objective/subjective assessments and usability studies show the feasibility of the proposed systems for rehabilitation in stroke patients with upper limb motor dysfunction.

M. Hoda (✉) · A. El Saddik
University of Ottawa, Ottawa, ON, Canada
e-mail: mhoda053@uottawa.ca; elsaddik@uottawa.ca

P. Phan
Division of Orthopaedic Surgery, University of Ottawa, Ottawa, ON, Canada
e-mail: pphan@toh.ca

E. Wai
The Ottawa General Hospital, Ottawa, ON, Canada
e-mail: ewai@toh.ca

© Springer Nature Switzerland AG 2020
T. McDaniel, S. Panchanathan (eds.), *Haptic Interfaces for Accessibility, Health, and Enhanced Quality of Life*, https://doi.org/10.1007/978-3-030-34230-2_5

1   Overview

Physical inactivity among large swaths of the population is contributing to many severe health issues [1, 2]. In fact, a recent study has shown that not exercising can be more harmful to your body than smoking, diabetes, or even heart disease [3]. Exergames, also known as active-play video games, have not only proven to be the bringer of great promise towards encouraging physical activity, but also played a major role in medical rehabilitation, especially post-stroke rehabilitation [4–6]. Recently, exergame rehabilitation systems combine recent rehabilitation approaches with efficient and affordable skeleton tracking input technology and multimodal interactive computer environments. Those systems provide real-time feedback to stroke patients, summarize the feedback after each session, and predict the overall recovery progress. In this chapter, we will state the benefits of exergames, and then describe the different types and techniques used in exergames and rehabilitation. This allows us to evaluate the ability to monitor and predict the progress of users, and compare the general performance of current approaches. Finally, we provide limitations and conclude with the potential for new techniques in exergames. Material from this chapter is partially from [7, 8].

2   Importance of Exergame

One of the main reasons that a group of patients with the same pathology responds differently to a rehabilitation program is motivation. Many previous researchers have proven that motivation is a significant factor in determining the patient's outcome in a rehabilitation program [9, 10]. Family and close friends are a source of motivation to stroke patients. The love and care stroke patients receive from their loved ones empowers them to fight back and motivates them to continue their treatment. In this context, by providing a multi-player gaming environment, exergames play a crucial role in engaging family and friends in motivating patients to adhere to the training program. Moreover, exergames have been customized to fit the patients' need for rehabilitation regardless of their medical status. They are cost-effective and provide the patients with safe, natural or real-life environments that minimize the risks during rehabilitation sessions. For all of the above-mentioned reasons, Exergames have been used to promote physical activities that fit healthy people and patients of all ages.

3   Exergames in Rehabilitation

Exergaming is the use of computer software and hardware to build a multi-sensory environment that simulates the real world. Exergames in rehabilitation are the use of exergame technology as a rehabilitation tool for patients.

Physiotherapists recommend that stroke patients perform a daily life activity, such as preparing a cup of coffee, in order to improve their upper limb movements. Since such tasks might put a patient with an arm injury at risk of peril, researchers have developed and implemented virtual home-based rehabilitation (VHR) frameworks that can prevent accidents. The effectiveness of using VHR over conventional rehabilitation therapy has been proven in many studies [11–15].

One of the early reviews in this field predicted that virtual reality would become an essential tool in stroke rehabilitation in the future [16]. Rose et al. [16] assessed the effect of using virtual reality on the major disabilities that could have been caused by stroke. Patients who suffered from executive dysfunction, the first disability assessed, showed a better performance in the sequencing and organization tasks with VR-based tests than those who did not use it. However, there was no improvement in memory impairment, the second disability assessed, in most of the studies they had reviewed. The assessment of the rest of the disabilities, namely spatial ability impairments, attention deficits, and unilateral visual neglect, suggested that the effectiveness of the VR on those impairments was limited but encouraging to continue researching using VR in rehabilitation.

Henderson et al. [17] reviewed twenty virtual reality rehabilitation papers to investigate whether the immersive and non-immersive VR-based stroke rehabilitation systems were better than conventional therapy or no therapy of the upper limb. The results showed that the patients who had immersive VR therapy significantly improved over those with conventional therapy or no therapy.

In [18], two studies used off-the-shelf commercially available VR gaming systems out of the nineteen papers that Laver et al. had explored. Again, the results were in line with the previous studies that had tackled the effectiveness of customized virtual reality exergames. Moreover, the improvement of the upper limb motor function was statistically significant as reported in seven other studies.

The previous review of [18] has been recently updated to investigate more studies and reach conclusive evidence on whether VR can improve the motor function of a stroke patient. The updated review of [18] was published in May 2015 and it included the results of analyzing thirty-seven papers with 1019 participants, almost double the number of papers (nineteen papers) and participants (565 participants) that had been included in the previous review. Although the number of participants was relatively small in each trail under study, the authors could reach conclusive evidence that VR could improve the motor function of the upper limb as well as the general activity of daily living (ADL) function of a stroke patient.

In conclusion, with the advances of VR technology, there is more evidence that VR-based rehabilitation systems can improve the motor function of stroke patients.

4 Types of Exergame Rehabilitation Frameworks

Exergame frameworks can be roughly divided into three categories: (1) exergame virtual reality (VR) frameworks, (2) exergame VR frameworks with haptic feedback, and (3) exergame VR frameworks with assistive devices (Fig. 5.1). VR takes

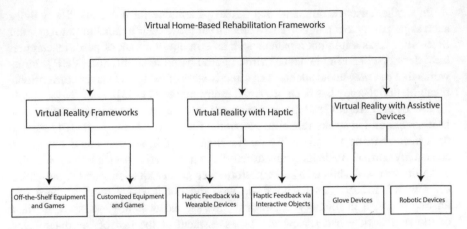

Fig. 5.1 Classification of virtual home-based rehabilitation

advantage of the advances in capturing and tracking sensor technology to build an artificial environment that enables users to interact with it. VR rehabilitation frameworks are divided into two more subcategories depending on whether they are off-the-shelf or customized. VR frameworks with haptic feedback enable the users to experience the sense of touch while they are interacting with virtual objects in the artificial environment. VR rehabilitation frameworks with assistive devices are the most popular among the other rehabilitation frameworks. We will explain the three categories in the next sections.

5 Exergame Virtual Reality Rehabilitation Frameworks

Virtual reality games are designed to help stroke patients gradually regain some of their lost motor functions. These games allow patients to interact with virtual objects in a safe and controlled environment. Users place head mounted displays on their eyes for simulation of 3D viewing or can interact with objects through special input devices similar to standard computer games. Sony PlayStation, Nintendo Wii, and Microsoft Kinect are examples of the video gaming systems that have been used for rehabilitation.

Perhaps one of the earliest uses of VR in stroke rehabilitation is by Holden et al. [19]. They recreated a "mailbox" scene used for stroke patients' training. Patients had to pick up an envelope and place it in the mailbox. A virtual world of the real hand movements was created on a computer screen using a special application prototype. Although the number of the subjects were very limited (two stroke patients), the patients showed a significant improvement in reaching ability in both virtual and real worlds. Such results encouraged other researchers to further investigate the feasibility and effectiveness of VR rehabilitation systems.

The usability of Microsoft Kinect as a tracking sensor in exergame rehabilitation exercises was studied by Chang et al. [20], Lange et al. [21], Obdrzalek et al. [22] and Clark et al. [23]. In [20], a rehabilitation system using Kinect (Kinerehab) was tested on two patients: Peter, suffering from severe cerebral palsy, and Sherry, from muscle atrophy. By the end of the study, results showed that the correctness of the performed exercises increased with the use of the proposed rehabilitation system. Moreover, the patients were motivated by the virtual environment provided by the rehabilitation system, and they showed an interest in contributing to their rehabilitation program using the Kinerehab system.

Later studies evaluated the correctness of the obtained 3D positions of the body joints from Kinect [22, 23]. The evaluation was done by comparing the values captured by Kinect to those captured by advanced multiple-camera 3D motion tracking. In [22], forty-three markers were placed on a human body and tracked by nine cameras. The obtained points were connected together to form a skeleton. Another skeleton was estimated by using the Kinect sensor. Results showed that Kinect could replace high-cost tracking systems for controlled body exercises. However, Kinect suffered from occlusions when the subjects were performing the exercises while they were sitting in a wheelchair. Clark et al. [23] have reached similar results when they tested Kinect pose estimation with a twelve-camera Vicon MX motion tracking system.

Stroke tele-rehabilitation was first proposed in [24] for the use of telemedicine in the form of video-teleconferencing to maximize the number of patients given effective acute stroke treatment. The system was not intended to support patients at home while they were conducting the rehabilitation exercises. In [25], Hoda et al. developed an intelligent, low-cost cloud-based rehabilitation gaming system using Microsoft Kinect sensor. They implemented two goal-oriented exercises that target the rehabilitation of patients with chronic upper limb motor dysfunction. While designing the games, they emphasized the collaboration between family/friends and the stroke patient in the post-stroke rehabilitation process. The games included moving a ball into a net (basketball game) in the vertical plane and reaching a cup (reaching game) in the horizontal plane. In addition, accurate performance measures were provided for each exercise to quantitatively evaluate the effectiveness of the treatment plan for each patient (Fig. 5.2). The implemented system increased user motivation to participate in the rehabilitation program, accelerated the recovery of the muscles, and offered the therapists the ability to assess the patients' progress.

6   Exergame Virtual Reality with Haptics

Although the new video game controllers are considered intuitive, they do not cater to a wide range of patients, especially those with upper limb impairments. In addition, the scores provided by the games are not adequate to evaluate the patients' progress [11]. Many researchers have reported that adding the sense of touch to virtual reality enhances the patient's sense of presence within the rehabilitation

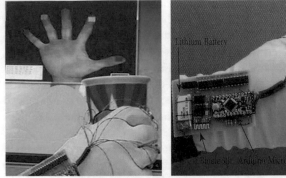
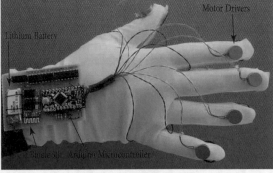
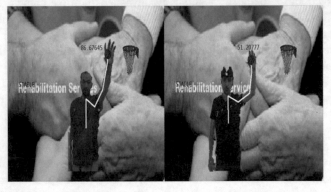

Fig. 5.2 Augmented reality rehabilitation exergames. The first picture (upper-left): A subject performing the holding and reaching cup game. The second picture (upper-right): The haptic glove for finger rehabilitation. The third picture (lower-middle): Two users play the basketball game. The screen on each terminal is split into two parts. The condition state of the hand (stable or unstable) is displayed during the running time of the game

program [26]. In addition, it provides the therapists with important data that helps them assess the rehabilitation status of the patients. In many studies, the Rutgers Master II-ND (RMII-ND) haptic glove has been used as a standard input haptic device for exergames [27–29]. The glove was proposed in 2002 as a replacement of another haptic robotic arm called PHANToM, the mostly used haptic interface at that time [30]. Perhaps the first use of RMII-ND was reported by Merians et al. [27] in which it was connected to a VR rehabilitation system and tested on three patients. In addition to the RMII-ND, the developed rehabilitation system consisted of another hand input device, namely, a CyberGlove and a personal computer. Four exergames were developed to improve the range of movement, the speed of movement, the fractionating of the fingers, and the strength of movement. All the developed exercises were targeting the upper limbs of stroke patients. In [28], Adamovich et al. used almost the same system (PC, CyberGlove, and RMII-ND) with a different set of games to help patients improve their range of motion, speed, independence, and strength of their affected fingers. The VR system was tested on eight subjects for two to two and one half hours each day over thirteen days. The same rehabilitation system was used by Boian et al. [29] with a set of exercises that targeted four finger injuries: range of motion of the finger, speed of motion of the finger, degree of independence of the finger, and the strength of the finger. The results of the three VR systems [27–29] were very encouraging. In fact, in [27], two of the three patients showed a significant improvement in their ADL tasks. In [28], six out of eight patients showed significant increase of their overall finger motor functions, while in [29] three of the four stroke patients showed a good degree of improvement. Jebsen Test of Hand Function (JTHF) was also used to evaluate the progress of stroke patients in the real-world. The results confirmed the fact that improvement of motor functions in the virtual world was reflected to the real-world. Another VR system with haptic feedback was developed by Reinkensmeyer [31]. A customized haptic joystick was used as a controller device for exergames. The joystick played a dual role during the rehabilitation exercises: it could physically assist or resist the movement of the patients while they were playing the game. The system was implemented using the Java programming language along with ASP and HTML. It consisted of two components: a client component, and a server component. The importance of such a VR system was that it was considered to be the first home-based rehabilitation system that connected stroke patients to their therapists through the internet. The system was tested on only one patient. The results showed that the subject improved on the Chedoke-McMaster Upper Extremity Scale, but showed no improvement on the Functional Test of the Upper extremity.

A new version of the PHANToM haptic device was used by Broeren et al. [32]. They developed a virtual reality environment to investigate the effect of using the sense of touch in a 3D environment in rehabilitation on stroke patients. The rehabilitation environment consisted of a computer game, haptic device, and stereoscopic shuttered glasses. Stereoscopy was used to give users an illusion of 3D virtual environments that they could interact with through the haptic device. The system was tested on five stroke patients over a period of five weeks. They collected three sets of measurements from each patient at three different periods. Results showed significant improvement of the patient's upper-limb kinematics; however, four of the five patients showed no improvement in performing their ADLs.

CyberGrasp is another important haptic device that has been used as an interface with virtual environment rehabilitation systems [33–36]. Almari et al. [33] used the CyberGrasp device with five virtual reality exergames to build an upper limb rehabilitation framework. The games were developed under the supervision of occupational therapists at the General Hospital of Ottawa and they were based on JTHF. Each game contained an exercise that was designed to help patients improve the strength and the range of motion of the affected upper limb. The proposed five exergames were evaluated by ten subjects and a usability study showed a great acceptance of the rehabilitation framework. Moreover, the analysis of the collected data helped the therapists to effectively evaluate the patients' medical status. In [34], McLaughlin et al. designed a set of virtual reality exergames for hand reaching and grasping exercises. CyberGrasp was used to capture the position and the orientation of the user's hand while the user was performing the tasks. The virtual environment could be displayed on a PC monitor or head mounted display. A very similar system was reported in [35]. A subjective assessment of the systems by two stroke patients showed an overall satisfaction of the whole rehabilitation experience. Another haptic/virtual environment rehabilitation system was proposed by Adamovich et al. [36]. One of the main advantages of the developed system is that it used intelligent algorithms to control the level of difficulty of the exergame and provide an adaptive haptic assistance to the patients. As a proof of concept, a virtual piano exergame was implemented to train an individual finger of the paralyzed hand at each exercise, and provide the user with auditory and visual feedback. The system was tested on four patients for ninety minutes per day in nine sessions. Two of the four patients showed improvement of their aggregate time in completing the task on the JTHF and three patients showed improvement in Wolf Motor Function test.

7 Exergame Virtual Reality with Assistive Devices

Many studies have suggested that repetitive training of the upper limb is very helpful in the rehabilitation course. However, stroke leaves the affected side of the body very weak in which training exercises become frustrating and tedious tasks. Many assistive devices have been introduced to help stroke patients interact with virtual reality during their rehabilitation regime [37–41]. Jack et al. [37] used CyberGlove with virtual reality exergames to train the fingers of the injured hand after stroke. CyberGlove is a commercial glove used for capturing finger movement data. It is mounted with eighteen piezo-resistive sensors to measure the angles of the fingers, palm arch, wrist flexion, and wrist abduction/adduction [42]. However, Jack et al. used the data captured by the two bend sensors on each finger to measure the angles of the thumb and fingers. Two virtual reality rehabilitation games have been developed which targeted the range-of-motion and speed-of-motion of the fingers. The system was tested on three stroke patients for nine days. All the three patients showed substantial changes on JTHF. In fact, one of the patients could button his shirt in the second week of the training program. Moreover, a subjective evaluation of the system showed positive feedback from all patients. All patients agreed that the program was effective; they felt that the movement of their fingers improved with the exercises. In [39], Connelly et al. designed and implemented a pneumatic glove (PneuGlove) for stroke rehabilitation that could be used for training grasp movements with real or virtual objects. One of the main advantages of the PneuGlove is that it can be used to assist finger extension while allowing the patients to fully move their affected arms. A head mounted display (HMD) is used to display the virtual environment that has been designed for stroke rehabilitation. The environment consists of a room in which the user can see his/her hand and the virtual objects inside it. Fourteen patients were recruited from the Rehabilitation Institute of Chicago Clinical Neuroscience Research Registry to test the proposed system. The subjects were randomly divided into two groups. The first group used the PneuGlove in their rehabilitation tasks while the second group used a shadow glove with the virtual environment. The rehabilitation program lasted for six weeks, three sessions a week with one hour training in each session. The Fugl-Meyer Assessment (FMA) test showed that there was an improvement of the clinical status of both groups, but even though the group using the PneuGlove showed a greater mean improvement than the second group, such improvement was not statistically significant. Morrow et al. designed and implemented two games that helped improve the velocity and range of motion of the fingers. The raw data captured from the glove finger sensors were calibrated to get valid data that could be used for further analysis. Although the system is significantly inexpensive compared to other systems, it is still unclear whether it would be beneficial for stroke rehabilitation since no real tests have been conducted on real patients to evaluate its effectiveness.

Robotic devices are garnering more interest for stroke rehabilitation to help patients improve motor recovery. Many studies have suggested that using robotics devices in rehabilitation have improved the overall motor function of patients [43–46]. More recently, robotic devices have been used in conjunction with virtual reality to provide patients with encouraging and entertaining environments that could help them continue their rehabilitation program [44].

The first interactive robotic device with 2-DOF was introduced by Hogan et al. [47] allowed therapists to monitor and control the rehabilitation of multiple patients without the need of the physical presence of the therapists. Several games were developed to motivate patients while they were conducting the exercises. Moreover, the patient's performance was recorded during the exercises for later evaluation. The proposed robotic device was evaluated on ninety-six stroke patients for two weeks, five days a week, and one hour per day. The exercises targeted improving the range of motion, force, direction and dexterity control of the patients. By the end of the study, patients showed a significant improvement in performing their daily life activities. Moreover, they gained strength in their shoulder and elbow muscles [48–50].

The effect of combining 3D video games with robotic devices on reaching in stroke patients was studied by Acosta et al. [46]. They used a 6 DOF robotic arm and developed computer software to track the positions of the arm, and hence find the maximum reaching point. Seven stroke patients participated in the experiment. They were asked to perform the reaching exercises in two different settings: while they were playing a video game, and while they were reaching for a specific target with visual feedback via a 3D avatar. Results showed reaching distances achieved by the patients in the avatar visual feedback setting were larger than those in the setting where they played the game. However, the results were not confirmed because, in certain conditions, the reaching distances in the game playing setting were much better than the avatar setting.

Recently, Sale et al. [51] conducted a study on fifteen stroke patients to determine long-term and short-term changes in their motor functions. They used the ReoGo robot device which had an arm with a platform to stabilize the patient's forearm. To motivate the patients, a user-friendly interface was developed with 3D objects displayed. The study was conducted on fifteen stroke patients over four weeks, five days a week for forty-minutes a session. The results showed that the patients had accepted the proposed robotic stroke therapy. Moreover, a statistically significant improvement of the patients' overall motor function was detected in the clinical assessment tests by the end of the study.

Robot-assisted stroke rehabilitation devices take advantage of low-cost tracking technologies to track real objects while patients perform their rehabilitation exercises. In [52], Loconsole et al. used Microsoft Kinect to identify and track generic objects to be moved by stroke patients. A robotic arm exoskeleton was used to assist patients to reach and move these objects. Although the proposed system (Kinect and a robotic arm) was not tested on stroke patients, the results showed that the system was robust to occlusions and light variations. Moreover, it was accurate in identifying and tracking generic objects. However, further studies are necessary to validate the correctness of the proposed system.

8   Performance Evaluation

There is no single standard measurement for the function of the upper extremities of patients with motor deficiencies. One of the main reasons is that therapists have not agreed on the definition of the word "function" [53]. Many researchers have developed quantitative tests in order to assess the improvement of patients. For each test there is a set of normative values, known as benchmarks, taken from normal people within different categories. Benchmarking is the basic step of understanding and evaluating the progress of patients during recovery from their disabilities. It is achieved by applying benchmarks, or simply markers, that form the basis of comparison between the data that has been gathered from stroke patients and the benchmarking data that has been drawn from healthy individuals. The importance of benchmarking has been shown in many studies [54–60]. Jebsen [57] provided the standard deviation, but not the norms, of seven tests used to assess the hand functions of two male and two female groups. The Australian version of the Jebsen test contains the norm and the standard deviation of eight tests for men and women within different aging groups, ranging from sixteen to ninety years [59]. Nagasaki [55] used an arm-rotator to find the kinematics measurements of human arm movements. The results of the benchmarks of four healthy men suggest that the asymmetric velocity and acceleration of the arm produced a minimum jerk and a smooth movement. In [61], a set of haptic-based hand exercises was created for use in stroke rehabilitation. The information gathered from five healthy subjects was used by occupational therapists to evaluate the progress of the patients under rehabilitation. Kim et al. [58] proposed a seven DOF robotic system (UL-EXO7) for stable post-stroke patients. Fifteen individuals were asked to play eight games that involved direct interaction with UL-EXO7. For each game, the angles of the joints were recorded and new conventional metrics were proposed to help therapists assess the patient's progress. A stress ball mounted with an accelerometer, gyroscope, and a magnetometer was used to find the angular tilting movements and their speeds, and the rotation around the z axis was used in [60]. The data was gathered from twelve healthy persons, and information about velocity, acceleration, jerk, and frequency were extracted. Such information is important for rehabilitation in virtual reality due to the lack of benchmarks for wrist kinematics.

9   Evaluated Parameters

We have provided a set of measurements of the most important kinematics parameters found in the literature of post stroke rehabilitation [54, 55, 58, 62]. These measurements will serve as ground truth therapists can depend on in their evaluation process. It is worth noting that this set of parameters changes depending on the rehabilitation exercises.

9.1   Direct Distance and Total Distance

The movement of the human arm is, in general, a point to point displacement. The trajectory is roughly a straight line, but in some cases, especially if there is a hidden deficiency in the arm, the trajectory will be a set of connected up and down lines. In such cases, the distance covered is not equal to the length of the straight line that connects the starting point of the experiment with the ending point, but rather a total distance that is the sum of all the short unsteady movements of the user. By comparing the values of these two parameters, the user can discover different impairments during the therapy in the early stages of the rehabilitation program.

9.2   Velocity

The velocity of the hand is an important factor in assessing its strength. Despite the fact that many performance evaluation tests, such as the Purdue Pegboard Test and the JHFT [57, 63], offer different normative data to accurately assess hand functions, they all agree on the significance of hand speed being a key factor in the evaluation process [64]. It has been shown in previous studies that the trajectory of the hand velocity in the real world tends to be bell-shaped [65]. The velocity of the upper hand in the x-direction is obtained by Eq. 5.1.

$$v = \frac{\Delta x}{\Delta t} \tag{5.1}$$

where $\Delta x = x_2 - x_1$ is the change in the displacement between t_2 and t_1, and $\Delta t = t_2 - t_1$.

9.3   Acceleration

Acceleration is defined as the rate of change of velocity (Eq. 5.2). It is well known that there is a strong correlation between the acceleration of the upper limb and the observed movement by an individual [65]. Moreover, therapists can use accelerometry measurements to distinguish between persons with normal or impaired upper limbs [66]. The acceleration is calculated in all three planes, but it is more sensitive to the x-y and y-z planes when the patient moves the affected hand in the horizontal and vertical planes, respectively. The magnitude of the acceleration vector is obtained by combining the values of accelerations at each axis and applying root-mean-square. The maximum acceleration is also extracted from the obtained data.

$$a = \frac{\Delta v}{\Delta t} \tag{5.2}$$

where $\Delta v = v_2 - v_1$ is the change in the velocity between t_2 and t_1, and $\Delta t = t_2 - t_1$.

9.4 Jerkiness

The jerkiness, or the change of acceleration with time, is used to measure the smoothness of movements. The smaller the value of jerk, the smoother the movement of the arm will be. In previous studies [55, 67], the jerkiness equation was studied to generate the smoothest trajectory between two points. In medicine, jerky movements can be caused by many diseases and conditions, namely, Angelman syndrome, movement disorders, primary dystonia, spastic paraparesis, and more. Researchers have found that "Jerk Cost," the time integral of the squared magnitude of hand jerkiness, is more effective at identifying the smoothness of the arm movement in space [68]. The cost of jerkiness is shown in Eq. 5.3:

$$J = \frac{1}{N} \sum_{i=1}^{N} \left(\frac{1}{2} \times \int_{0}^{T} a'^2(t) dt \right) \tag{5.3}$$

Where J is the jerk cost, N is the number of times the user has performed the movement, T is the time interval, and $a'(t)$ is the rate of change of acceleration. The zero-line cross is taken into consideration as another indication of the smoothness of hand motion. When the motion of the upper limb is controlled, both jerkiness and zero-line cross should be their optimum values.

9.5 Joints Angles

The joints create angles that need to be calculated in real-time when the user is moving his or her hands, wrists, elbows, and shoulders. To calculate the angle between the three joints representing essentially three vectors, the performance-efficient Eq. 5.4, which uses only one $acos()$ method call, is employed.

$$uv = P_{j1} - P_{j2}$$
$$wv = P_{j2} - P_{j3}$$

$$\theta = 0.5 * acos\left(2 * \left(\frac{\frac{(uv.wv)^2}{uv^2}}{wv^2}\right) - 1\right) \tag{5.4}$$

Equation 5.4 can be obtained by a simple derivation of the dot product formula between two vectors. In general, the angle between two vectors is equal to:

$$cos\theta = \frac{\overline{uv}.\overline{wv}}{|\overline{uv}| \cdot |\overline{wv}|} \tag{5.5}$$

squaring both side and substitute $cos^2\theta$ by $((1 + cos\ 2\theta)/2)$

$$\frac{(1 + cos\ 2\theta)}{2} = \left(\frac{\overline{uv}^2.\overline{wv}^2}{|\overline{uv}|^2.|\overline{wv}|^2}\right)$$

Now through cross multiplication and inverse of the cosine, Eq. 5.5 will be obtained.

10   Model Matching

For the evaluation of a patient's recovery progress, one requires a time series algorithm that computes the similarity between two time series regardless of their lengths. The time complexity of the algorithm should be reasonable in which the output could be obtained at real-time when the size of the data input is small.

10.1   Existing Model Matching Algorithms

Researchers have developed a variety of algorithms to find the distance between two time series. Most of these algorithms are based on dynamic programming in which the tackled problems are divided into simpler sub-problems. Dynamic Time Warping (DTW) [69] and Levenshtein Distance are the most popular algorithms that have been used to find the distance between two signals.

Dynamic Time Warping (DTW) is a well-known algorithm that is widely used to find the similarities between two temporal sequences. It was introduced in the sixties [70], and extensively used for data mining and speech recognition after that [71, 72]. In modern work, DTW is used in a variety of domains, including computer vision [73], gesture recognition [74], DNA sequence alignment [75], and much more.

Consider two time series sequences X and Y of lengths n and m respectively.

$$X = x_1, x_2, \ldots, x_i, \ldots x_n$$

Fig. 5.3 Optimal path
between the two sequences

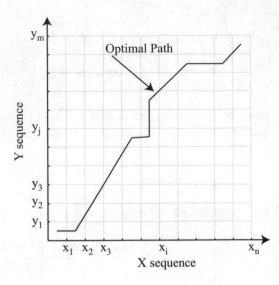

and

$$Y = y_1, y_2, \ldots, y_j, \ldots y_m$$

The smaller the distance between two different points x_i and y_j, the more similar they are.

Let c be defined as the local cost distance (Euclidian distance) between x_i and y_j such that:

$$c : X * Y \rightarrow \mathbb{R}$$

To find the alignment path, we first build a cost matrix in which each element in the matrix represents the distance between two different points from the two series. Then we find the shortest path that connects the starting and ending points of the minor diagonal of the matrix (Fig. 5.3). However, such an alignment path should satisfy the following boundary conditions:

1. Monotonicity: $x_i \leq x_{i+1}$ and $y_i \leq y_{i+1}$.
2. Continuity: The alignment path should advance one step at a time. This step could be taken in the right, up, or diagonal directions.
3. Boundaries: The alignment path starts at the bottom left point and ends at the upper right point of the matrix.

Taking into consideration the three previous constraints, the total cost P of the alignment path can be defined as:

$$P(n, m) = c(x_n, y_n) + \min \{ P(i-1, j), P(j-1, j-1), P(i, j-1) \}$$

When considering the movement pattern of a healthy subject (e.g., replacement, velocity, acceleration, jerkiness) as a template, the variance between the patient's movement pattern and this template indicates the recovery status of the patient. We can include the angles, the angular velocities, and other parameters in our template depending on the parameters captured during the exercises. Given a healthy subject's movement pattern (H):

$$H = \begin{cases} R_{healthy} = \left[R_{healthy,1} \cdots R_{healthy,m} \right]^T \\ V_{healthy} = \left[V_{healthy,1} \cdots V_{healthy,m} \right]^T \\ A_{healthy} = \left[A_{healthy,1} \cdots A_{healthy,m} \right]^T \\ J_{healthy} = \left[J_{healthy,1} \cdots J_{healthy,m} \right]^T \end{cases}$$

and the patient's movement pattern (P):

$$P = \begin{cases} R_{patient} = \left[R_{patient,1} \cdots R_{patient,m} \right]^T \\ V_{patient} = \left[V_{patient,1} \cdots V_{patient,m} \right]^T \\ A_{patient} = \left[A_{patient,1} \cdots A_{patient,m} \right]^T \\ J_{patient} = \left[J_{patient,1} \cdots J_{patient,m} \right]^T \end{cases}$$

the task is to find a correspondence matching between these two patterns:

$$\begin{aligned} \Theta \left(Healthy\ Subject's\ Pattern,\ Patient's\ Pattern \right) \\ = \left[\Theta \left(R_{healthy}, R_{patient} \right) \Theta \left(V_{healthy}, V_{patient} \right) \right. \\ \left. \Theta \left(A_{healthy}, A_{patient} \right) \Theta \left(J_{healthy}, J_{patient} \right) \right]^T \end{aligned}$$

where $R_{healthy}$, $V_{healthy}$, $A_{healthy}$, and $J_{healthy}$ are the replacement, velocity, acceleration, and jerkiness of the healthy subject. $R_{patient}$, $V_{patient}$, $A_{patient}$, and $J_{patient}$ are the replacement, velocity, acceleration and jerkiness of the patient subject. T is the transpose operator. The optimized correspondence relation $\hat{\Theta}$ is the minimum distance between the corresponding patterns [76]. In other words, the smaller the distance between two corresponding patterns, the more similarities they have in common.

$$\hat{\Theta} = argMin \begin{bmatrix} \sum_{i=1,j=1}^{i=m,j=n} \dfrac{d\left(R_{healthy,i}, R_{patient,j} \right) p_{i,j}}{\sum_{i=1,j=1}^{i=m,j=n} p_{i,j}} \\ \sum_{i=1,j=1}^{i=m,j=n} \dfrac{d\left(V_{healthy,i}, V_{patient,j} \right) q_{i,j}}{\sum_{i=1,j=1}^{i=m,j=n} q_{i,j}} \\ \sum_{i=1,j=1}^{i=m,j=n} \dfrac{d\left(A_{healthy,i}, A_{patient,j} \right) r_{i,j}}{\sum_{i=1,j=1}^{i=m,j=n} r_{i,j}} \\ \sum_{i=1,j=1}^{i=m,j=n} \dfrac{d\left(J_{healthy,i}, J_{patient,j} \right) k_{i,j}}{\sum_{i=1,j=1}^{i=m,j=n} k_{i,j}} \end{bmatrix}$$

By using the above optimization criteria, we conclude that we need to minimize the cumulative distance representing the unmatching part $D_{unmatch}$ which can be stated as:

$$D_{unmatch} = \begin{bmatrix} \sum_{i=1,j=1}^{i=m,j=n} \frac{d\left(R_{healthy,i},R_{patient,j}\right)p_{i,j}}{\sum_{i=1,j=1}^{i=m,j=n} p_{i,j}} \\ \sum_{i=1,j=1}^{i=m,j=n} \frac{d\left(V_{healthy,i},V_{patient,j}\right)q_{i,j}}{\sum_{i=1,j=1}^{i=m,j=n} q_{i,j}} \\ \sum_{i=1,j=1}^{i=m,j=n} \frac{d\left(A_{healthy,i},A_{patient,j}\right)r_{i,j}}{\sum_{i=1,j=1}^{i=m,j=n} r_{i,j}} \\ \sum_{i=1,j=1}^{i=m,j=n} \frac{d\left(J_{healthy,i},J_{patient,j}\right)k_{i,j}}{\sum_{i=1,j=1}^{i=m,j=n} k_{i,j}} \end{bmatrix}$$

Then the patient's rehabilitation status is modeled as a combination of the elements of $D_{unmatch}$, as shown in Eq. 5.6:

$$\begin{cases} S_{rehab} = w_1.D_{unmatch}[1] + w_2.D_{unmatch}[2] + w_3.D_{unmatch}[3] + w_4.D_{unmatch}[4] \\ \qquad\qquad w_1 + w_2 + w_3 + w_4 = 1 \end{cases}$$

$$(5.6)$$

11   Exergame Examples

Unlike a typical video game, designing exergames for rehabilitation requires close coordination and cooperation between therapists and game developers. For this reason, we have had brainstorming sessions with a therapist to design simple, yet effective games. The games need to be easy to play as stroke patients have physical limitations and may not be able to stand up for the entire game session. Moreover, users should be able to perform the exercises while they are standing up or sitting down, though for capturing more accurate data, standing is better than sitting. Two exergames have been designed and implemented: the Basketball game, and the Touching Cup game.

11.1   Basketball Game

The Basketball game is designed to measure the mobility of the upper limb of the patient in the vertical direction. The goal of this game is to train the patient on a simple repetitive exercise that does not require any cognitive challenge. Users see themselves standing in front of a ball and net. A user must grab a ball and move it vertically to the net, as shown in Fig. 5.4. The ball is placed at 45 degrees below the horizontal plane while the net is placed at 45 degrees above the plane, giving a total movement of 90 degrees. The timer starts the moment the user touches the ball and

Fig. 5.4 A subject playing the Basketball game using the Kinect cameras

stops when the patient reaches the net. During this period, we calculate the values of different parameters such as: direct distance, total distance traversed, velocity, angular velocity, acceleration, and jerkiness of the hand/arm. The game provides users with real-time feedback about the stability of their hand movements.

11.2 Touching Cup Game

The Touching Cup game is designed to help stroke patients restore the function of their upper limbs so they can perform their daily activities. The game is divided into several levels that correspond to essential physical tasks (horizontal and vertical movement) for stroke patients, such as attempting to reach for an object, opening a pill box to retrieve pills, and so forth. The first level includes the challenge of reaching the pill bottle. In the second level, the patients must flex the fingers of the opened arm and grasp the box. Each level contains additional challenges, so by the end of all levels the user will be able to take his/her medications independently. In the Touching Cup game, a multi-level trajectory line connects a real cup with a virtual one, and the patient is asked to connect the real cup with the virtual one by tracking the line. The line increases its slope by 15 degrees every time a level is completed successfully until it reaches 90 degrees with the horizontal. A timer is set to start the minute the patient touches the real cup and stops when the real cup reaches the virtual cup. A thorough description of the game implementation is presented in [25].

12 Developed Algorithms

Capturing accurate and reliable data at real-time are two key features for successful performance evaluation of the rehabilitation status of stroke patients. For this purpose, two algorithms are developed to capture, at real-time, the time needed to complete the task, the displacement, the velocities, the accelerations, the jerkiness, and the angular measurements of the paralysed upper limb of stroke patients. The extracted values are to be used in further analysis such as providing the appropriate feedback to users, evaluating their progress, and recommending new levels of rehabilitation exercises.

12.1 Angles Calculation

Algorithm 5.1 calculates the angles between two sequential bones in real-time. The importance of extracting the angles is two-fold: first, the angular measurements are used in advanced levels of the games, along with the other parameters, to determine the rehabilitation status of the patients; and second, these measurements are used to modify the level of difficulty of the game when the patient exceeds pre-set

Algorithm 5.1 Finding the Angle between two consecutive bones

---
Algorithm 1 Overview of the angles calculation using the Kinect camera.

Input:

 U_M : $Object\ U_M\ (User\ Model)\ that\ represents\ the\ person's\ skeleton\ standing\ in\ front\ of\ the\ kinect\ camera$

Output:

 α : $The\ angle\ between\ any\ two\ bones\ in\ the\ upper\ limb.$

1: $Loop$

2:   $q_u \leftarrow Skeleton().IsTracking(U_M)$

3:   $if\ q_u is\ true\ then$

4:    $Find\ S_B \leftarrow SkeletonJointPosition(U_M, J_i, N)$

    $//J_i\ is\ the\ set\ of\ joints\ to\ be\ tracked.$

    $//N\ is\ the\ number\ of\ tracked\ joints.$

    $//0 \leq i \leq N.$

5:    $\forall p, q\ in\ J_i,\ construct\ the\ set\ of\ lines\ L_{pq}\ between\ every\ consecutive\ joints\ by\ using\ the\ calculated\ positions\ S_B\ of\ these\ joints.$

6:    $\forall\ l_k, l_r \epsilon L\ (set\ of\ constructed\ lines)$

    $if\ l_k\ and\ l_r have\ a\ common\ joint\ then$

7:     $find\ the\ dot\ product\ of\ these\ two\ vectors$

     $//l_1.l_2 = |\ l_1\ | * |\ l_2\ | * \cos\alpha$

8:     $deduce\ and\ update\ the\ angle\ \alpha\ (inverse\ cosine)$

9:    $end\ if$

10:   $end\ if$

11: $end\ loop\ (when\ exit\ button\ pressed).$

12: $return\ \alpha.\ //End\ of\ Algorithm\ 1.$

---

angle threshold values. To describe the working process of the algorithm, the joint detection and joint angle determination data are presented. Every three consecutive joints of the upper limb form two vectors with the middle joint as the intersection point between them. A simple, yet effective way to find the cosine angle between the two vectors is by using their dot product. In our framework, we have calculated in real-time the angles of the wrist, the elbow, and the shoulder when the users moves his or her upper limb. An analysis of these angular results is further mapped with the game interaction commands.

12.2   Kinematics Calculation

Algorithm 5.2 calculates the basic parameters used to indicate the rehabilitation status of the patient. It extracts the kinematics by finding the coordinates of the tracked joints every 35 milliseconds which we found sufficient to capture the joints' coordination. After that, all that is required is to apply the basic kinematics functions to determine the velocity, acceleration, and jerkiness.

13   Games Sequence Flow

In this section, we will describe the flow diagram of the Touching Cup game (Fig. 5.5). We will also describe in detail various chart states along with their connectivity with the whole system.

13.1   Initialize Variables/Load Devices

The first basic step of the Touching Cup flowchart is to initialize different variables that the game needs for its operation. These variables may include the threshold time of the exercises (T_s), the rehabilitation status (R_s), and the game difficulty level. After the initialization of the variables, the framework starts identifying and loading the connected devices. Since all exercises require the Kinect camera to be connected to the computer, the framework checks whether the camera is connected. In case the camera is not connected, a message is displayed on the screen asking the user to connect the camera to operate the rehabilitation tasks. The framework continues loading the drivers of other assistive devices if they are connected to the rehabilitation framework. Moreover, the patient's profile is retrieved from the database at this stage of the initialization process.

Algorithm 5.2 Finding kinematics of the tracked joints

Algorithm 2 Finding the kinematics of a joint using kinect.

Input:

U_M : Object U_M (User Model) represents the skeleton
of the person standing infront of the kinect camera.

Output:

k : The kinematics measurements of the joints
in the upper limb.

1: Loop
2:     $q_u \leftarrow Skeleton().IsTracking(U_M)$
3:     if $q_u$is true then
4:       $g_s \leftarrow Game.IsStart( )$
5:       if $g_s$is true then
6:         Loop
7:           Find $S_{BI} \leftarrow SkeletonJointPosition(U_M, J, N)$
             //S_{BI} is the initial positions of the joints.
             //J is the set of joints to be tracked.
             //N is the number of tracked joints.
             //$0 \leq I \leq N$.
8:             $partialTime \leftarrow currentTime - initialTime$
           //$initialTime$ is the time at the start of the game.
           //It is rest every 35 milliseconds.
9:             if $partialTime \geq 50$ milliseconds then
10:          Find $S_{BC} \leftarrow SkeletonJointPosition(U_M, J, N)$
             //S_{BC} is the current positions of the joints.
11:    $\forall j_r$in J, and j_r is tracked, find kinematics of j_r
           //Velocity : $V_{j_r} = (S_{BCr} - S_{BIr})/partialTime$.
         //Acceleration; $A_{j_r} = (V_{BCr} - V_{BIr})/partialTime$.
         //Jerkiness : $J_{j_r} = (A_{BCr} - A_{BIr})/partialTime$.
         //Where S_{BCr} and S_{BIr} are the positions
         //of j_r at the end and at the beginning
         //of the time interval respectively

12:        end if
13:      end loop
14:      end if
15:    end if
16: end loop (when exit button pressed).
12: return kinematics. //End of Algorithm 2.

13.2   Start/Track Timer

The timer starts when the user touches the red ball which indicates the beginning of the exercise. The program tracks the timer and checks whether the user has touched the cup. The positions of the joints are recorded every 35 ms. The timer stops when the patient reaches the cup. To motivate the patient, the time is displayed on the screen while the patient is performing the exercise.

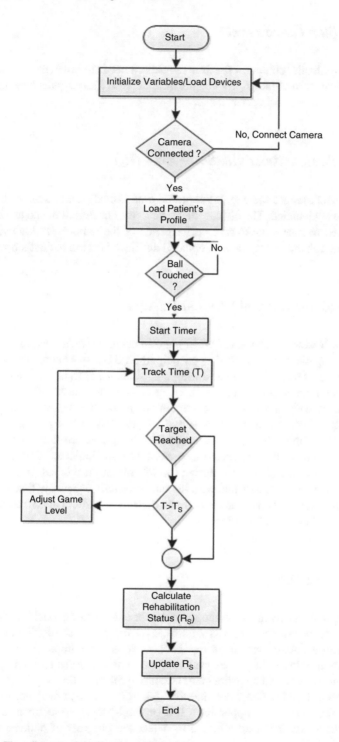

Fig. 5.5 A Flow diagram of the Touching Cup game

13.3   Adjust Game Level

The game controller compares the time the patient needs to finish the exercise with a predefined threshold time (T_s). When the time exceeds T_s, the game level is adjusted to meet the new patient's performance.

13.4   Calculate Rehabilitation Status (R_s)

When the user reaches the cup, the captured data is analyzed and the rehabilitation status (R_s) is calculated. The displacement, velocity, acceleration, and jerkiness may then be used to assess the overall performance of the patient. By the end of each exercise, the rehabilitation status is updated and saved on the patient's profile.

14   Experimental and Clinical Results

The Action Research Arm Test (ARAT), a clinical assessment test, has been taken by all stroke patients who agreed to participate in this experiment. To evaluate the recovery process of stroke patients during rehabilitation exercises, a low-cost virtual environment rehabilitation glove system has been introduced. This new system helps patients with upper limb impairments to reduce the degree of dysfunction. The novelty glove, which acts as a natural robotic arm, is designed to be worn on the unaffected limb. The FSR sensors of the glove measure the force exerted by the injured limb on the unaffected one. Results have indicated that the patients' limb movements have improved during the rehabilitation process, and the patients themselves were greatly satisfied with the entire system, especially that wearing the glove on the unaffected limb not only made their lives easier, but also allowed them to enjoy the rehabilitation session.

15   Clinical Test

Prior to these experiments, an orthopedic doctor and a professional physiotherapist evaluated the severity and impact of stroke on the patients. ARAT is considered a strong and accurate assessment test because it assesses upper limb functioning through observation unlike other available outcome measures that tend to rely on questionnaires answered by patients. ARAT has excellent intra-rater (ICC = 0.0989) and inter-rater (ICC = 0.0995) reliability [63, 76], and great evidence of criterion validity compared to the upper limb test of Fugl-Meyer Assessment and Motor Assessment Scale. It is also efficient to detect the progress of patients over time

[77]. Due to those advanced characteristics, professional association groups, such as Stroke Taskforce (StrokEdge) for chronic stroke, have adopted it as a highly-recommended measure outcome.

The test was divided into four subtests: grasp, grip, pinch, and gross movement. Each subtest contained ordered items graded from 0 to 3: "3", the highest grade means the subject has performed the task normally; "2" means the subject has completed the task but with some difficulties; "1" that the subject could not complete all the parts of the task; and "0" the subject has failed to complete any part of the task. The total number of items of the test is nineteen with a total score of fifty-seven, which indicates that the arm is functioning normally.

16 Rehabilitation Protocol

The rehabilitation exercises took place mainly in rehabilitation centers with exceptions for some severely injured patients. In the following, descriptions of the rehabilitation protocol are illustrated.

16.1 General Requirements

1. A Kinect Camera.
2. A laptop (at least 4 g RAM).
3. A room with minimum area of $4 \times 4\,\mathrm{m}^2$.
4. A cup.
5. Small set of shelves (three shelves of height 1 m, 1.4 m and 2.0 m).
6. A consent form to be completed by the subject.

16.2 Qualitative Test Protocols

1. The test subject should see a pre-recorded video that describes the tasks he/she is going to perform.
2. The minimum distance between the test subject and the camera should be 1.2 m, while the maximum distance should be 3.5 m.
3. Allow several minutes for the test subject to use the system before starting the official task. During this preparatory stage, it would be recommendable to talk to the subject about the importance of his/her cooperation and about the steps of the test, and to demonstrate some of the exercises.
4. If at any time during the test, the subject feels tired or uncomfortable, the test should be stopped immediately, but repeated sometime later.
5. The test subject should be facing the camera at all times.

16.3 Quantitative Test Protocols

1. The average set of kinematics (speed, acceleration, jerkiness, etc.) of each joint under testing is determined by taking the average kinematics of all the trials attempted on each day of the rehabilitation period.
2. The overall average is then determined by taking the average of the final data of each day.
3. The angles of the body joints under testing should be tracked at real-time and saved to a database every 35 milliseconds.
4. A statistical model should be provided by the end of the test showing the performance of the patient during the rehabilitation exercises.

17 Experiment

17.1 Subjects

Three post-stroke patients of average age 68.5 years were recruited to take part in this experiment. The patients who had not undergone any rehabilitation treatment other than the one provided by our rehabilitation system and suffered from chronic strokes (<10 months) were included in the study. The patients who had serious cognitive problems, able to move the injured limb without the support of the unaffected one, or fully dependent upon wheelchair were excluded from the study.

Patient One, Mr. S, 79 years old, suffered from a left cerebral artery (MCA) ischemic stroke four months prior to the experiment. The stroke hit him while drinking his tea by the fireplace on a cold winter night of 2014. It caused numbness to his face and right arm, and he was hit by whole body dizziness. He tried to stand up, but he instantly fell on the floor. He also suffered from hypertension and had been on medication ever since. Mr. S was admitted to hospital, wherein he was diagnosed with ischemic stroke on the right side of his body. The right arm became very weak to the extent that he could neither move nor control it. The stroke did not affect his cognitive abilities. Mr. S had not experienced a stroke previously.

Patient two, Mr. M, 67 years old, suffered from a right cerebral artery (MCA) ischemic stroke four months prior to the experiment. The stroke hit him while exercising. He had a sudden severe headache the minute he bowed his head down. Mr. M had never had headaches before. A few hours later, he felt numbness on the left side of his body. Not taking the symptoms seriously, Mr. M woke up the next morning with a completely paralyzed left side. His speech and vision were impaired, yet he did not lose consciousness. He was hospitalized for three months.

Patient three, Mrs. F, 60 years old, suffered from a right middle cerebral artery (MCA) ischemic stroke six months prior to the experiment. Mrs. F was diabetic and a heavy smoker as well. At night, she felt dizzy every time she tried to walk. At 4:00 AM the next morning, and while getting up, Mrs. F fell on the floor. She was rushed to hospital and was diagnosed with a severe ischemic stroke. Mrs. F had not been able to move her left upper limb prior to the experiment.

17.2   Virtual Environment System

The virtual environment system, which incorporates recent rehabilitation approaches, and combines efficient and affordable skeleton tracking input technologies with a multimodal interactive computer environment, consists of three main components: the upper limb tracking unit, motion analysis unit, and multimodal feedback unit. The Kinect, an affordable and control-free skeleton tracking sensor from Microsoft, is used as a 3D motion sensor to capture and track the movement of the upper limbs. The motion analysis unit evaluates the data it receives and compares it with that previously collected. At this stage, therapists are able to assess the patient's progress and determine the treatment paradigm effectiveness. The motion analysis unit sends requests to the game controller system to upgrade the difficulty level of the exergame as needed. The feedback unit provides the users with both visual and auditory feedback. The visual feedback, displayed in the progress bar on the left bottom of the screen, shows the progress of the patient depending on the real time data sent by the glove worn by the patient: red means the unaffected limb is performing the task, while green means that the injured limb is performing the task without the support of the unaffected limb. On the other hand, the auditory feedback reflects the overall performance of the user. The events that take place during the exercise and the data measurements received from the glove are all recorded. At the end of the task performed, the system plays the overall performance pre-recorded clip.

17.3   Method

During each rehabilitation task, the kinematics of the upper limb have been captured. By kinematics, this work refers to velocity, acceleration, and jerkiness. Moreover, the force exerted by the affected hand on the unaffected one has been measured. The rehabilitation session lasted for six weeks in which each patient had four sessions per week. The patients were asked to play the moving cup game under the supervision of a certified physiotherapist. A more detailed description of the design, approach and implementation of this game is provided in our previous work [25]. The ARAT clinical assessment has also been used to evaluate the motor functions of both the affected and unaffected limbs regardless of the severity of the stroke.

17.4   Results and Discussion

In week one of the study, all the work was done by the unaffected hand. Such a conclusion could be drawn by the fact that the force exerted by the affected hand on the unaffected hand was almost constant. However, the velocity curve had a bell-shape which indicated that the cup movement was smooth and steady. By the end of the experiment, results showed that the patients were able to partially depend on their affected hand. In fact, the force exerted by the unaffected hand on the affected hand had significantly decreased. Moreover, the velocity curve had more peaks than the first week. The clinical assessment (ARAT) score at the end of this study showed an improvement of the hands, the affected and the unaffected.

Our results are compatible with previous studies [14, 78, 79]. In [14], a virtual environment rehabilitation system along with conventional treatment was used to train and assess the performance of one group of stroke patients, while the second group used only the conventional treatment. At the end of the study, stroke patients who received virtual environment rehabilitation recovered faster than those who received only the conventional therapy. In [78], the time needed for stroke patients to complete a rehabilitation exercise was longer, by a significant amount, than that of non-stroke subjects. Additionally, the measurement values of the stroke patient velocities included a larger degree of variation than those of healthy subjects. Furthermore, the clinical test had revealed a significant correlation between stroke severity and affected hand kinematics.

References

1. Deckelbaum, R.J., Williams, C.L.: *Childhood obesity: the health issue.* Obesity research **9**(S11), 239S–243S (2001)
2. Manson, J.E., Skerret, P.J., Greenland, P., VanItallie, T.B.: *The escalating pandemics of obesity and sedentary lifestyle: a call to action for clinicians.* Archives of internal medicine **164**(3), 249–258 (2004)
3. Mandsager, K., Harb, S., Cremer, P., Phelan, D., Nissen, S.E., Jaber, W.: *Association of cardiorespiratory fitness with long-term mortality among adults undergoing exercise treadmill testing.* JAMA network open **1**(6), e183605-e183605 (2018)
4. Toscos, T., Faber, A., An, S., Gandhi, M.P.: *Chick clique: persuasive technology to motivate teenage girls to exercise.* In: *CHI'06 extended abstracts on Human factors in computing systems,* ACM (2006)
5. Maitland, J., Sherwood, S., Barkhuus, L., Anderson, I., Hall, M., Brown, B., Chalmers, M., Muller, H.: *Increasing the awareness of daily activity levels with pervasive computing.* In: *2006 Pervasive Health Conference and Workshops,* IEEE (2006)
6. Finin, T., Joshi, A., Kagal, L., Ratsimore, O., Korolev, V., Chen, H.: *Information agents for mobile and embedded devices.* In: *International Workshop on Cooperative Information Agents,* Springer (2001)
7. Hoda, M.: *SHECARE: Shared Haptic Environment on the Cloud for Arm Rehabilitation Exercises.* Doctoral dissertation, Université d'sOttawa/University of Ottawa (2016)
8. Hoda, M., Hoda, Y., Hage, A., Alelaiwi, A. El Saddik, A.: Cloud-based rehabilitation and recovery prediction system for stroke patients. *Cluster Computing, 18*(2), 803–815 (2015)

9. Maclean, N., Pound, P., Wolfe, C., Rudd, A.: *The concept of patient motivation: a qualitative analysis of stroke professionals' attitudes.* Stroke **33**(2), 444–448 (2002)
10. Choy, S., Wong, B., Simon, G., Rosenberg, C.: *The brewing storm in cloud gaming: A measurement study on cloud to end-user latency.* In: *Proceedings of the 11th annual workshop on network and systems support for games*, IEEE (2012)
11. Deutsch, J.E., Brettler, A., Smith, C., Welsh, J., John, R., Guarrera-Bowlby, P., Kafri, M.: *Nintendo wii sports and wii fit game analysis, validation, and application to stroke rehabilitation.* Topics in stroke rehabilitation **18**(6), 701–719 (2011)
12. Viau, A., Feldman, A.G., McFadyen, B.J., Levin, M.F.: *Reaching in reality and virtual reality: a comparison of movement kinematics in healthy subjects and in adults with hemiparesis.* Journal of neuroengineering and rehabilitation **1**(1), 11 (2004)
13. Keshner, E.A.: *Virtual reality and physical rehabilitation: a new toy or a new research and rehabilitation tool?* Journal of NeuroEngineering and Rehabilitation **1**(1), 8 (2004)
14. Turolla, A., Dam, M., Ventura, L., Tonin, P., Agostini, M., Zucconi, C., Kiper, P., Cagnin, A., Piron, L.: *Virtual reality for the rehabilitation of the upper limb motor function after stroke: a prospective controlled trial.* Journal of neuroengineering and rehabilitation **10**(1), 85 (2013)
15. Adamovich, S.V., Fluet, G.G., Tunik, E., Merians, A.S.: *Sensorimotor training in virtual reality: a review.* NeuroRehabilitation **25**(1), 29–44 (2009)
16. Rose, F.D., Brooks, B.M., Rizzo, A.A.: *Virtual reality in brain damage rehabilitation.* Cyberpsychology & behavior **8**(3), 241–262 (2005)
17. Henderson, A., Korner-Bitensky, N., Levin, M.: *Virtual reality in stroke rehabilitation: a systematic review of its effectiveness for upper limb motor recovery.* Topics in stroke rehabilitation **14**(2), 52–61 (2007)
18. Laver, K.E., Lange, B., George, S., Deutsch, J.E., Saposnik, G., Crotty, M.: *Virtual reality for stroke rehabilitation.* Cochrane database of systematic reviews (11), CD008349 (2017)
19. Holden, M., Todorov, E., Callahan, J., Bizzi, E.: *Virtual environment training improves motor performance in two patients with stroke: case report.* Journal of neurologic physical therapy **23**(2), 57–67 (1999)
20. Chang, Y.J., Chen, S.F., Huang, J.D., *A Kinect-based system for physical rehabilitation: A pilot study for young adults with motor disabilities.* Research in developmental disabilities **32**(6) 2566–2570 (2011)
21. Lange, B., Chang, C.Y., Suma, E., Newman, B., Rizzo, A.S., Bolas, M.: *Development and evaluation of low cost game-based balance rehabilitation tool using the Microsoft Kinect sensor.* In: *2011 Annual International Conference of the IEEE Engineering in Medicine and Biology Society*, IEEE (2011)
22. Obdržálek, Š., Kurillo, G., Ofli, F., Bajcsy, R., Seto, E., Jimison, H., Pavel, M.: *Accuracy and robustness of Kinect pose estimation in the context of coaching of elderly population.* In: *2012 Annual International Conference of the IEEE Engineering in Medicine and Biology Society*, IEEE (2012)
23. Clark, R.A., Pua, Y.H., Fortin, K., Ritchie, C., Webster, K.E., Denehy, L., Bryant, A.L.: *Validity of the Microsoft Kinect for assessment of postural control.* Gait & posture **36**(3), 372–377 (2012)
24. Levine, S.R., Gorman, M.: *"Telestroke" The Application of Telemedicine for Stroke.* Stroke **30**(2), 464–469 (1999)
25. Hoda, M., Dong, H., Ahmed, D., El Saddik, A.: *Cloud-based rehabilitation exergames system.* In: *2014 IEEE International Conference on Multimedia and Expo Workshops (ICMEW)*, IEEE (2014)
26. Durlach, P.J., Fowlkes, J., Metevier, C.J.: *Effect of variations in sensory feedback on performance in a virtual reaching task.* Presence: Teleoperators & Virtual Environments, **14**(4), 450–462 (2005)
27. Merians, A.S., Jack, D., Boian, R., Tremaine, M., Burdea, G.C., Adamovich, S.V., Recce, M., Poizner, H.: *Virtual reality–augmented rehabilitation for patients following stroke.* Physical therapy **82**(9), 898–915 (2002)

28. Adamovich, S.V., Merians, A.S., Boian, R., Lewis, J.A., Tremaine, M., Burdea, G.S., Recce, M., Poizner, H.: *A virtual reality—based exercise system for hand rehabilitation post-stroke.* Presence: Teleoperators & Virtual Environments **14**(2), 161–174 (2005)
29. Boian, R., Sharma, A., Han, C., Merians, A., Burdea, G., Adamovich, S., Recce, M., Tremaine, M., Poizner, H.: *Virtual reality-based post-stroke hand rehabilitation.* Studies in health technology and informatics 85, 64–70 (2002)
30. Bouzit, M., Burdea, G., Popescu, G., Boian, R.: *The Rutgers Master II-new design force-feedback glove.* IEEE/ASME Transactions on mechatronics **7**(2), 256–263 (2002)
31. Reinkensmeyer, D.J., Pang, C.T., Nessler, J.A., Painter, C.C., *Web-based telerehabilitation for the upper extremity after stroke.* IEEE transactions on neural systems and rehabilitation engineering **10**(2), 102–108 (2002)
32. Broeren, J., Rydmark, M., Björkdahl, A., Sunnerhagen, K.S.: *Assessment and training in a 3-dimensional virtual environment with haptics: a report on 5 cases of motor rehabilitation in the chronic stage after stroke.* Neurorehabilitation and neural repair **21**(2), 180–189 (2007)
33. Alamri, A., Iglesias, R., Eid, M., El Saddik, A., Shirmohammadi, S., Lemaire, E.: *Haptic exercises for measuring improvement of post-stroke rehabilitation patients.* In: *2007 IEEE International Workshop on Medical Measurement and Applications,* IEEE (2007)
34. McLaughlin, M., Rizzo, A., Jung, Y., Peng, W., Yeh, S., Zhu, W.: *Haptics-enhanced virtual environments for stroke rehabilitation.* In: Proc. IPSI (2005)
35. Rizzo, A.A., McLaughlin, M., Jung, Y., Peng, W., Yeh, S.C., Zhu, W.: *Virtual Therapeutic Environments with Haptics: An Interdisciplinary Approach for Developing Post-Stroke Rehabilitation Systems.* CPSN **5**, 70–76 (2005)
36. Adamovich, S.V., Fluet, G.G., Mathai, A., Qiu, Q., Lewis, J., Merians, A.S.: *Design of a complex virtual reality simulation to train finger motion for persons with hemiparesis: a proof of concept study.* Journal of neuroengineering and rehabilitation **6**(1), 28 (2009)
37. Jack, D., Boian, R., Merians, A.S., Tremaine, M., Burdea, G.C., Adamovich, S.V., Recce, M., Poizner, H.: *Virtual reality-enhanced stroke rehabilitation.* IEEE transactions on neural systems and rehabilitation engineering **9**(3), 308–318 (2001)
38. Khor, K.X., Chin, P.J.H., Rahman, H.A., Yeong, C.F., Su, E.L.M., Narayanan, A.L.T.: *A novel haptic interface and control algorithm for robotic rehabilitation of stoke patients.* In: *2014 IEEE Haptics Symposium (HAPTICS),* IEEE (2014)
39. Massie, T.H., Salisbury, J.K.: *The phantom haptic interface: A device for probing virtual objects.* In: *Proceedings of the ASME winter annual meeting, symposium on haptic interfaces for virtual environment and teleoperator systems,* Citeseer (1994)
40. Morrow, K., Docan, C., Burdea, G., Merians, A.: *Low-cost virtual rehabilitation of the hand for patients post-stroke.* In: *2006 International Workshop on Virtual Rehabilitation,* IEEE (2006)
41. Luo, X., Kline, T., Fischer, H.C., Stubblefield, K.A., Kenyon, R.V., Kamper, D.G.: *Integration of augmented reality and assistive devices for post-stroke hand opening rehabilitation.* In: *IEEE Engineering in Medicine and Biology 27th Annual Conference,* IEEE (2006)
42. Dipietro, L., Sabatini, A.M., Dario, P.: *A survey of glove-based systems and their applications.* IEEE Transactions on Systems, Man, and Cybernetics, Part C (Applications and Reviews) **38**(4), 461–482 (2008)
43. Coote, S., Murphy, B., Harwin, W., Stokes, E.: *The effect of the GENTLE/s robot-mediated therapy system on arm function after stroke.* Clinical rehabilitation **22**(5), 395–405 (2008)
44. Marchal-Crespo, L., Reinkensmeyer, D.J.: *Review of control strategies for robotic movement training after neurologic injury.* Journal of neuroengineering and rehabilitation **6**(1), 20 (2009)
45. Hesse, S., Schmidt, H., Werner, C., Bardeleben, A.: *Upper and lower extremity robotic devices for rehabilitation and for studying motor control.* Current opinion in neurology **16**(6), 705–710 (2003)
46. Acosta, A.M., Dewald, H.A., Dewald, J.P.: *Pilot study to test effectiveness of video game on reaching performance in stroke.* Journal of rehabilitation research and development **48**(4), 431 (2011)
47. Hogan, N., Krebs, H.I., Sharon, A., Charnnarong, J.: *Interactive robotic therapist.* Google Patents (1995)

48. Aisen, M.L., Krebs, H.I., Hogan, N., McDowell, F., Volpe, B.T.: *The effect of robot-assisted therapy and rehabilitative training on motor recovery following stroke.* Archives of neurology **54**(4), 443–446 (1997)
49. Volpe, B., Krebs, H., Hogan, N., Edelsteinn, L., Diels, C., Aisen, M.: *Robot training enhanced motor outcome in patients with stroke maintained over 3 years.* Neurology **53**(8), 1874–1874 (1999)
50. Volpe, B.T., Krebs, H.I., Hogan, N.: *Is robot-aided sensorimotor training in stroke rehabilitation a realistic option?* Current opinion in neurology **14**(6), 745–752 (2001)
51. Sale, P., Bovolenta, F., Agosti, M., Clerici, P., Franceschini, M.: *Short-term and long-term outcomes of serial robotic training for improving upper limb function in chronic stroke.* International Journal of Rehabilitation Research **37**(1), 67–73 (2014)
52. Loconsole, C., Banno, F., Frisoli, A., Bergamasco, M.: *A new Kinect-based guidance mode for upper limb robot-aided neurorehabilitation.* In: *2012 IEEE/RSJ International Conference on Intelligent Robots and Systems,* IEEE (2012)
53. Jayson, M.I.: *Quantification of Disability: Methods of Clinical Measurement and the Approach to the Problems of Disability.* Journal of the Royal Society of Medicine **67**(5), 400–401 (1974)
54. Alamri, A., Cha, J., El Saddik, A.: *AR-REHAB: An augmented reality framework for poststroke-patient rehabilitation.* IEEE Transactions on Instrumentation and Measurement **59**(10), 2554–2563 (2010)
55. Nagasaki, H.: *Asymmetric velocity and acceleration profiles of human arm movements.* Experimental brain research **74**(2), 319–326 (1989)
56. Beggs, W., Howarth, C.: *The movement of the hand towards a target.* Quarterly Journal of Experimental Psychology **24**(4), 448–453 (1972)
57. Jebsen, R.H., Taylor, N., Trieschmann, R., Trotter, M.J., Howard, L.A.: *An objective and standardized test of hand function.* Archives of physical medicine and rehabilitation **50**(6), 311–319 (1969)
58. Kim, H., Miller, L.M., Fedulow, I., Simkins, M., Abrams, G.M., Byl, N., Rosen, J.: *Kinematic data analysis for post-stroke patients following bilateral versus unilateral rehabilitation with an upper limb wearable robotic system.* IEEE transactions on neural systems and rehabilitation engineering **21**(2), 153–164 (2012)
59. Agnew, P., Maas, F.: *An interim Australian version of the Jebsen test of hand function.* Australian Journal of Physiotherapy **28**(2), 23–29 (1982)
60. Karime, A., Eid, M., Gueaieb, W., El Saddik, A.: *Determining wrist reference kinematics using a sensory-mounted stress ball.* In: *2012 IEEE International Symposium on Robotic and Sensors Environments Proceedings,* IEEE (2012)
61. Shakra, I., Orozco, M., El Saddik, A., Shirmohammadi, S., Lemaire, E.: *VR-based hand rehabilitation using a haptic-based framework.* In: *2006 IEEE Instrumentation and Measurement Technology Conference Proceedings,* IEEE (2006)
62. Pang, M.Y., Harris, J.E., Eng, J.J.: *A community-based upper-extremity group exercise program improves motor function and performance of functional activities in chronic stroke: a randomized controlled trial.* Archives of physical medicine and rehabilitation **87**(1), 1–9 (2006)
63. Lyle, R.C.: *A performance test for assessment of upper limb function in physical rehabilitation treatment and research.* International journal of rehabilitation research **4**(4), 483–492 (1981)
64. Hardin, M.: *Assessment of hand function and fine motor coordination in the geriatric population.* Topics in Geriatric Rehabilitation **18**(2), 18–27 (2002)
65. Morasso, P.: *Spatial control of arm movements.* Experimental brain research **42**(2), 223–227 (1981)
66. Uswatte, G., Giuliani, C., Winstein, C., Zeringue, A., Hobbs, L., Wolf, S.L.: *Validity of accelerometry for monitoring real-world arm activity in patients with subacute stroke: evidence from the extremity constraint-induced therapy evaluation trial.* Archives of physical medicine and rehabilitation **87**(10), 1340–1345 (2006)
67. Flash, T., Hogan, N.: *The coordination of arm movements: an experimentally confirmed mathematical model.* Journal of neuroscience **5**(7), 1688–1703 (1985)

68. Hogan, N.: *Control and coordination of voluntary arm movements.* in *1982 American Control Conference*, IEEE (1982)
69. Berndt, D.J., Clifford, J.: *Using dynamic time warping to find patterns in time series.* In: *KDD workshop*, Seattle, WA, USA (1994)
70. Bellman, R. Kalaba, R.: *On adaptive control processes.* IRE Transactions on Automatic Control **4**(2), 1–9 (1959)
71. Keogh, E.J. Pazzani, M.J.: *Scaling up dynamic time warping for datamining applications.* In: *Proceedings of the sixth ACM SIGKDD international conference on Knowledge discovery and data mining*, ACM (2000)
72. Furtună, T.F.: *Dynamic programming algorithms in speech recognition.* Revista Informatica Economică nr, **2**(46), 94 (2008)
73. Rath, T.M., Manmatha, R.: *Word image matching using dynamic time warping.* In: *2003 IEEE Computer Society Conference on Computer Vision and Pattern Recognition, 2003. Proceedings*, IEEE (2003)
74. Corradini, A: *Dynamic time warping for off-line recognition of a small gesture vocabulary.* In: *Proceedings IEEE ICCV workshop on recognition, analysis, and tracking of faces and gestures in real-time systems*, IEEE (2001)
75. Aach, J., Church, G.M., *Aligning gene expression time series with time warping algorithms.* Bioinformatics **17**(6), 495–508 (2001)
76. Yozbatiran, N., Der-Yeghiaian, L., Cramer, S.C.: *A standardized approach to performing the action research arm test.* Neurorehabilitation and neural repair **22**(1), 78–90 (2008)
77. Hsieh, Y.W., Wu, C.Y., Lin, K.C., Chang, Y.F., Chen, C.L., Liu, J.S.: *Responsiveness and validity of three outcome measures of motor function after stroke rehabilitation.* Stroke **40**(4), 1386–1391 (2009)
78. Cirstea, M. Levin, M.F.: *Compensatory strategies for reaching in stroke.* Brain **123**(5), 940–953 (2000)
79. Lambercy, O., Dovat, L., Yun, H., Wee, S.K., Kuah, C., Chua, K., Gassert, R., Milner, T., Teo, C.L., Burdet, E.: *Rehabilitation of grasping and forearm pronation/supination with the Haptic Knob.* In: *2009 IEEE International Conference on Rehabilitation Robotics*, IEEE (2009)

Chapter 6
Therapeutic Haptics for Mental Health and Wellbeing

Troy McDaniel and Sethuraman Panchanathan

Abstract Preliminary results in the literature have demonstrated how haptic technology can support mental health and wellbeing by addressing associated risks including abnormal levels of stress/anxiety as well as social isolation/exclusion. Given the importance of mental health on our personal, social, and professional lives, the aim of this chapter is to review therapeutic haptic technologies and their results for mental health and wellbeing to garner attention and interest from the community to investigate this topic more deeply. Included in this survey are commercial products, research prototypes, paradigms, and ideas, organized as follows: games, toys, and play; emotion regulation; stimulation therapy; distributed touch therapy; and haptics for social wellness. The final section, haptics for social wellness, presents a review of technologies to support and enrich social interactions, including devices to enable social touch from a distance (i.e., haptic interpersonal communication) as well as social assistive aids for individuals who are blind. Given the innovation seen within haptics for social wellness, this section is given the widest coverage, and surveys technologies, including those for therapy, play, and general communication, from a historical perspective. Finally, guidelines for designing mediators for haptic interpersonal communication are discussed.

1 Introduction

The World Health Organization (WHO) defines mental health as a state of wellbeing in which people can be resilient to reasonable amounts of stress, contribute to their communities, work productively, and realize their potential [1]. Risks to mental health are varied, complex, and person-specific, depending on biological,

T. McDaniel (✉)
The Polytechnic School, Arizona State University, Mesa, AZ, USA
e-mail: troy.mcdaniel@asu.edu

S. Panchanathan
Arizona State University, Tempe, AZ, USA
e-mail: panch@asu.edu

© Springer Nature Switzerland AG 2020
T. McDaniel, S. Panchanathan (eds.), *Haptic Interfaces for Accessibility, Health, and Enhanced Quality of Life*, https://doi.org/10.1007/978-3-030-34230-2_6

psychological, and personality factors [1]. A few reviews of technology to promote mental health are available. For example, Woodward et al. [2] present a review of technologies for supporting mental health and wellness including traditional tools, mobile health (mHealth) applications such as chatbots, tangible user interfaces, virtual reality, augmented reality, biofeedback, and advancements related to sensing emotions using smart phones and watches. Sanches et al. [3] conducted a systematic analysis of literature on human-computer interaction technologies and methods addressing depression and anxiety within the last ten years. Another survey by Eid et al. [4] reviews affective computing for health and wellness. While these survey efforts showcase the speed at which this field is growing, no thorough review of haptic technology for mental health exists. Given that preliminary research has demonstrated the potential of haptics to help individuals cope with and address a subset of risks to mental health including experiences of abnormal stress/anxiety and social isolation/exclusion, this chapter aims to fill a gap by surveying therapeutic haptic technology for mental health.

Haptic technology to support mental health and wellbeing has seen much less exploration compared to haptics for physical health and therapy (for surveys, see the following chapters in this book: Kurita et al. "*Assistive Soft Exoskeletons with Pneumatic Artificial Muscles*", Tadayon et al. "*Haptics for Accessibility in Hardware for Rehabilitation*", and Fong et al. "*Intelligent Robotics and Immersive Displays for Enhancing Haptic Interaction in Physical Rehabilitation Environments*"). This is surprising given the implications of poor mental health on personal, social, and professional life. There is thus a pronounced need for research in therapeutic haptics for mental health to design and evaluate technologies to promote psychological, emotional, and social wellness. This chapter surveys and organizes related work including research ideas, prototypes, findings, and commercial products into the following categories: games, toys, and play; emotion regulation; stimulation therapy; distributed touch therapy; and haptics for social wellness.

Section 2 Games, Toys, and Play provides a survey of serious games, toys, and environments supporting play where haptics is the central interaction paradigm. While serious games for mental health have been extensively explored within recent years, there has been little exploration of using haptics within purposeful games. This section presents technologies and ideas on robot-assisted therapy and activities, immersive augmented reality experiences, gamification of gardening, and interactive multisensory environments for children with autism.

Section 3 Emotion Regulation presents technologies to regulate (alter) users' emotional states using haptics. The most common approach within this category is guided breathing exercises via mobile applications, wearables, and immersive multisensory environments. Other examples include strategies to increase users' awareness of stress, and devices and environments to support meditation.

Section 4 Stimulation Therapy reviews therapies, treatments and technologies using haptic stimulation to target specific areas of the brain for therapeutic effects. Techniques and associated technologies covered include whole body vibrations, vibroacoustic therapy, bilateral alternating stimulation in tactile, deep pressure, thermal stimulation, and safe, controlled pain for sensory grounding. This section aims to present an overview of approaches rather than a comprehensive review and history of studies for each treatment type.

Section 5 Distributed Touch Therapy presents a review of technologies to mediate touch therapy (e.g., acupressure and massage). *Section 6 Haptics for Social Wellness* presents a survey of haptic technologies to support and enrich social interactions. In Sect. 6, the history of technologies for mediated interpersonal communication (i.e., social touch from a distance such as holding hands or hugging a long-distance partner, or social assistive aids for individuals who are blind) is presented. From a purely technological lens, of the aforementioned categories, haptic mediators for interpersonal communication have seen the most innovation, with research going back to the 1980s. Section 6's historical perspective offers insights into how both technology and our needs as a society have advanced.

2 Games, Toys, and Play

Over the last couple of decades, a variety of serious games for mental health have been proposed; yet more research is needed to explore the role of haptics in these games. This section focuses on play where touch is targeted as the key design aspect to elicit therapeutic outcomes. In this context, therapeutic haptics using serious games has seen little exploration compared to the much broader use of serious games for mental wellness.

Inspired by the well-known therapeutic benefits of animal-assisted therapy (AAT) and activities (AAA), researchers are exploring robots simulating animals in this context, i.e., robot-assisted therapy (RAT) and activities (RAA). Wada et al. [5] propose *Paro*, depicted in Fig. 6.1, a robot seal toy embedded with sensors and actuators to respond and react to stimuli to encourage shared play and activities. To enhance realism, *Paro* is capable of not only reactive behavior, but also proactive behavior (based on internal states of desire and rhythm) and physiological behavior

Fig. 6.1 Artist's rendition of *Paro*, a robot seal toy capable of responding and reacting to its environment

(based on basic needs such as sleep). *Paro* is designed to support mental health and well being, and such robots are known as *mental commit robots*. In their previous work, Wada et al. explored the use of *Paro* in pediatric therapy, and found increased communication and activities among this population. In [5], Paro was deployed in a day service center catering to elderly people to assess the impact of RAA on this population. Participants interacted with Paro one, two, or three days per week for five weeks. *Paro* was placed at a dedicated table, allowing up to eight participants to interact with it as a group for about twenty minutes at a time. Data assessing mood was collected before and after interacting with *Paro*. Mood was shown to improve after interactions with *Paro*, and participants felt vigor during their interactions. The nursing staff was also evaluated for burnout. They were asked to complete a burnout scale once per week beginning a week before *Paro* was introduced. The findings revealed a reduction in mental impoverishment of the nursing staff, perhaps by the improved moods of the patients, which in turn Wada et al. hypothesized reduced their dependency on the nursing staff.

Roo et al. [6] present *Inner Garden*, an ambient, augmented sandbox toy consisting of polymeric "wet" sand, an overhead projector to overlay graphics onto the sandbox, and physiological sensors to measure breathing and mental state. Users begin by "shaping" the world; manipulating and forming the sand to bring the world to life (e.g., day/night cycle, weather), which slowly grows over time (e.g., water, vegetation, creatures). Growth is based on users' mental state, detected by physiological sensors: if stress is high, the world "dries out"; if users remain calm (e.g., through breathing exercises), the world thrives. Two physiological signals are monitored: Breathing, using a stretch sensor; and brain activity, using EEG. These signals are linked to different aspects of the world; e.g., breathing is linked to sea level and day/night cycle; frustration level is linked to weather. This genre of game is known as the god game genre because users are world building. *Inner Garden* is aimed at supporting contemplation, mental awareness, and meditation. The toy combines the peace of a Zen garden with the play and exploration of a sandbox to create a relaxing experience.

In their follow-up work [7], Roo et al. add a virtual reality experience to *Inner Garden*, enabling users to go "into" the sandbox world. This was done by having the user place a miniature avatar at a location within the sandbox, and putting on a virtual reality headset to find him or herself at that location, sitting next to a campfire. The fire interacts with the user: as the user breathes out, the glow of the fire intensifies. This latest version of *Inner Garden* also maps physiological signals differently based on feedback from meditation experts and medical professionals familiar with meditation: breathing is mapped to water level and wind strength, and cardiac coherence is mapped to the overall "health" of the world (e.g., good health reduces clouds, increases the growth of vegetation, etc.). When an "unhealthy" state is detected, there are no explicit, obvious outcomes to avoid judgment, e.g., trees do not die; instead, more subtle indicators are used, e.g., animals hide. A big difference between the first and latest version of *Inner Garden* is that EEG is no longer used due to inaccuracies in recording. *Inner Garden* was tested with meditation experts of varying background and experience, who found the system useful for contemplation and mindfulness.

The therapeutic effects of gardening and tending to plants was exploited by Park et al. [8] who developed three plant-based games with touch-enabled interactions. In particular, plants were explored as controller interfaces, targeting their use in supporting mental health and wellness. Three games were designed around the concept of a plant pet. The game controls involve tapping, pinching, and patting plants. The capacitive touch sensing hardware used an electrode attached to the plant. The capacitive value read correlates with the surface area and strength of the touch. A user study was run to explore the impact of plant-based games on reducing anxiety level. Participants were recruited with higher than normal anxiety levels using the State-Trait Anxiety Inventory (STAI) survey. The control condition used a pressure pad instead of a plant. The experimental condition, but not the control condition, showed a decrease in anxiety level after playing the plant-based games.

Haptics for therapeutic play among children with autism spectrum disorders (ASD) has seen some exploration, largely between multisensory environments (MSE) and tangible user interfaces. For the remainder of this section, we present two MSE approaches, although the coverage here is not meant to be exhaustive. Parés et al. [9] propose *Mediate*, a MSE using multimodal stimulation to enable children with ASD and without verbal communication to express themselves. Their goal is to create a fun and engaging environment to help children with ASD gain a sense of agency/control and to feel comfortable. Their implementation avoids body-worn sensors, opting for all external sensors. Children interact with the system using simple movements and gestures (e.g., moving lateral, forward/backward, touching the screen, etc.). They developed a suite of interactive games, all involving manipulation of "particles" on a screen. The game that seemed to hold the most promise based on preliminary feedback was *Ta-to-mo*, a game where particles (on-screen tiles) grew/shrunk as the user moves closer/farther from the screen, respectively, eventually forming into a silhouette of the user. The tiles could be interacted with when touched, creating playful multisensory experiences.

Zalapa and Tentori [10] present *SensoryPaint*, an interactive multisensory environment to support sensory therapy for children with ASD. The interactive environment combines physical and digital elements to allow users to use tangible and body-based interactions to control an audiovisual painting application and game to enhance body awareness. A child interacts with balls of different colors/sizes, moving them in front of his or her body to paint on a wall via a projection. The child can also throw or kick the ball at the wall to create a "splash" effect. The child can change the music being played using his or her body. The system setup consists of two Microsoft Kinect devices to detect movement, a multimedia projector, speakers, and software for video streaming, tracking (OpenCV), rendering (OpenGL), and synthesizing (openAL).

In their follow-up work [11], functionality is expanded to include three modes: free-form, coloring book, and hitting moving targets. Another addition includes changing the color of the projection (e.g., limbs of the user) based on proximity to the screen. Two studies were conducted. The first study was a lab-based, one-hour study enrolling fifteen children with autism or other neurodevelopment disorders. Each participant experienced each mode, and could select one of the three modes to play again. Post-experiment interviews focused on impressions of the experience and suggestions. The second study deployed *SensoryPaint* within a rehabilitation clinic where four children with autism used the technology daily as part of their sensory therapy; but a much larger set of children also tried *SensoryPaint* during its installment at the clinic. The first phase involved participants using their normal therapy procedures (two weeks). The second phase used *SensoryPaint* (three weeks). Overall, through observations and interviews, *SensoryPaint* could effectively balance attention between body and stimuli, be easily integrated into existing therapies, and overall, participants found the interactive environment easy to use and fun.

3 Emotion Regulation

This section reviews ideas and technologies that use haptics to directly alter or regulate a user's current emotion. Some examples include guided vibrotactile breathing exercises or haptic cues to create awareness of current stress levels. Approaches falling under the umbrella of emotion regulation have garnered much research and commercial interest over the last few years. For example, the *Breathe* app on the latest versions of Apple's WatchOS reminds users to take a break to relax and focus on their breathing. Frequency of breaks and number of deep breaths per one-minute break are personalizable. On-screen instructions as well as haptic feedback walk users through the guided breathing exercises. Haptic feedback allows users to close their eyes during exercise. The remainder of this section surveys research prototypes investigating different techniques to regular emotion.

Using their custom-built device, *CalmMeNow*, Paredes and Chang [12] investigated guided acupressure and breathing using vibrotactile stimulation. For guided acupressure, a vibrating bracelet stimulates two acupressure points: the wrists, activated by the form factor itself, and the sternum, activated by the user holding his or her wrist against the body site. For guided breathing, the same vibrating bracelet communicates breathing rhythm, which the user mimics for a calming effect. Paredes and Chang conducted a study comparing these approaches to other intervention techniques including social networks and playing games. They found similarities between these intervention techniques and commonly used stress reduction techniques of positive thinking and visualization. This result demonstrates the potential for haptics as a method to reduce stress and provide relief.

Dijk and Weffers [13] present *Breathe with the Ocean*, shown in Fig. 6.2, a breathing guidance system using a multimodal experience of visual, auditory, and haptic stimuli. Users experience being at the shore of the ocean; they experience the sounds of the waves, see ambient lighting reminiscent of being at the beach (i.e., colors of yellow, orange, and blue), and feel the waves up and down their body, meant to indicate breathing rate. Haptic stimuli are delivered using the *Touch Blanket*, a blanket embedded with over one hundred vibration motors arranged in a two-dimensional grid. Different versions of the system were implemented and

Fig. 6.2 Conceptual sketch of *Breathe with the Ocean*, depicting the audio and haptic aspects of the user experience

tested. The first version is a non-adaptive, open-loop design in which the stimuli does not change based on the user's state. In the second version (closed loop), the stimuli follow the user's breathing pattern, i.e., the stimuli are adjusted to mimic the user's breathing as detected by a respiration belt. Whereas the first version used static lighting, in version 2.0, the intensity of the lighting changes based on the user's breathing. Version 3.0 is a closed-loop implementation that incorporates an adaptive, guided breathing exercise designed to achieve optimal breathing. The guided breathing is driven by physiological signals of respiration rate (measured using a respiration belt) and heart signal (blood-volume pulse, measured using a clip-on finger sensor). In an informal, preliminary pilot test, participants tried the three different versions of the system. The closed-loop outperformed open-loop implementation: users often found it difficult and uncomfortable to adapt their breathing to what was considered "normal" to the system. Everyone has their own unique breathing patterns, and the lack of person-centeredness and adaptation made version 1.0 difficult to use. Closed-loop systems were easier to use, and participants appreciated the multimodal stimuli. They found the haptic stimuli to be particularly pleasant, even more so when combined with audio.

Alonso et al. [14] propose three tangibles, each embedded with sensors to detect stress-related behaviors (e.g., fidgeting) and actuators to provide multimodal feedback to increase awareness of and reduce stress. *Squeeze It* is a handheld cylindrical object, which uses pressure sensors to detect squeezing of greater than normal force and/or frequency, triggering vibrotactile and visual feedback respectively. A blue light indicates a state of relaxation. *Marmoro* is a handheld object consisting of a row of rollable marbles, which a user can easily roll with his or her thumb. Sensors beneath the row of marbles detect rolling. On both sides of each marble are differently colored LEDs. When rolling is detected, solenoids on the back of the device are actuated to give the sensation of pressure on the palm. When stress-related behavior is detected (e.g., fast rolling), the red LEDs light up. When normal behavior is detected, the green LEDs blink. During the aforementioned stress-related behavior, blue LEDs are used to guide the user toward normal direction and speed of movement. If normal rolling is detected, indicating a relaxed state, the device vibrates. The final tangible is *Wigo*, a bulb-shaped, handheld device with a knob at the top where the user places his or her thumb and applies pressure to the side of the knob to rotate it. Sensors detect the distance, speed, duration, and frequency of the movement. Stress-related behavior (e.g., short, fast movements) cause the knob to be harder to rotate. The user is forced to slow down rotation, which in turn slowly reduces the friction of the knob, assisting the user at reaching a more relaxed state. A blinking blue LED communicates the desired frequency of the rotation needed to reach a relaxed state. Once the user achieves this relaxed state, the LED turns green. A baseline (stressless) state is captured when the user slowly rotates the knob, then pushes the knob down (the knob also functions as a button). *Wigo* was found to have the most sufficient feedback based on preliminary evaluations given its close coupling between movement and feedback.

Hernandez et al. [15] explored how vehicles may be equipped with low-cost sensors for stress monitoring, and how novel multimodal interactions (e.g., thermal stimulation, visualizations, and music) may be used to regulate emotion when stress is detected. While preliminary, the ideas show potential for haptics to be used in driving scenarios to reduce stress and anxiety.

SWARM, by Williams et al. [16], is a fashionable scarf embedded with actuators for emotional awareness and regulation. Their final prototype utilizes vibrotactile, thermal (heating), and auditory (music) stimulation to augment, not replace, users' existing coping strategies. A study found that users viewed warmth and music as useful intervention methods to change mood, whereas vibrations were viewed as a useful indicator of a change in emotion. *SWARM* was designed with individuals with disabilities in mind, and as such, user studies engaged this population. *SWARM* was also designed to support group emotion communication to help support the exchange and sharing of emotional states, especially for those with disabilities.

Kelling et al. [17] present *Good Vibes*, an athletic sleeve embedded with two C2 actuators to provide a dynamic vibration pattern simulating a calming touch sensation. A heart rate monitor is used to recognize changes in heart rate, which drives the actuators, delivering vibrotactile stimulation up and down the arm. A smartwatch app allows users to adjust the vibrations and monitor heart rate. Study results show potential for reducing heartrate in stressful situations, but a larger sample size is needed.

The *Soma Mat* and *Breathing Light*, developed by Ståhl et al. [18], support meditation and bodily awareness. The *Soma Mat*, see Fig. 6.3, is made of memory foam material and uses heat pads to deliver thermal stimulation to different body sites for directing users' attention to different body parts. With built-in speakers, *Soma Mat* can be used for guided meditation and other training sessions to enhance bodily awareness and relaxation. While laying on *Soma Mat*, users are enclosed in a peaceful, private environment by the string curtain of the *Breathing Light*. The light displayed by the *Breathing Light* dims in response to users' own breathing, measured by a proximity sensor estimating the displacement of the rise and fall of

Fig. 6.3 Artist's sketch of the *Soma Mat*. Thermal stimulation directs users' attention to different body sites to enhance meditation and bodily awareness

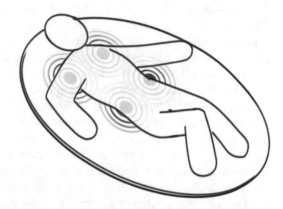

the chest. Speakers in the lamp can play any background sounds or music of the user's choice. The *Breathing Light* helps enhance the focus of one's own breathing for meditation and calmness.

Costa et al. [19] present *EmotionCheck*, a wrist-worn device, much like a watch, that simulates the sensation of a heartbeat using subtle vibrations. The device simulates a slow heartbeat to help users feel less anxious during anxiety and stress-inducing episodes. Key aspects of *EmotionCheck* are subtlety (not requiring much attention) and focus on regulating anxiety. An experiment was conducted involving three experimental conditions and one control condition. In the vibration experimental condition, subjects felt vibration pulses on their wrist at 60 bpm, and were not informed of their purpose. In the slow heart rate experimental condition, participants were told that the vibrations indicated current heartrate, but they were simulating a slow heartrate (60 bpm). In the real heart rate experimental condition, participants were told that the vibrations indicated current heart rate, and did adapt to match the subject's current heartrate. In the control condition, the device delivered no vibrotactile stimulation. The study design was between-subject. The experiment consisted of a baseline phase, then the stress test, consisting of a two-part process: a preparation phase followed by a presentation phase, for which the subject prepares then presents for their dream job as part of a simulated interview that is video recorded. The study results revealed: lower anxiety scores, collected using the State-Trait Anxiety Inventory pre and post intervention, for the slow heart rate experimental condition compared to the control group; and no difference between the vibration experimental group and the control group, indicating that vibrotactile stimulation alone as a distractor did not lower anxiety. While the real heart rate experimental group did not have higher anxiety scores compared to the control group, subjective feedback indicated subjects felt more stressed, anxious and nervous when they felt their heart rate increase. Participants did not find the vibrations distracting, and were able to notice them in the background while fully focused on the task at hand. The results of Costa et al.'s study have important implications: truthful biofeedback may increase anxiety during stressful situations.

The *Breathing Friend* [20] is a tangible, portable device to help users cope with stress using mindful breathing. The design is aimed at reducing technostress by appearing nothing like a computerized device; it takes the form of an embryonic shape with silicon skin. Inside, actuators provide the tactile "breathing" stimulation.

Miri et al. [21] introduce *HapLand*, a research platform for investigating haptic wearables for emotion regulation using a scalable biofeedback testbed. *HapLand* enables researchers to create and visualize haptic effects based on biofeedback or non-biofeedback, and conduct experimental studies leveraging the resources and infrastructure of the platform. Muri et al. use haptic stimulation given the evidence showing haptics can be used to elicit emotions. Muri et al. propose three ways in which haptics can support emotion regulation, all of which would be driven by physiological signals. First, Muri et al. propose that haptics can be used to gamify tasks (e.g., counting patterns) to redirect attention (or distract) away from a particular situation that may be negatively impacting emotion. Second, Muri et al. propose that haptic cues can be learned and associated with specific ideas and thoughts to help users focus on redefining the meaning of certain situations, e.g., a stressful situation helps build experience and character. Third, Muri et al. propose haptics for assisting users with altering their physiological and behavioral responses by either becoming aware of their current state (e.g., elevated heartrate, and therefore, taking steps to become calm) or the ideal state he or she could work toward (e.g., slow heartrate, calm and steady breathing). The platform consists of the following components: (i) physical hardware devices and sensors to capture physiological signals including GSR and ECG; (ii) logic driven by biosignals to decide actuator type, body site, and the dimensions of the stimulation pattern to activate; (iii) wireless wearables that receive commands from the logic component; (iv) an experimental design component; and (v) a visualization component. In its current implementation, *HapLand* supports two types of actuators (ERM or LRA) and four actuators of each type. Simple effects (e.g., pulses, ramps, etc.) are pre-defined. These simple effects may be combined into more complex effects.

Just Breathe [22] is an augmented in-car driving experience that provides therapeutic guided breathing exercises through vibrotactile feedback via the driver's seat. The work compares vibrotactile-guided with voice-guided exercises. The back rest of the driver's seat is embedded with a two-dimensional grid of over forty LRA vibrating motors. A variety of spatiotemporal vibrotactile patterns were investigated to explore their naturalness at communicating the stages of breathing (e.g., breath in, hold breath, breath out). A simple upward or downward "swipe" to communicate breath in and breath out, respectively, were found to be the most intuitive. No vibration indicated hold breath. Swipes were simulated using the perceptual illusion apparent tactile motion. The voice-guided exercise used simple voice commands "breathe in" and "breathe out". A driving simulator was used in the experiment. The experiment showed that while both inventions effectively reduced breathing rate and arousal level, participants preferred haptic feedback as they felt it was less obtrusive compared to spoken commands as well as more natural for communicating breathing pace give its continuous presentation versus the discrete nature of voice-commands.

4   Stimulation Therapy

Stimulation therapies and treatment programs focus on basic stimuli to target areas of the brain to achieve therapeutic goals. This section focuses on techniques using vibrotactile and pressure stimulation with some coverage of thermal stimulation.

Vaucelle et al. [23], inspired by touch therapy procedures found in sensory integration and sensory grounding therapies, propose four novel haptic devices: *Touch Me*, *Squeeze Me*, *Hurt Me*, and *Cool Me Down*. *Touch Me* consists of a large felt wrap embedded with an array of vibration motors, driven by a music keyboard such that intensity and body site of the vibrotactile stimulation can be varied. *Touch Me* allows patients and caregivers to conduct touch therapy sessions anytime over a distance. *Squeeze Me* consists of a vest of inflatable pneumatic chambers distributed around the torso and activated with the push of a button. *Squeeze Me* provides distributed stimulation to simulate the pressure of weighted vests/blankets and therapeutic holding. *Hurt Me* consists of a wearable bracelet of inflatable pneumatic bladders that push embedded plastic, rounded "teeth" into the skin for controlled, safe pain. *Cool Me Down* consists of a computerized wrap that may be applied to any part of the body for thermal stimulation at temperature variations that are safe. Technologies for "on-the-go" thermal stimulation are beginning to be explored and commercialized: *The Embr Wave*, by Embr labs[1], is a wrist-worn device for "anytime, anywhere" thermal stimulation for temperature comfort.

Vibroacoustic therapy (VAT), first proposed in the 1960s, combines music and low-frequency vibrotactile stimulation (typically 20–120 Hz) for relaxation and meditation. The health benefits of the multimodality of VAT, i.e., compared to music alone, have been well demonstrated. In a study by Koike et al. [24] involving fifteen elderly nursing home residents with depression, VAT was investigated for its therapeutic benefits for this population. Participants used the device for thirty minutes per day for two weeks (excluding weekends). Classical music was played through speakers embedded in the mattress near the head. Six vibration motors were embedded in the mattress near the upper body (shoulders) and lower body (waist and femoral region). Survey instruments (self-ratings), physiological signals (e.g., heart rate, tympanic temperature), and measures related to sleep duration and quality were collected. After VAT treatment, statistically significant reductions in heart rate and tympanic temperature were found. Further, statistically significant increases in waking hours during the day were found. Finally, survey instruments revealed statistically significant improvements in sadness and depression, indicating that VAT helped with relaxation. For interested readers, Punkanen and Ala-Ruona [25] present a detailed contemporary survey of VAT. The accumulated results of VAT demonstrate its potential for reducing tension, spasticity, pain, stress, and even challenging behaviors present in developmental disabilities and ASD. Punkanen and Ala-Ruona outline hypotheses around its therapeutic benefits including: (i)

---

[1] https://embrlabs.com/

relaxation response, by matching frequencies between VAT and the patient's own natural resonate frequency, targeting specific muscles; (ii) inhibition of pain through stimulation of the Pacinian corpuscles; and (iii) use of vibration to facilitate a cellular cleansing process.

Serin et al. [26] investigated the therapeutic effects of bilateral alternating stimulation in tactile (BLAST), which has been shown to reduce stress and lessen the physical sensations of stress by modulating electrical activity in brain areas mediating stress response. They conducted a study exploring whether BLAST can reduce both emotional stress and bodily stress using a novel implementation of the BLAST principle called *Touchpoints*. Over one thousand participants took part in the study. The ages of these participants ranged from twenty-one to forty-seven. All participants had purchased the commercially-available *Touchpoints* system. Using an app, participants were asked to rate their emotional stress and bodily distress on a scale from zero to ten before and after each thirty-second BLAST treatment. Results showed a significant reduction in both ratings after the treatment. While the study is not without limitations, e.g., it was difficult to exercise control over experimental procedures since data was collected in the wild without experimenter supervision, the results still demonstrate potential for BLAST as a therapeutic treatment for stress and anxiety. Much more research is needed in this space, particularly around the residual effects of BLAST treatment.

Another type of stimulation therapy uses deep pressure, such as weighted vests (Stephenson et al. [27]) and compression jackets[2], and joint compression, particularly as part of sensory integration therapy for children with ASD, ADHD, cerebral palsy, and other developmental disorders. Such activities and stimulation are believed to help calm subjects and reduce stereotypical behaviors by changing how sensory information is processed in the brain. Individuals with ASD and other developmental disorders often suffer from hyposensitivity and/or hypersensitivity to stimuli, and so repeated exposure to stimuli as part of sensory integration therapy and "sensory diets" is believed to be helpful. Finally, another stimulation therapy is the use of whole body vibration (WBV) to enhance cognitive capacity, especially attention in adults with ADHD [28].

5   Distributed Touch Therapy

Inspired by the therapeutic benefits of touch therapy (e.g., massage, acupressure) and social touch, researchers have extensively explored technology-mediated touch therapy for the last few decades, especially devices to mediate *interpersonal haptic communication* between remote partners. The next section presents a history of mediators for interpersonal communication, particularly social touch, as well as background information about social wellness and evidence of the therapeutic

---

[2]https://www.mytjacket.com/

Fig. 6.4 Artist's rendition of *TapTap*, a wearable felt scarf embedded with actuators to deliver therapeutic touch stimulation

benefits and importance of social touch. In this section, devices for distributed touch therapy are presented; that is, mediators that facilitate touch therapy from a distance where personal associations are largely removed, or are not the focus. These technologies are less concerned about connecting people for social engagement, and more focused on delivery of touch stimulation for treatment.

Bonanni et al. [29] propose *TapTap*, shown in Fig. 6.4, a wearable haptic device for recording and asynchronously distributing therapeutic touch. The main motivation behind *TapTap* is to provide therapists with a platform capable of recording therapeutic touch stimulation for broadcast to wide array of patients for use anytime, anyplace. The form factor is that of a felt scarf with six pockets for which up to six actuator sheets (i.e., actuation circuits mounted on thin plastic sheets) may be inserted. Users decide the type of actuation circuits, e.g., vibration, solenoid, etc., based on preference and need.

Stressed OutSourced (*SOS*) by Chung et al. [30] is a peer-to-peer network for mediated massage therapy to reduce stress. The system consists of embed-dable/attachable wearable modules for both signal sensing and actuation. The signal sensing modules communicate to *SOS* users an incoming *SOS* signal through a subtle stimulus, i.e., an ambient vibrotactile, thermal, or auditory cue that an anonymous user is requesting touch therapy. The receiver responds by touching the signal module's receptive surface, translating the touch input into a massage output, delivered by the actuator module. Actuators are arranged based on geographical location: actuators closer to the spine are dedicated to responding to touch therapy from users within a 10-mile radius, whereas those within the same country or international locations use actuators farther from the spine, respectively. *SOS* also provides a companion website to users, allowing access to *SOS* responses while concealing the identities of users. The sensing and actuation modules of *SOS* are depicted in Fig. 6.5.

Tang et al. [31] present a tactile sleeve for use in conjunction with virtual reality to help acclimate individuals with ASD to social touch. The sleeve is

Fig. 6.5 Conceptual sketch
of *SOS*, depicting one
possible instantiation of the
sensing module (top) and
actuation module (bottom)

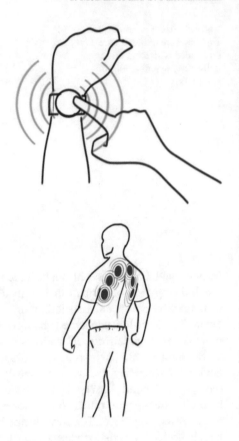

embedded with a four-by-three array of LRA vibrating motors to simulate social touch gestures using perceptual illusions of apparent tactile motion and phantom sensations. Besides being compatible with virtual reality, other design requirements of the sleeve include low cost to build and capable of reproducing "simple" touch cues (e.g., a tap) to "dynamic" touch cues (e.g., repetitive movements like rubbing). The following four touch gestures were implemented using the sleeve: finger poke, hand pat, stroke up, and stroke down. The work of Tang et al. focuses on comparing LRA and ERM motors to realistically convey simple and dynamic touch patterns through a user study involving non-ASD participants. The results of their study found smoother sensations of movements when the LRA motors were used to convey touch gestures.

Neidlinger et al. [32] introduce *AWElectric*, a wearable device aimed at producing and sharing the sensation of awe, often described as experiencing thrills, chills, and goosebumps; an experience often desired to be shared with others. Experiencing awe is known to have therapeutic benefits, both mentally and physically. *AWElectric* aims to create amplified sensations of awe and facilitate the sharing of such experiences with others. Both the sender and receiver wear a garment equipped with sensors and actuators. In the sender's wearable, biosensors detect feelings of awe, triggering the inflation of 3D printed fabric pieces (i.e., *Goosebump Fractals*) to mimic piloerection (hair standing on end) often accompanied by feelings of awe. The receiver's *Goosebump Fractals* are also activated as is their *AudioTactile fabric* using a modified speaker to create inaudible, low-frequency vibrations to induce goosebumps. While having many applications for mediated interpersonal haptic communication, this work is included in this section because it is easy to imagine such a technology being used for distributed touch therapy by enabling asynchronous sharing of awe-inducing stimulation.

Haritaipan et al. [33] present a mediated haptic interpersonal communication device for facilitating massage therapy. The device consists of two parts: A "sender" device that is human shoulder-shaped, intended to be massaged by the sender, the tactile interactions of which are recorded by force sensitive resistors; and a "receiver" device consisting of shoulder-worn "hands" embedded with vibration motors to deliver the massage. Only the duration of pressure is delivered by turning on/off the vibration motors; the strength of the massage is not conveyed. A user study revealed a stronger physical and emotional connection between remote participants (couples) while using the device during a video call compared to not using the device during a video call. Haritaipan et al.'s device has potential as a mediator for haptic interpersonal communication, but given its aim to facilitate massage therapy, their device is included in this section.

The Humming Wall [34–35] is a responsive urban installation in the form of a curved wall with places to sit. Accompanying *The Humming Wall* is a vibrotactile vest, the wearer of which can experience, through actuators, touch stimulation delivered to the wall (e.g., knocks and strokes) as visitors interact with their environment. The wall has two interactive zones: one where visitors see and hear their heartbeat, and another where visitors feel and hear their breath. The vibrotactile vest consists of thirty-two adjustable actuators, and sensors for reading heart rate and breath rate. The wall is equipped with actuators and sensors to provide multimodal (visual, haptic, and audio feedback) and recognize touches (knocks and swipes), respectively. The wall is divided into different parts, each varying functionally to create an exploratory interactive environment. Two parts of the wall provide biofeedback of heart rate and breath. Three parts of the wall relay different touch inputs to the vest. The wall was placed at Utzon Park in Aalborg, Denmark. *The Humming Wall* also gives off ambient humming and vibration to entice visitors to explore. In user studies, participants found their experiences engaging, sociable, and pleasant, with some subjects commenting that their experience was therapeutic, soothing, and calming.

6   Haptics for Social Wellness

Social wellness relates to the quality of our personal and professional relationships, and our effectiveness at building and maintaining such connections over time for psychological needs (e.g., meaningful relationships) and self-fulfilling needs (e.g., career goals). Healthy interpersonal relationships thrive on social touch and effective communication of emotions. Animal studies have found that nurturing touch is paramount to the survival of offspring, supporting mental and physical development and growth including reducing stress and fear, building immunity, increasing weight, and increasing physical activity, with similar findings for humans [36]. Through laboratory experiments, social touch conveying implicit, emotional content, e.g., hugging or holding hands, has been shown to have therapeutic benefits including lowering cortisol (stress), heart rate, and blood pressure, and increasing the love hormone, oxytocin [37]. Not surprisingly, long-distance romantic relationships without social touch may struggle to develop and grow [37]. Effective communication of intentions and emotions relies on access to not only verbal cues (i.e., speech), but as importantly, non-verbal cues (e.g., facial expressions, hand gestures, body language, and eye gaze). The majority of information exchanged during a social interaction is non-verbal (approximately 65% [38]), which puts individuals who are blind at a social disadvantage compared to their sighted peers. Social miscommunications may lead to embarrassing situations, and eventually social avoidance and isolation, which can cause psychological issues including social anxiety and depression [39].

Mediators for haptic interpersonal communication designed to facilitate the exchange of social touch cues from a distance have the potential to support intimate interactions and feelings of co-presence and togetherness between remote partners; and, more broadly, mediators for interpersonal communication between people at any distance (remote or colocated), such as social assistive aids for individuals who are blind, are paving way for accessible and enriched social interactions for individuals with disabilities. The term *haptic interpersonal communication* refers to any exchange of information between people through the sense of touch. When the communication channel is mediated, this information exchange happens through a *mediator*, or a device, system, or protocol facilitating the transfer of information between persons. Multimodal mediators engage more than one sense; e.g., touch, vision, hearing, and/or olfactory. This section focuses on mediators for social touch, and presents a historical perspective of the progress of this subfield. We also provide some coverage of mediators for interpersonal communication, focusing on research prototypes and technologies to enhance access to non-verbal cues for individuals who are blind.

Haptic mediators may be divided into three broad categories: *therapy*, *play*, and *general communication*. These categories are not distinct; often a technology will be applicable to two or more categories whether or not this was intended by the designer. Here, technologies are categorized based on their intended use. Games and playful activities involving touch (direct or indirect) normally require participants to be colocated, but through mediators, participants may be remote. The most common examples are video games played using controllers with haptic-feedback capabilities. Here, the focus is on technologies encouraging more connected, personable play. Social assistive aids as well as devices for mediated touch for which the intended purpose is implicit or nonverbal communication without a specific use beyond communication, are gathered under the broad category of general communication. As communication tools, these devices provide a medium for exchanging concrete to abstract information, and often, new forms of expression develop through use of these alternative channels of expression. Often, these technologies augment existing communication systems, e.g., instant messaging platforms, to provide redundant or complementary information.

6.1   Early Innovations in Haptics for Social Wellness (1980s–Early 2000s)

The majority of early mediators for maintenance of social wellness were aimed at intimate communication within a family or couple. Some of the earliest technologies were designed by Strong and Gaver [40]: *Feather, Scent* and *Shaker. Feather* is a uni-directional, two-device communication system that converts touch input into visual output between two long-distance partners. The traveling partner is in possession of a smart picture frame with a photo of his or her loved one. The picture frame is a portable device of sentimental value that can transmit a signal to a fixed device that the other partner has installed at home. The fixed device is a large structure, treated as furniture, in the shape of a transparent cone with a feather inside and fan at its base. When the picture frame is picked up and manipulated, the fan turns on, lifting the feather into the air, while the cone prevents the feather from escaping. When the partner at home notices the floating feather, he or she is reminded that the traveling partner is thinking of him or her. *Scent* is similar, but engages the sense of smell. Manipulating the picture frame activates a heating element at the bottom of a bowl, which vaporizes essential oils, producing a scent that fills the house. *Scent* is more pervasive than *Feather*, but both provide an abstract communication link that is emotionally meaningful. *Shaker* consists of two handheld devices each containing a solenoid, and is intended for play. When one device is shaken, a current is produced by the solenoid and sent to the other device, which activates the solenoid, causing the device to shake. Although there is not a one-to-one correspondence between haptic signals generated and subsequently felt, there is a correspondence between timing parameters and magnitudes.

Chris Dodge's *The Bed* [41], depicted in Fig. 6.6, was inspired by the rich communication channel afforded by a real bed when someone is resting next to you: you can feel your partner's movement, hear your partner's breath and whispers, and feel your partner's body warmth; all these sensory experiences combine to create a feeling of togetherness and intimacy. When couples are spatially separated, this sensory-rich experience is lost. *The Bed* attempts to recreate this experience of togetherness and intimacy for remote couples using two augmented beds equipped with a myriad of sensors and actuators. For example, when holding a body pillow, your partner feels your body warmth through a heating blanket inside his or her own body pillow. In the head pillow, a microphone captures your breathing, which is communicated to your partner by curtain sway, created by activating a fan that blows a curtain surrounding the bed.

Fig. 6.6 Sketch of user's experience in *The Bed* as user feels the body warmth and heartbeat of his or her remotely located partner

The concept of virtual hugs, or hugging from a distance, has been of interest to researchers since the early 2000s given the myriad of health benefits of social touch. One of the earliest devices is *The Hug* [42]: *The Hug* connects remotely-located persons using abstract anthropomorphic forms, each consisting of (i) sensors on its back to detect strokes, rubs and pressure, and a microphone to capture voice messages; and (ii) actuators including vibration motors and thermal fibers in and around its belly to convey captured tactile interactions as well as a speaker to deliver recorded voice messages. Other early concepts and devices that promote intimate communication from a distance include the *LoveBomb* [43] – emotional communication between strangers to make public spaces more personal; and *LumiTouch* [44] – communication of emotional and sentimental feelings through interactive picture frames.

The *Telephonic Arm-Wrestling device*[3] is one of the earliest known instances of mediated haptic interpersonal communication for play. The system connects two robotic arms over a telephone line, which allowed players in Toronto to arm wrestle opponents in Salerno or Paris. While latency limited the realism of the experience, the work was novel for its time. Besides *Shaker*, described earlier, another early entertainment device is *HandJive* [45], which consists of two interconnected spheres each held in a separate hand. The spheres can be shifted from their upright position either forward or background, together or separately, allowing nine possible combinations. Shifting a sphere causes the corresponding sphere to shift on your partner's *HandJive*, but from side to side rather than forward or backward. This protocol prevents users from fighting for control, i.e., users can create and send *HandJive* haptic signals while still receiving haptic signals from their partners without interference or interruption. Further, *HandJive* is proposed as a general haptic communication tool using the *Tactilese* language. Essentially, the smallest

---

[3]https://v2.nl/archive/works/telephonic-arm-wrestling

Fig. 6.7 Conceptual
rendering of *InTouch*,
demonstrating a user's
interaction with the three-pin
rollers

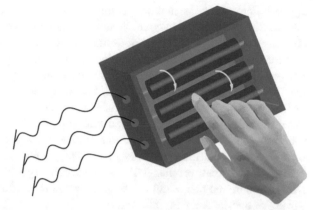

units (position of the spheres) are used to create simple movements (patterns), and
in turn, combined to create complex movements (routines).

One of the earliest mediated haptic interpersonal communication devices for
general communication is *InTouch* [46], depicted in Fig. 6.7. The system consists
of two three-pin rollers, each controlled by one partner. To communicate, a user
moves the rollers with his or her hand, and his or her partner's roller changes
accordingly. The communication channel is bi-directional, and its input and output
signals are mapped to the same channel (symmetric I/O mapping), operating on the
principle of a shared object, i.e., it is as if there is only one object being manipulated.
InTouch enables two types of interactions: passive, where the user's hand is placed
on the device to feel what his or her partner is communicating; or active, where
both users manipulate the object, and perceive its output, simultaneously. No formal
user testing was completed, but a pilot test revealed user interest and better use of
the device to convey abstract, subtle communication cues, such as those found in
intimate communication, as opposed to communication in general.

Another early system is Chang et al.'s *ComTouch* [47]: a vibrotactile glove for
complementing verbal information exchanged during a phone conversation. The
device is bi-directional and its I/O mapping is asymmetric. When a user, e.g., user
A, applies pressure through use of the glove, a vibrotactile signal is sent to his or
her partner, e.g., user B, where the intensity is proportional to the amount of applied
pressure. User B feels the vibration at his or her index metacarpophalangeal joint
(MPJ). User A also feels a vibration, but on his or her index proximal interphalangeal
joint (PIJ), and in the form of a feedback signal, enabling a way to assess the
intensity of the signal being sent to readjust the pressure as needed. To evaluate
the usefulness of *ComTouch*, two user studies were conducted by Chang et al.
In the first study, subjects participated in an audio conversation while using the
glove, but they were not instructed on how to use it – that was left for them
to decide. Experimenters observed that subjects created their own novel tactile
gestures, usually to (i) emphasize what they were saying by applying pressure while
saying certain words or phrases; (ii) indicate turn-taking by sending a vibratory

signal before speaking; and/or (iii) mimic the other user by exchanging the same vibrotactile pattern, which could be used to indicate presence or acknowledgment. Although simple in its conceptual design, *ComTouch*'s addition of a tactile channel provided a powerful form of expression, complementing the auditory channel with nonverbal information. In their second user study, participants had to perform the Desert Survival Negotiation Task, but using only the tactile channel afforded by the device. In this task, subjects had to sort a list of items based on their importance. Experimenters observed that subjects developed their own encoding scheme, deciding on tactile gestures indicating agreement and numbers. This user study showed the possibility of *ComTouch* for sensory substitution wherein information normally exchanged through speech might be exchanged through touch. Such a system may be useful in situations when the auditory channel is unavailable for use or overloaded with information.

Early work has also explored augmenting instant messaging (IM) with haptic signals to communicate nonverbal cues. For the *Haptic Instant Messenger*, Rovers and van Essen [48] implemented a platform to enrich instant messages (textual content and emoticons) with customizable haptic effects and programmable forces captured and delivered by devices of users' choosing. *Contact IM* [49] augments instant messages with force-feedback using a Phantom haptic device to enable users to chat while throwing around a virtual ball, providing a familiar, yet subtle and abstract form of expression. Other early innovations in mediated nonverbal and general communication include *HyperMirror* [50], *RobotPHONE* [51], *Hover* [52], and *FeelLight* [53].

6.2   Recent Work in Haptics for Social Wellness (2005–2019)

Within the last ten to fifteen years, interest in technology to support social wellness (non-verbal communication, co-presence, togetherness, and long-distance intimate communication) has steadily grown. Mediated haptic interpersonal communication has enjoyed much attention, particularly around topics such as simulating social touch [54–55] and facilitating intimate expressions [56]. Technological advances have aided this progress, enabling the development of mediators that provide more engaging, immersive, and/or veridical experiences. This section begins by surveying technologies supporting intimate communication, followed by play, and then finally, general communication.

Hug over a Distance [57] utilizes a koala bear plush toy with a PDA touch screen incorporated in its belly. When the touch screen is rubbed, a virtual hug is sent to the user's partner who wears a vest outfitted with air compartments that fill with air to simulate a hug. Other related devices for mediating virtual hugs or kisses include *HaptiHug* [58], *HugMe* [59], *Huggy Pajama* [60], and *Kissenger* [61]. The *Ring*U device [62] leverages the symbolism of a ring-shaped form factor to represent relationship promise, commitment, and connection between two people. *Ring*U uses visual and tactile effects for intimate, non-verbal communication. When a user squeezes his or her *Ring*U*, the paired device vibrates and shines based on fourteen different combinations of RGB values (specified through the sender's smartphone app) and squeeze pressure. *MemoryReel* [63] helps loved ones and family stay connected by supporting capture, persistence, and reminiscence of meaningful and sentimental memories within our digital social lives. The device consists of two tangible, anthropomorphic forms for which interpersonal distance sets the message type: far (text message), middle (voice message), and close (digital projection).

Other recent technologies to facilitate intimate communication and connectedness include *Patches* [64] – literal Facebook pokes through haptic wearables to support nonverbal and emotion communication or even play; and the *Tactile Sleeve for Social Touch (TaSST)* [65] – a touch-sensitive wearable sleeve enabling synchronous communication of a user's social touches (e.g., a poke, squeeze, stroke, etc.) to a paired *TaSST* device and delivered using vibration motors such that the stimulated body site and pressure intensity are kept consistent between input and output. For interested readers, see *Lover's Cups* [66], *Keep in Touch* [67], and *Implicit Emotion Communication* [68]. While outside the scope of this chapter, mediators to enable remote interactions between humans and pets is also building interest; for example, *Poultry.Internet* [69] mediates haptic interpersonal communication between humans and poultry.

Ambient displays for communicating and sharing physiological biofeedback (e.g., breathing, heart rate) have garnered much interest recently. *BreathingFrame* devices [70] are inflatable picture frames to support intimate communication, co-presence, and play between remote partners. Each user wears a belt embedded with a stretch sensor to detect breathing. The inhalation/exhalation of a user is mapped, in real-time, to the deflation/inflation of the picture frame of his or her partner. The partner may view and touch the picture frame, which is made of an inflatable latex surface, actuated using an air pump. Other contributions include *Breeze* [71] – a pendant, worn around the neck, that measures breathing and shares captured biofeedback through visual, auditory, and haptic modalities with both the user and potentially friends, family, and loved ones to increase connectedness and empathy; *CoupleVibe* [72] – a mobile application to connect long-distance couples using vibrotactile cues to implicitly alert a user about a change in his or her partner's predefined locations (e.g., at home, at work, going on lunch break, etc); *United-Pulse* [73] – communication of heartbeat pulses between partners; and *imPulse* [74] – enables two people to remotely share heartbeats using a wireless, lap-based device capable of sensing a pulse (sender places his or her hands on the device) and communicating partner's pulse (using vibrations and lights).

Intimate interaction and communication between humans and robots, see playful illustration of Fig. 6.8, has grown as an active area of research. Block et al. [75–76] investigated how robots should hug, i.e., how to design emotionally desirable robot hugs. They conducted a study to assess whether a robot hug similar to that of a human (similar in pressure, duration, warmth, and softness) would provide an emotionally satisfying hug. The Willow Garage Personal Robot 2 (PR2) was

Fig. 6.8 A playful depiction
of a robot sharing a gentle
hug with a human. Intimate
human-robot interaction is
garnering much research
interest

used. Three physical variations were evaluated: *Hard-cold condition: no padding*;
Soft-cold condition: additional padding (layers of foam, cotton and fabric); and
Soft-warm condition: heated (similar to soft-cold condition, but with a heating
element). Various behavioral conditions were also tested: Varying pressure levels
of the robot hug (appropriate pressure, loose, or tight) and varying timings of the
robot hug (short, release when human releases, or long, driven by the touch-sensitive
capabilities of the robot). Post-experiment survey responses suggest a preference
toward hugs of similar duration to the participants' own, and that hugs were more
comforting in the soft-warm condition. Overall, in soft conditions, the robot was
viewed as being safer.

   Mole Messenger [77] is reminiscent of the whack-a-mole game, aimed at
supporting sibling and parent-child remote communication, connectedness, and
play. *Mole Messengers* are paired communication devices, supporting equal communication through a one-to-one ratio of interaction, i.e., when one mole is pushed
down, the other pops up, and vice versa, detected by a force sensor at the base
of the mole. A force sensor on the head of the mole detects tactile gestures used
to communicate mood by color changes. This simple communication paradigm
uses an identifiable and fun anthropomorphic figure to allow children to focus on
play and shared interactions rather than complex textual communication. *PlayPals*
[78], depicted in Fig. 6.9, is a system to promote co-play between remotely-
located children. *PlayPals* are networked dolls, together with communication and
multimedia tools, wherein moving a limb on one doll causes the corresponding limb
to move accordingly on the other. Tangible tokens add functionality to *PlayPals*
such as the capability to capture and share multimedia content including audio,
images, and video. For example, a "walkie-talkie" token transforms *PlayPals* into
a voice communication system. A pilot test showed promising results wherein
children displayed interest in using the system, and new forms of communication
and expression were observed. *Toys Keeping in Touch* [79] is a network of
connected dollhouses enabling a child to interact with his or her playmates through

Fig. 6.9 Concept of *PlayPals* wherein remotely-located children interact through connected dolls; in this example, moving the arm of one teddy bear causes the arm of the other teddy bear to move

communication technologies (voice, text messages, and scanned images) built into the dollhouses and associated props.

Shake2Talk [80] is a cell-phone based system to support general communication through the use of gestural inputs, such as strokes and taps, to create audio-tactile messages. For example, a tapping gesture may generate the sound of gentle tapping and the sensation of someone tapping; such a message may be interpreted as the caller asking the recipient to call back soon. In a user study involving six couples [81], some couples developed a vocabulary, assigning meanings to certain messages; the majority of couples used the multimodal messages for coordination, e.g., "I'm on my way over", but the messages were also used for awareness/reassurance, play and social touch. Research continues to explore how to augment instant messages; interested readers should see *iFeel_IM!* [82], *Touch Emoticons* [83] and *FootIO chatter* [84].

Since the late 2000s, researchers have explored sensory substitution algorithms to convert visual non-verbal cues (e.g., facial expressions), extracted using computer vision algorithms, into haptic representations for accessible display to individuals who are blind. These algorithms are employed within social assistive aids – wearable or table-top systems to provide access to visual information exchanged during a social interaction. Researchers have mainly targeted non-verbal cues of facial expressions, emotions, interpersonal distance, and eye gaze.

Very little work has focused on improving access to eye gaze information for people who are blind. Qiu et al. [85] propose a headband embedded with vibration motors to map an interaction partner's eye gaze to sites around a user's head. Gaze dwell time (length of fixation) is mapped to tactile patterns: a short visual glance is presented as a quick vibrotactile burst, whereas a longer gaze is presented as a repeating vibration.

McDaniel et al. have explored enhancing social situational awareness for individuals who are blind by providing access to the relative direction and interpersonal distance of interaction partners. In [86], they present the *Haptic Belt*, a waist-worn belt embedded with equidistantly-spaced vibration motors, to convey direction and distance information of those around a user. The system is wearable and uses a discreetly embedded video camera in a pair of sunglasses to extract frames of a user's surroundings, which are then analyzed using computer vision algorithms to detect faces. Face detection drives the belt's actuators to communicate direction, delivered using body site and based on where faces appear in a frame; and interpersonal distances of intimate, personal, social, and public, communicated using vibration duration and based on the size of the detected face. Subsequent work by McDaniel et al. [87–88] investigated tactile rhythms for improving recognition of interpersonal distances.

Most research in this space has targeted enriching access to facial expressions and emotions. Using an office chair, the back of which was embedded with vibration motors arranged along three axes, Réhman et al. [89] investigated a one-to-one mapping of basic emotion to one of the three axes where emotion intensity was represented based on where the stimulation occurred along the axis. Buimer et al. [90] explored mapping basic emotions to the sides of a user's torso using a waist-worn vibrotactile belt. Krishna et al. [91] propose the *VibroGlove*, a glove embedded with vibration motors on the back of the hand to communicate basic emotions (happy, sad, surprise, anger, fear, and disgust) using spatiotemporal vibrotactile patterns inspired by emoticons. For example, a frown is felt as an upside-down U, with the leftmost motor actuated first, and the rightmost motor actuated last.

To expand the versatility of these systems beyond the six basic emotions, researchers have explored how facial action units (universal facial movements representing the building blocks of facial expressions [92]) can be mapped to vibrotactile representations [93–95] for use in social assistive aids. A plethora of software now exists for reliably extracting facial action units in real-time from video [96–98]. Most recently, McDaniel et al. [99] conducted a comparative study between individuals who are blind and sighted to better understand how well people who are blind can learn and recognize visual facial action units when presented through touch. Vibrotactile facial action units, refined over the course of a few studies, were presented to participants using the *Haptic Chair*, depicted in Fig. 6.10, an office chair embedded with a six-by-eight array of vibration motors on the back. Study results revealed individuals who are blind can recognize tactile facial action units as well as their sighted counterparts, hinting at the potential for this technology as an assistive aid for social wellness.

Of equal importance are machine learning and computer vision challenges toward extracting non-verbal cues from video for use in social assistive aids. Panchanathan et al. [100] present the *Social Interaction Assistant* (SIA) – a wearable system for vision-based extraction and haptic delivery of non-verbal information using active learning, conformal predictions, and topic models. The SIA performs person and facial expression recognition, and conveys this information to its wearer through audio and haptic devices including the previously described *Haptic Belt* [86], *VibroGlove* [91], and *Haptic Chair* [99].

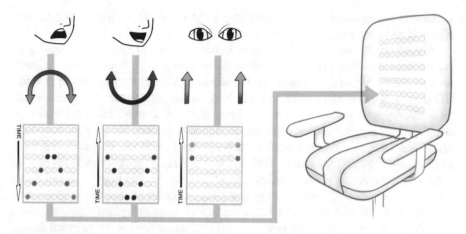

Fig. 6.10 Visual depiction of how three different facial action units would be display on the *Haptic Chair* (from left to right): "Lip Corner Depressor", "Lip Corner Puller", and "Upper Lid Raiser"

6.3 Key Design Considerations

This section summarizes helpful considerations when designing mediators for haptic interpersonal communication. Here, the theory behind designing such devices is explored through two facets: (1) design space, i.e., key parameters that will ultimately influence how a technology is used; and (2) associated design issues.

Chang et al. [101] propose a mediator design space consisting of four variables: *data direction*, *data transfer*, *I/O mapping*, and *data content*. Data direction refers to how information can be exchanged. If the information exchange is one-way, i.e., one device is dedicated to sending information, and another to receiving information, then the information flow of these devices is uni-directional. On the other hand, if both devices can send and receive signals, then information exchange is two-way, and the information flow of these devices is bi-directional. In general, bi-directional information exchange is more useful as it enables both interaction partners to exchange information. It also provides for a more rich and interesting communication channel. However, with respect to mediated communication, a uni-directional information exchange may still be very useful in certain contexts. In the previous survey of the field, examples of uni-directional (*Feather*, *Scent*) and bi-directional (*HandJive*, *InTouch*) devices were presented.

With bi-directional information exchange, the communication protocol becomes more complex; for example, should devices send and receive information at the same time with the possibility of interference and interruption (i.e., *asynchronous data transfer*)? On the other hand, with *synchronous data transfer*, a user could not send information if a signal is being received, and likewise, he or she could not receive information if a signal is being sent. From our everyday experiences with cell phones and email communication, it is obvious how useful asynchronous data transfer is for communication. It provides a communication channel that is rich, dynamic and spontaneous; see, for example, *ComTouch*.

Another complexity of bi-directional information exchange is the mapping of input and output channels. In a *symmetric I/O mapping*, the input and output signals are mapped to the same channel of communication. As one might expect, symmetric I/O mappings cause communication problems; in particular, users fight for control. In a simple, exploratory prototype by Fogg et al. [45], two fluid-filled balloons, connected by a tube, allowed two communication partners to interact with each other by squeezing a balloon to fill the other partner's balloon. However, this form of interaction caused fights for control, and in one case, competition, where one user attempted to make her partner's balloon pop. To alleviate what Fogg et al. call *direct resistance* encountered by communication partners, an alternate form of interaction was pursued wherein input and output channels were separated. The described balloon prototype was a preliminary concept evaluation toward the eventual design and development of *HandJive*, described earlier.

The last design variable to consider is data content. Data content may either be *discrete* or *analog*. Analog communication provides for infinite variations in the signal, enabling rich and interesting information exchange (e.g., see *ComTouch*). Discrete communication, however, may provide a limited channel for information exchange in that only a finite number of signals can be communicated; but that is not to say discrete data content is not useful. For example, although the *Haptic Instant Messenger* may not provide a rich communication channel on its own, it may be useful for enriching textual communication by engaging users' sense of touch, thereby creating a new channel for expression.

Chang et al. also suggested design issues that warrant further consideration: *feedback*, *flow of control*, and *language of touch*. Designers must decide whether input channels should provide users with feedback to gauge certain characteristics (e.g., quality) of the information being sent. The importance of feedback should be evident by how much we rely on it during both co-located and mediated social interactions; feedback helps guide our communication to ensure effective message delivery. Feedback may be particularly helpful when information is converted from one modality to another. For example, in *ComTouch*, based on the intensity and timing of the feedback signal, the sender will learn the mapping between pressure and vibration, and can adjust his or her messages accordingly. On the other hand, without feedback, it may be difficult to gauge how hard or soft to squeeze given the unfamiliar mapping between pressure and vibration.

Another important design issue is flow of control, i.e., the communication protocol, which concerns how control is obtained and transferred during a communication; or perhaps there is no concept of control such as is largely the situation during asynchronous communication. Flow of control should be carefully considered to ensure a smooth, intuitive communication protocol. At no point in the communication process should users be confused about whose turn it is to communicate, or how control is transferred between interaction partners.

Lastly, the language of touch should be considered. Haptic mediators provide a new channel for information exchange through the sense of touch. As with any type of communication signal, careful consideration must be given to how information is exchanged. This may be particularly true here given that touch is rarely used, outside of social touching during social interactions, for communication by the general population. The first design choice pertains to the kind of information that needs to be communicated. For example, the information may be alphanumeric (letters and numbers), or conceptual, i.e., concepts that evoke certain thoughts or emotions (e.g., a hug or pat on the back). Next, the mapping between information and the haptic signal must be decided. At one end of the spectrum, the mapping may be *literal*, i.e., the mapping is intuitive and familiar in that no training is required to learn it; on the other end, it may be *symbolic*, i.e., meanings are arbitrarily mapped to symbols, such as in languages, and therefore, this mapping must be learned and continually recalled.

There are more facets to the language of touch to be considered: the *form factor* through which the haptic signal will be delivered; and the *implementation* of the haptic signal, which raises many important questions: which *modality* of touch (pressure, vibration, temperature, etc.) will be engaged; what *dimensions* of the signal will be used (e.g., a vibratory signal has many dimensions that can encode information, such as amplitude, frequency, duration, etc.); what *encoding scheme* will be used to map meaning to the haptic signal; and lastly, will data content be discrete or analog.

Acknowledgments The authors thank Josh Chang and Megan McGroarty for their assistance with designing the graphics contained within this chapter. The authors thank Arizona State University and the National Science Foundation for their funding support. The preparation of this chapter is supported by the National Science Foundation under Grant No. 1828010.

References

1. World Health Organization: Mental health: Strengthening our response. https://www.who.int/news-room/fact-sheets/detail/mental-health-strengthening-our-response (2018).
2. Woodward, K., Kanjo, E., Brown, D., McGinnity, T.M., Inkster, B., Macintyre, D.J., Tsanas, A.: Beyond mobile apps: A survey of technologies for mental well-being. arXiv pre-print. 1905.00288 (2019).
3. Sanches, P., Janson, A., Karpashevich, P., Nadal, C., Qu, C., Daudén Roquet, C., Umair, M., Windlin, C., Doherty, G., Höök, K., Sas, C.: HCI and Affective Health: Taking Stock of a Decade of Studies and Charting Future Research Directions. In: Proceedings of the 2019 CHI Conference on Human Factors in Computing Systems. pp. 245:1–245:17. ACM, New York, NY, USA (2019). https://doi.org/10.1145/3290605.3300475.
4. Eid, M.A., Osman, H.A.: Affective Haptics: Current Research and Future Directions. IEEE Access. 4, 26–40 (2016). https://doi.org/10.1109/ACCESS.2015.2497316.
5. Wada, K., Shibata, T., Saito, T., Tanie, K.: Effects of robot-assisted activity for elderly people and nurses at a day service center. Proceedings of the IEEE. 92 (11), 1780–1788 (2004). https://doi.org/10.1109/JPROC.2004.835378.
6. Roo, J.S., Gervais, R., Hachet, M.: Inner Garden: An Augmented Sandbox Designed for Self-Reflection. In: Proceedings of the TEI '16: Tenth International Conference on Tangible, Embedded, and Embodied Interaction. pp. 570–576. ACM, New York, NY, USA (2016). https://doi.org/10.1145/2839462.2856532.

7. Roo, J.S., Gervais, R., Frey, J., Hachet, M.: Inner Garden: Connecting Inner States to a Mixed Reality Sandbox for Mindfulness. In: Proceedings of the 2017 CHI Conference on Human Factors in Computing Systems. pp. 1459–1470. ACM, New York, NY, USA (2017). https://doi.org/10.1145/3025453.3025743.
8. Park, T., Hu, T., Huh, J.: Plant-based Games for Anxiety Reduction. In: Proceedings of the 2016 Annual Symposium on Computer-Human Interaction in Play. pp. 199–204. ACM, New York, NY, USA (2016). https://doi.org/10.1145/2967934.2968094.
9. Parés, N., Carreras, A., Durany, J., Ferrer, J., Freixa, P., Gómez, D., Kruglanski, O., Parés, R., Ribas, J.I., Soler, M., Sanjurjo, À.: MEDIATE: An interactive multisensory environment for children with severe autism and no verbal communication. In: Third International Workshop on Virtual Rehabilitation, pp. 43–52 (2004).
10. Zalapa, R., Tentori, M.: Movement-Based and Tangible Interactions to Offer Body Awareness to Children with Autism. In: Urzaiz, G., Ochoa, S.F., Bravo, J., Chen, L.L., and Oliveira, J. (eds.) Ubiquitous Computing and Ambient Intelligence. Context-Awareness and Context-Driven Interaction. pp. 127–134. Springer International Publishing (2013).
11. Ringland, K.E., Zalapa, R., Neal, M., Escobedo, L., Tentori, M., Hayes, G.R.: SensoryPaint: A Multimodal Sensory Intervention for Children with Neurodevelopmental Disorders. In: Proceedings of the 2014 ACM International Joint Conference on Pervasive and Ubiquitous Computing. pp. 873–884. ACM, New York, NY, USA (2014). https://doi.org/10.1145/2632048.2632065.
12. Paredes, P., Chan, M.: CalmMeNow: Exploratory Research and Design of Stress Mitigating Mobile Interventions. In: CHI '11 Extended Abstracts on Human Factors in Computing Systems. pp. 1699–1704. ACM, New York, NY, USA (2011). https://doi.org/10.1145/1979742.1979831.
13. Dijk, E.O., Weffers, A.: Breathe with the Ocean: A system for relaxation using audio, haptic and visual stimuli. In: EuroHaptics 2010, pp. 47–60 (2010).
14. Alonso, M.B., Varkevisser, M., Keyson, D.V.: Expressive stress relievers. In: 7th Nordic Conference on Human-Computer Interaction: Making Sense Through Design, pp. 761–764 (2012).
15. Hernandez, J., McDuff, D., Benavides, X., Amores, J., Maes, P., Picard, R.: AutoEmotive: Bringing Empathy to the Driving Experience to Manage Stress. In: Proceedings of the 2014 Companion Publication on Designing Interactive Systems. pp. 53–56. ACM, New York, NY, USA (2014). https://doi.org/10.1145/2598784.2602780.
16. Williams, M.A., Roseway, A., O'dowd, C., Czerwinski, M., Morris, M.R.: SWARM: an actuated wearable for mediating affect. In: Proceedings of the Ninth International Conference on Tangible, Embedded, and Embodied Interaction. pp. 293–300. ACM (2015).
17. Kelling, C., Pitaro, D., Rantala, J.: Good Vibes: The Impact of Haptic Patterns on Stress Levels. In: Proceedings of the 20th International Academic Mindtrek Conference. pp. 130–136. ACM, New York, NY, USA (2016). https://doi.org/10.1145/2994310.2994368.
18. Ståhl, A., Jonsson, M., Mercurio, J., Karlsson, A., Höök, K., Banka Johnson, E.-C.: The Soma Mat and Breathing Light. In: Proceedings of the 2016 CHI Conference Extended Abstracts on Human Factors in Computing Systems. pp. 305–308. ACM, New York, NY, USA (2016). https://doi.org/10.1145/2851581.2889464.
19. Costa, J., Adams, A.T., Jung, M.F., Guimbretière, F., Choudhury, T.: EmotionCheck: Leveraging Bodily Signals and False Feedback to Regulate Our Emotions. In: Proceedings of the 2016 ACM International Joint Conference on Pervasive and Ubiquitous Computing. pp. 758–769. ACM, New York, NY, USA (2016). https://doi.org/10.1145/2971648.2971752.
20. Macik, M., Prazakova, K., Kutikova, A., Mikovec, Z., Adolf, J., Havlik, J., Jilekova, I.: Breathing Friend: Tackling Stress Through Portable Tangible Breathing Artifact. In: Bernhaupt, R., Dalvi, G., Joshi, A., K. Balkrishan, D., O'Neill, J., and Winckler, M. (eds.) Human-Computer Interaction – INTERACT 2017. pp. 106–115. Springer International Publishing (2017).
21. Miri, P., Flory, R., Uusberg, A., Uusberg, H., Gross, J.J., Isbister, K.: HapLand: A Scalable Robust Emotion Regulation Haptic System Testbed. In: Proceedings of the 2017 CHI

Conference Extended Abstracts on Human Factors in Computing Systems. pp. 1916–1923. ACM, New York, NY, USA (2017). https://doi.org/10.1145/3027063.3053147.

22. Paredes, P.E., Zhou, Y., Hamdan, N.A.-H., Balters, S., Murnane, E., Ju, W., Landay, J.A.: Just Breathe: In-Car Interventions for Guided Slow Breathing. Proceedings of the ACM on Interactive, Mobile, Wearable and Ubiquitous Technologies 2(1), 28:1–28:23 (2018). https://doi.org/10.1145/3191760.

23. Vaucelle, C., Bonanni, L., Ishii, H.: Design of Haptic Interfaces for Therapy. In: Proceedings of the SIGCHI Conference on Human Factors in Computing Systems. pp. 467–470. ACM, New York, NY, USA (2009). https://doi.org/10.1145/1518701.1518776.

24. Koike, Y., Hoshitani, M., Tabata, Y., Seki, K., Nishimura, R., Kano, Y.: Effects of Vibroacoustic Therapy on Elderly Nursing Home Residents with Depression. Journal of Physical Therapy Science. 24(3), 291–294 (2012). https://doi.org/10.1589/jpts.24.291.

25. Punkanen, M., Ala-Ruona, E.: Contemporary Vibroacoustic Therapy: Perspectives on Clinical Practice, Research, and Training. Music and Medicine. 4, 128–135 (2012).

26. Serin, A., Hageman, N.S., Kade, E.: The Therapeutic Effect of Bilateral Alternating Stimulation Tactile Form Technology on the Stress Response. JBBS. 1(2), 42–47 (2018). https://doi.org/10.14302/issn.2576-6694.jbbs-18-1887.

27. Stephenson, J., Carter, M.: The Use of Weighted Vests with Children with Autism Spectrum Disorders and Other Disabilities. J Autism Dev Disord. 39, 105 (2008). https://doi.org/10.1007/s10803-008-0605-3.

28. Fuermaier, A.B.M., Tucha, L., Koerts, J., van Heuvelen, M.J.G., van der Zee, E.A., Lange, K.W., Tucha, O.: Good Vibrations – Effects of Whole Body Vibration on Attention in Healthy Individuals and Individuals with ADHD. PLOS ONE. 9(2), e90747 (2014). https://doi.org/10.1371/journal.pone.0090747.

29. Bonanni, L., Vaucelle, C., Lieberman, J., Zuckerman, O.: TapTap: A Haptic Wearable for Asynchronous Distributed Touch Therapy. In: CHI '06 Extended Abstracts on Human Factors in Computing Systems. pp. 580–585. ACM, New York, NY, USA (2006). https://doi.org/10.1145/1125451.1125573.

30. Chung, K., Chiu, C., Xiao, X., Chi, P.-Y.: Stress Outsourced: A Haptic Social Network via Crowdsourcing. In: CHI '09 Extended Abstracts on Human Factors in Computing Systems. pp. 2439–2448. ACM, New York, NY, USA (2009). https://doi.org/10.1145/1520340.1520346.

31. Tang, F., McMahan, R.P., Allen, T.T.: Development of a low-cost tactile sleeve for autism intervention. In: 2014 IEEE International Symposium on Haptic, Audio and Visual Environments and Games (HAVE) Proceedings. pp. 35–40 (2014). https://doi.org/10.1109/HAVE.2014.6954328.

32. Neidlinger, K., Truong, K.P., Telfair, C., Feijs, L., Dertien, E., Evers, V.: AWElectric: That Gave Me Goosebumps, Did You Feel It Too? In: Proceedings of the Eleventh International Conference on Tangible, Embedded, and Embodied Interaction. pp. 315–324. ACM, New York, NY, USA (2017). https://doi.org/10.1145/3024969.3025004.

33. Haritaipan, L., Hayashi, M., Mougenot, C.: Design of a Massage-Inspired Haptic Device for Interpersonal Connection in Long-Distance Communication. Advances in Human-Computer Interaction. 2018, 5853474 (2018). https://doi.org/10.1155/2018/5853474.

34. Morrison, A., Manresa-Yee, C., Knoche, H.: Vibrotactile Vest and The Humming Wall: "I Like the Hand Down My Spine." In: Proceedings of the XVI International Conference on Human Computer Interaction. pp. 3:1–3:8. ACM, New York, NY, USA (2015). https://doi.org/10.1145/2829875.2829898.

35. Morrison, A., Manresa-Yee, C., Knoche, H.: Vibrotactile and vibroacoustic interventions into health and well-being. Universal Access in the Information Society 17(1), 5–20 (2018). https://doi.org/10.1007/s10209-016-0516-6.

36. Field, T.: Touch. MIT Press (2003).

37. Field, T.: Touch for socioemotional and physical well-being: A review. Developmental Review. 30(4), 367–383 (2010). https://doi.org/10.1016/j.dr.2011.01.001.

38. Knapp, M.L.: Nonverbal Communication in Human Interaction. Harcourt College, San Diego, CA, USA (1996).
39. Segrin, C., Flora, J.: Poor Social Skills Are a Vulnerability Factor in the Development of Psychosocial Problems. Human Communication Research 26(3), 489–514 (2000). https://doi.org/10.1111/j.1468-2958.2000.tb00766.x.
40. Strong, R., Gaver, B.: Feather, Scent, and Shaker: Supporting simple intimacy. In: ACM Conference on Computer-Supported Cooperative Work (1996).
41. Dodge, C.: The Bed: A Medium for Intimate Communication. In: CHI '97 Extended Abstracts on Human Factors in Computing Systems. pp. 371–372. ACM, New York, NY, USA (1997). https://doi.org/10.1145/1120212.1120439.
42. DiSalvo, C., Gemperle, F., Forlizzi, J., Montgomery, E.: The Hug: An exploration of robotic form for intimate communication. In: 2003 IEEE International Workshop on Robot and Human Interactive Communication. pp. 403–408. (2003).
43. Hansson, R., Skog, T.: The LoveBomb: Encouraging the Communication of Emotions in Public Spaces. In: CHI '01 Extended Abstracts on Human Factors in Computing Systems. pp. 433–434. ACM, New York, NY, USA (2001). https://doi.org/10.1145/634067.634319.
44. Chang, A., Resner, B., Koerner, B., Wang, X., Ishii, H.: LumiTouch: An Emotional Communication Device. In: CHI '01 Extended Abstracts on Human Factors in Computing Systems. pp. 313–314. ACM, New York, NY, USA (2001). https://doi.org/10.1145/634067.634252.
45. Fogg, B., Cutler, L.D., Arnold, P., Eisbach, C.: HandJive: A Device for Interpersonal Haptic Entertainment. In: Proceedings of the SIGCHI Conference on Human Factors in Computing Systems. pp. 57–64. ACM Press/Addison-Wesley Publishing Co., New York, NY, USA (1998). https://doi.org/10.1145/274644.274653.
46. Brave, S., Dahley, A.: in Touch: A Medium for Haptic Interpersonal Communication. In: CHI '97 Extended Abstracts on Human Factors in Computing Systems. pp. 363–364. ACM, New York, NY, USA (1997). https://doi.org/10.1145/1120212.1120435.
47. Chang, A., O'Modhrain, S., Jacob, R., Gunther, E., Ishii, H.: ComTouch: Design of a Vibrotactile Communication Device. In: Proceedings of the 4th Conference on Designing Interactive Systems: Processes, Practices, Methods, and Techniques. pp. 312–320. ACM, New York, NY, USA (2002). https://doi.org/10.1145/778712.778755.
48. Rovers, A.F., van Essen, H.A.: HIM: A Framework for Haptic Instant Messaging. In: CHI '04 Extended Abstracts on Human Factors in Computing Systems. pp. 1313–1316. ACM, New York, NY, USA (2004). https://doi.org/10.1145/985921.986052.
49. Oakley, I., O'Modhrain, S.: Contact IM: Exploring asynchronous touch over distance. In: ACM Conference on Computer-Supported Cooperative Work (2002).
50. Morikawa, O., Yamashita, J., Fukui, Y.: The Sense of Physically Crossing Paths: Creating a Soft Initiation in HyperMirror Communication. In: CHI '00 Extended Abstracts on Human Factors in Computing Systems. pp. 183–184. ACM, New York, NY, USA (2000). https://doi.org/10.1145/633292.633393.
51. Sekiguchi, D., Inami, M., Tachi, S.: RobotPHONE: RUI for Interpersonal Communication. In: CHI '01 Extended Abstracts on Human Factors in Computing Systems. pp. 277–278. ACM, New York, NY, USA (2001). https://doi.org/10.1145/634067.634231.
52. Maynes-Aminzade, D., Tan, B.-K., Goulding, K., Vaucelle, C.: Hover: Conveying Remote Presence. In: ACM SIGGRAPH 2002 Conference Abstracts and Applications. pp. 194–194. ACM, New York, NY, USA (2002). https://doi.org/10.1145/1242073.1242207.
53. Suzuki, K., Hashimoto, S.: Feellight: A Communication Device for Distant Nonverbal Exchange. In: Proceedings of the 2004 ACM SIGMM Workshop on Effective Telepresence. pp. 40–44. ACM, New York, NY, USA (2004). https://doi.org/10.1145/1026776.1026786.
54. Haans, A., IJsselsteijn, W.: Mediated social touch: a review of current research and future directions. Virtual Reality. 9(2), 149–159 (2006). https://doi.org/10.1007/s10055-005-0014-2.
55. Huisman, G.: Social Touch Technology: A Survey of Haptic Technology for Social Touch. IEEE Transactions on Haptics. 10(3), 391–408 (2017). https://doi.org/10.1109/TOH.2017.2650221.

56. Hassenzahl, M., Heidecker, S., Eckoldt, K., Diefenbach, S., Hillmann, U.: All You Need is Love: Current Strategies of Mediating Intimate Relationships Through Technology. ACM Transactions on Computer-Human Interaction 19(4), 30:1–30:19 (2012). https://doi.org/10.1145/2395131.2395137.
57. Mueller, F. "Floyd," Vetere, F., Gibbs, M.R., Kjeldskov, J., Pedell, S., Howard, S.: Hug over a Distance. In: CHI '05 Extended Abstracts on Human Factors in Computing Systems. pp. 1673–1676. ACM, New York, NY, USA (2005). https://doi.org/10.1145/1056808.1056994.
58. Tsetserukou, D.: HaptiHug: A Novel Haptic Display for Communication of Hug over a Distance. In: Kappers, A.M.L., van Erp, J.B.F., Bergmann Tiest, W.M., and van der Helm, F.C.T. (eds.) Haptics: Generating and Perceiving Tangible Sensations, EuroHaptics '10, LNCS 6191. pp. 340–347. Springer Berlin Heidelberg (2010).
59. Cha, J., Eid, M., Rahal, L., Saddik, A.E.: HugMe: An interpersonal haptic communication system. In: 2008 IEEE International Workshop on Haptic Audio Visual Environments and Games. pp. 99–102 (2008). https://doi.org/10.1109/HAVE.2008.4685306.
60. Cheok, A.D., Zhang, E.Y.: Huggy Pajama: Remote Hug System for Family Communication. In: Cheok, A.D. and Zhang, E.Y. (eds.) Human–Robot Intimate Relationships. pp. 33–75. Springer International Publishing, Cham (2019). https://doi.org/10.1007/978-3-319-94730-3_3.
61. Cheok, A.D., Zhang, E.Y.: Kissenger: Transmitting Kiss Through the Internet. In: Cheok, A.D. and Zhang, E.Y. (eds.) Human–Robot Intimate Relationships. pp. 77–97. Springer International Publishing, Cham (2019). https://doi.org/10.1007/978-3-319-94730-3_4.
62. Choi, Y., Tewell, J., Morisawa, Y., Pradana, G.A., Cheok, A.D.: Ring*U: A Wearable System for Intimate Communication Using Tactile Lighting Expressions. In: Proceedings of the 11th Conference on Advances in Computer Entertainment Technology. pp. 63:1–63:4. ACM, New York, NY, USA (2014). https://doi.org/10.1145/2663806.2663814.
63. Wei, H., Hua, D., Blevis, E., Zhang, Z.: MemoryReel: A Purpose-designed Device for Recording Digitally Connected Special Moments for Later Recall and Reminiscence. In: Proceedings of the Thirteenth International Conference on Tangible, Embedded, and Embodied Interaction. pp. 135–144. ACM, New York, NY, USA (2019). https://doi.org/10.1145/3294109.3295649.
64. He, Y., Schiphorst, T.: Designing a Wearable Social Network. In: CHI '09 Extended Abstracts on Human Factors in Computing Systems. pp. 3353–3358. ACM, New York, NY, USA (2009). https://doi.org/10.1145/1520340.1520485.
65. Huisman, G., Frederiks, A.D., Dijk, B.V., Hevlen, D., Kröse, B.: The TaSSt: Tactile sleeve for social touch. In: 2013 World Haptics Conference (WHC). pp. 211–216 (2013). https://doi.org/10.1109/WHC.2013.6548410.
66. Chung, H., Lee, C.-H.J., Selker, T.: Lover's Cups: Drinking Interfaces As New Communication Channels. In: CHI '06 Extended Abstracts on Human Factors in Computing Systems. pp. 375–380. ACM, New York, NY, USA (2006). https://doi.org/10.1145/1125451.1125532.
67. Motamedi, N.: Keep in Touch: A Tactile-vision Intimate Interface. In: Proceedings of the 1st International Conference on Tangible and Embedded Interaction. pp. 21–22. ACM, New York, NY, USA (2007). https://doi.org/10.1145/1226969.1226974.
68. Ceballos, R., Ionascu, B., Park, W., Eid, M.: Implicit Emotion Communication: EEG Classification and Haptic Feedback. ACM Transactions on Multimedia Computing, Communications, and Applications 14(1), 3:1–3:18 (2017). https://doi.org/10.1145/3152128.
69. Teh, K.S., Lee, S.P., Cheok, A.D.: Poultry.Internet: A Remote Human-pet Interaction System. In: CHI '06 Extended Abstracts on Human Factors in Computing Systems. pp. 251–254. ACM, New York, NY, USA (2006). https://doi.org/10.1145/1125451.1125505.
70. Kim, J., Park, Y.-W., Nam, T.-J.: BreathingFrame: An Inflatable Frame for Remote Breath Signal Sharing. In: Proceedings of the Ninth International Conference on Tangible, Embedded, and Embodied Interaction. pp. 109–112. ACM, New York, NY, USA (2015). https://doi.org/10.1145/2677199.2680606.

71. Frey, J., Grabli, M., Slyper, R., Cauchard, J.R.: Breeze: Sharing Biofeedback Through Wearable Technologies. In: Proceedings of the 2018 CHI Conference on Human Factors in Computing Systems. pp. 645:1–645:12. ACM, New York, NY, USA (2018). https://doi.org/10.1145/3173574.3174219.

72. Bales, E., Li, K.A., Griwsold, W.: CoupleVIBE: Mobile Implicit Communication to Improve Awareness for (Long-distance) Couples. In: Proceedings of the ACM 2011 Conference on Computer Supported Cooperative Work. pp. 65–74. ACM, New York, NY, USA (2011). https://doi.org/10.1145/1958824.1958835.

73. Werner, J., Wettach, R., Hornecker, E.: United-pulse: Feeling Your Partner's Pulse. In: Proceedings of the 10th International Conference on Human Computer Interaction with Mobile Devices and Services. pp. 535–538. ACM, New York, NY, USA (2008). https://doi.org/10.1145/1409240.1409338.

74. Lotan, G., Croft, C.: imPulse. In: CHI '07 Extended Abstracts on Human Factors in Computing Systems. pp. 1983–1988. ACM, New York, NY, USA (2007). https://doi.org/10.1145/1240866.1240936.

75. Block, A.E., Kuchenbecker, K.J.: Emotionally Supporting Humans Through Robot Hugs. In: Companion of the 2018 ACM/IEEE International Conference on Human-Robot Interaction. pp. 293–294. ACM, New York, NY, USA (2018). https://doi.org/10.1145/3173386.3176905.

76. Block, A.E., Kuchenbecker, K.J.: Softness, Warmth, and Responsiveness Improve Robot Hugs. Int J of Soc Robotics. 11(1), 49–64 (2019). https://doi.org/10.1007/s12369-018-0495-2.

77. Shen, X., George, M., Hernandez, S., Park, A., Liu, Y., Ishii, H.: Mole Messenger: Pushable Interfaces for Connecting Family at a Distance. In: Proceedings of the Thirteenth International Conference on Tangible, Embedded, and Embodied Interaction. pp. 269–274. ACM, New York, NY, USA (2019). https://doi.org/10.1145/3294109.3300990.

78. Bonanni, L., Vaucelle, C., Lieberman, J., Zuckerman, O.: PlayPals: Tangible Interfaces for Remote Communication and Play. In: CHI '06 Extended Abstracts on Human Factors in Computing Systems. pp. 574–579. ACM, New York, NY, USA (2006). https://doi.org/10.1145/1125451.1125572.

79. Freed, N.: Toys Keeping in Touch: Technologies for Distance Play. In: Proceedings of the Fourth International Conference on Tangible, Embedded, and Embodied Interaction. pp. 315–316. ACM, New York, NY, USA (2010). https://doi.org/10.1145/1709886.1709960.

80. Brown, L.M., Williamson, J.: Shake2Talk: Multimodal Messaging for Interpersonal Communication. In: Oakley, I. and Brewster, S. (eds.) Haptic and Audio Interaction Design, LNCS 4813. pp. 44–55. Springer Berlin Heidelberg (2007).

81. Brown, L.M., Sellen, A., Krishna, R., Harper, R.: Exploring the Potential of Audio-tactile Messaging for Remote Interpersonal Communication. In: Proceedings of the SIGCHI Conference on Human Factors in Computing Systems. pp. 1527–1530. ACM, New York, NY, USA (2009). https://doi.org/10.1145/1518701.1518934.

82. Tsetserukou, D., Neviarouskaya, A.: iFeel_IM!: Augmenting Emotions during Online Communication. IEEE Computer Graphics and Applications. 30(5), 72–80 (2010). https://doi.org/10.1109/MCG.2010.88.

83. Shin, H., Lee, J., Park, J., Kim, Y., Oh, H., Lee, T.: A Tactile Emotional Interface for Instant Messenger Chat. In: Smith, M.J. and Salvendy, G. (eds.) Human Interface and the Management of Information. Interacting in Information Environments, LNCS 4558. pp. 166–175. Springer Berlin Heidelberg (2007).

84. Rovers, A.F., van Essen, H.A.: FootIO – Design and evaluation of a device to enable foot interaction over a computer network. In: First Joint Eurohaptics Conference and Symposium on Haptic Interfaces for Virtual Environment and Teleoperator Systems (2005).

85. Qiu, S., Rauterberg, M., Hu, J.: Designing and Evaluating a Wearable Device for Accessing Gaze Signals from the Sighted. In: Antona, M. and Stephanidis, C. (eds.) Universal Access in Human-Computer Interaction. Methods, Techniques, and Best Practices. 9737, pp. 454–464. Springer International Publishing (2016).

86. McDaniel, T., Krishna, S., Balasubramanian, V., Colbry, D., Panchanathan, S.: Using a haptic belt to convey non-verbal communication cues during social interactions to individuals who are blind. In: 2008 IEEE International Workshop on Haptic Audio visual Environments and Games. pp. 13–18 (2008). https://doi.org/10.1109/HAVE.2008.4685291.
87. McDaniel, T.L., Krishna, S., Colbry, D., Panchanathan, S.: Using Tactile Rhythm to Convey Interpersonal Distances to Individuals Who Are Blind. In: CHI '09 Extended Abstracts on Human Factors in Computing Systems. pp. 4669–4674. ACM, New York, NY, USA (2009). https://doi.org/10.1145/1520340.1520718.
88. McDaniel, T.L., Villanueva, D., Krishna, S., Colbry, D., Panchanathan, S.: Heartbeats: A Methodology to Convey Interpersonal Distance Through Touch. In: CHI '10 Extended Abstracts on Human Factors in Computing Systems. pp. 3985–3990. ACM, New York, NY, USA (2010). https://doi.org/10.1145/1753846.1754090.
89. Réhman, S.U., Liu, L.: Vibrotactile rendering of human emotions on the manifold of facial expressions. Journal of Multimedia. 3, 18–25 (2008).
90. Buimer, H.P., Bittner, M., Kostelijk, T., van der Geest, T.M., van Wezel, R.J.A., Zhao, Y.: Enhancing Emotion Recognition in VIPs with Haptic Feedback. In: Stephanidis, C. (ed.) HCI International 2016 – Posters' Extended Abstracts. 618, pp. 157–163. Springer International Publishing (2016).
91. Krishna, S., Bala, S., McDaniel, T., McGuire, S., Panchanathan, S.: VibroGlove: An Assistive Technology Aid for Conveying Facial Expressions. In: CHI '10 Extended Abstracts on Human Factors in Computing Systems. pp. 3637–3642. ACM, New York, NY, USA (2010). https://doi.org/10.1145/1753846.1754031.
92. Paul Ekman Group: Facial Action Coding System. https://www.paulekman.com/ (2019).
93. Bala, S., McDaniel, T., Panchanathan, S.: Visual-to-tactile mapping of facial movements for enriched social interactions. In: 2014 IEEE International Symposium on Haptic, Audio and Visual Environments and Games (HAVE) Proceedings. pp. 82–87 (2014). https://doi.org/10.1109/HAVE.2014.6954336.
94. McDaniel, T., Devkota, S., Tadayon, R., Duarte, B., Fakhri, B., Panchanathan, S.: Tactile Facial Action Units Toward Enriching Social Interactions for Individuals Who Are Blind. In: Basu, A. and Berretti, S. (eds.) Smart Multimedia. pp. 3–14. Springer International Publishing (2018).
95. McDaniel, T., Tran, D., Devkota, S., DiLorenzo, K., Fakhri, B., Panchanathan, S.: Tactile Facial Expressions and Associated Emotions Toward Accessible Social Interactions for Individuals Who Are Blind. In: Proceedings of the 2018 Workshop on Multimedia for Accessible Human Computer Interface. pp. 25–32. ACM, New York, NY, USA (2018). https://doi.org/10.1145/3264856.3264860.
96. Seeing Machines. https://www.seeingmachines.com (2019).
97. IMOTIONS. https://imotions.com (2019).
98. Torre, F.D. la, Chu, W., Xiong, X., Vicente, F., Ding, X., Cohn, J.: IntraFace. In: 2015 11th IEEE International Conference and Workshops on Automatic Face and Gesture Recognition (FG). pp. 1–8 (2015). https://doi.org/10.1109/FG.2015.7163082.
99. McDaniel, T., Tran, D., Chowdhury, A., Fakhri, B., Panchanathan, S.: Recognition of Tactile Facial Action Units by Individuals Who Are Blind and Sighted: A Comparative Study. Multimodal Technologies and Interaction. 3(2), 32 (2019). https://doi.org/10.3390/mti3020032.
100. Panchanathan, S., Chakraborty, S., McDaniel, T.: Social Interaction Assistant: A Person-Centered Approach to Enrich Social Interactions for Individuals With Visual Impairments. IEEE Journal of Selected Topics in Signal Processing. 10(5), 942–951 (2016). https://doi.org/10.1109/JSTSP.2016.2543681.
101. Chang, A., Kanji, Z., Ishii, H.: Designing touch-based communication devices. In: Workshop No. 14: Universal Design: Towards Universal Access in the Information Society, organized in the context of CHI '01 (2001).

Chapter 7
Applications of Haptics in Medicine

Angel R. Licona, Fei Liu, David Pinzon, Ali Torabi, Pierre Boulanger, Arnaud Lelevé, Richard Moreau, Minh Tu Pham, and Mahdi Tavakoli

Abstract Touch is one of the most important sensory inputs during the performance of surgery. However, the literature on kinesthetic and tactile feedback both called haptics in surgical training remains rudimentary. This rudimentary knowledge is partial since that haptic feedback is difficult to describe, as well as record and playback. This chapter aims at focusing on the use of haptics in the training of medical staff and also as a complementary tool for robotized and remote procedures. It provides an overview of the various available technologies to perform haptic feedback and details on how haptic guidance can enhance surgical skill acquisition. A critical review of available haptic interfaces vis-a-vis medical interventions to be performed is provided. The chapter ends with an illustration merging the advantages of usual supervised hands-on training and the ones offered by computer-based training: dual-user training simulators.

1 Introduction

Imagine yourself as a surgeon reaching deep inside a patient's body, pushing aside organs and trying to stop bleeding from a ruptured spleen. Before performing such dexterous manipulation, initial and continuing hands-on training is necessary. Continuing training as the technologies and procedures evolve, which requires

A. R. Licona · F. Liu · A. Lelevé (✉) · R. Moreau · M. T. Pham
Laboratoire Ampère (UMR 5005), INSA Lyon, University of Lyon, Lyon, France
e-mail: angel-ricardo.licona-rodriguez@insa-lyon.fr; arnaud.leleve@insa-lyon.fr;
richard.moreau@insa-lyon.fr; minh-tu.pham@insa-lyon.fr

D. Pinzon · P. Boulanger
Department of Computing Science, University of Alberta, Edmonton, AB, Canada
e-mail: dpinzon@ualberta.ca; pierreb@ualberta.ca

A. Torabi · M. Tavakoli
Department of Electrical and Computer Engineering, University of Alberta, Edmonton, AB, Canada
e-mail: torabipa@ualberta.ca; mahdi.tavakoli@ualberta.ca

© Springer Nature Switzerland AG 2020
T. McDaniel, S. Panchanathan (eds.), *Haptic Interfaces for Accessibility, Health, and Enhanced Quality of Life*, https://doi.org/10.1007/978-3-030-34230-2_7

surgeons to stay up to date. Analog Medical simulators such as cadavers and animals have been a convenient way to learn surgical skills for decades in medical universities. However, due to the growing cost of providing them, non-continuous availability problems and ethical issues, phantoms are more and more used. Yet, phantoms provide a limited set of common cases to practice on. Providing a larger set of lab setups imposes to own a great variety of available phantoms, which becomes rapidly expensive. Nowadays, medical universities still offer too few opportunities for over-crowded medical student populations to perform hands-on training during their curriculum. It has become essential to provide cost-efficient solutions they could use during supervised sessions but also autonomously to sufficiently train themselves during their curriculum.

Over the last decade, Virtual Reality (VR) simulators (see [97] for a recent review about this topic) have been designed to overcome the limitations mentioned previously. With patient models that can be selected online, trainees are provided with an infinite set of medical cases and further to provide difficulty levels that fit to a learning curve. VR simulators have been progressively improved to offer a more realistic environment in 2D and more recently in 3D (with one example [18]). With haptic training simulators, the additional force feedback provides a realistic tool behavior, which leads to efficient training for advanced tasks [63]. More recently, the interest of haptic feedback in robotic-assisted surgery has been proven by Talasaz et al. in the case of suturing applications [81]. It is demonstrated that force feedback plays a key role in helping users to manipulate the instruments better and to reduce tissue stress, tissue damage, and accidental hits.

Also, nowadays, surgical procedures are being transformed by robots entering the operating rooms. Robots are enhancing surgical techniques and expanding surgeons' capabilities. Surgical robots present the patient to the surgeon through a haptic interface and replicate the surgeons' forces and motions on the patient.

This chapter illustrates the current state-of-the-art of how haptic feedback can be used for training. Section 2 describes the various available technologies while Sect. 3 proposes a critical review of available haptic interfaces vis-a-vis various medical interventions to be performed. Section 4 details how haptic guidance can enhance surgical skill acquisition, and at last, Sect. 5 illustrates, in this training context, how the advantages of usual supervised hands-on training can be mixed with the ones offered by computer-based training.

2 Involved Technologies

2.1 Introduction

Haptics started its growing development in the 1990s. Researchers and private companies developed force feedback technology for different applications. As components costs are reduced, new designs and various sizes of haptic interfaces are now available. Concerning sensors, the critical requirement is their high resolution. To limit perturbances with haptic feedback, sensors must be contactless. Therefore, optical encoders and non-contact potentiometers are required to avoid compromising the back-drivability of haptic devices.

The computational performance of modern platforms facilitate analog and digital transformations with higher resolutions. They also ensure real-time computations which help to obtain sufficient haptic rendering (typically around 1kHz) to provide greater realism. The progress of embedded systems also allows the design of lighter haptic interfaces, of which some can be portable.

Technological advances are also an essential criterion for a quality haptic interface. Thus new actuators, such as piezoelectric polymers, electroactive, etc., allow a new approach in the design of a haptic interface. Small and light mechanisms have allowed the emergence of portable haptic interfaces. However, more conventional actuators such as electric motors, hydraulic and pneumatic cylinders, are still often used for actual haptic interfaces due to the knowledge of their characteristics and their associated control laws.

In this section, a review of the different technologies available will be carried out. We will deal not only with commercialized solutions but also solutions developed in laboratories.

2.2 Actuation

Actuators in haptic interfaces aim to produce force or tactile sensations to the user in order to ensure realism during simulation. An efficient control law is a guarantee of their performance.

To be efficient, haptic interfaces must be as transparent as possible. A transparent interface means that when there are no physical interactions during the simulation no force should be applied to the user. It is, therefore, necessary to follow the user's motion without any opposition. All friction forces in actuators, gears, and transmissions used in the interface should be compensated so that the user does not feel them. Thus, low friction and inertia in the chosen actuators are necessary or they should at least include an efficient control law to compensate them.

As haptic interfaces interact with the user, user safety is also a crucial concern. High-voltage, high-current, or high-pressure fluids should be avoided. The noise generated by actuators can also disturb users and should be avoided as much as possible.

The next subsection is dedicated to the description of the standard actuators used in most haptic interfaces today (electrical motors, hydraulic, and pneumatic cylinders). Following this, a non-exhaustive list of several novel actuators is introduced. A brief description of their characteristics is carried out. Finally, a comparison between the different haptic actuation technologies is carried out.

2.3   Classical Actuators

2.3.1   Electrical Actuators

Recently, most haptic actuators are based on electrical actuators. This is due to several reasons: ease of implementation since they only require an electric supply, ease of control since they are very common, cleanliness because there is no leakage and non-disturbance due to their low noise level. Figure 7.1 indicates some examples of commercialized haptic interfaces.

2.3.2   Pneumatic Actuators

Pneumatic actuators can be used as a haptic interface. The force produced by these actuators depends on the pressure of the power source. The power source is generally a tank linked to a compressor. Their advantages are:

- Compared to hydraulic ones, they are lighter and cheaper.
- Air is inflammable, cheap, and clean. Air can be exhausted outside (no need for a return pipe unlike hydraulic actuators).
- Compared to electrical motors, pneumatic actuators have a higher mass-to-force ratio and can generate more significant force without any reduction mechanism. This leads to a higher power-to-weight ratio compared to electrical motors.

Pneumatic actuators also have several drawbacks. The first is due to air compressibility. This leads to a lower bandwidth than hydraulic actuators. A second issue concerns friction forces. Indeed, since air is not self-lubricating, this may cause the existence of friction forces (viscous and dry). These forces may mask desired small feedback forces during interaction with soft virtual objects. Concerning their control, they are often considered as more complex to control than electrical actuators but this is mainly due to the fact that they are less used.

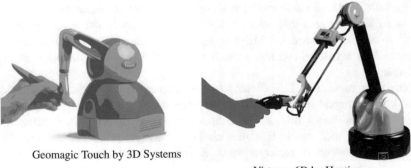

Geomagic Touch by 3D Systems

Virtuose 6D by Haption

Fig. 7.1   Various examples of commercialized haptic interfaces

There exist two main types of pneumatic actuators: cylinders and pneumatic artificial muscles.

Cylinders produce a force along their main axis with a high range of displacement. Their lifetime is quite long compared to electrical actuators. However, their static friction force is generally not negligible. They are quite cumbersome and heavier compared to a pneumatic artificial muscle, which could lead to complications during the design of a haptic interface.

Air compressibility is usually a drawback of air cylinders. When they are used in a haptic interface, however, this drawback may become an advantage. Indeed, this phenomenon can help to reproduce human tissue behavior [33]. For instance, a haptic interface has been developed to help obstetricians and midwives during childbirth [34]. This simulator is based on two pneumatic cylinders in order to reproduce realistic childbirth to offer practicians a risk-free learning tool. Figure 7.2 shows a CAD view of the childbirth simulator using two pneumatic cylinders.

Pneumatic artificial muscles have a higher power-to-mass ratio and are lighter than air cylinders. Among different types of pneumatic artificial muscles, one could distinguish braided muscles and pleated muscles. Braided muscles are also known as McKibben muscles. McKibben used them as an orthotic actuator in the late 1950s. Pleated muscles are another variant of pneumatic artificial muscles. Whatever their nature, pneumatic artificial muscles have some drawbacks compared to air cylinders, including very long dynamics, non-negligible hysteresis [88], limited displacement [16], and limited lifetime due to material fatigue. Moreover, they are quite complex to model and to control. Due to their likeness to human muscles,

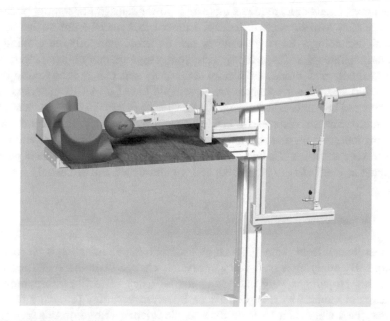

Fig. 7.2 Childbirth simulator actuated by two pneumatic cylinders

they are mostly used as actuators for the exoskeleton. In [10], they proposed a lower extremity exoskeleton based on pleated pneumatic artificial muscles.

2.3.3 Hydraulic Actuators

Concerning hydraulic actuators, the incompressibility of hydraulic fluid (oil) leads to obtaining a higher bandwidth than pneumatic actuators. They can produce a tremendous force which results in a very high power-to-weight ratio. The incompressibility of oil ensures a high stiffness. Moreover, as oil is self-lubricating, friction forces are negligible.

Concerning their drawbacks, hydraulic actuators are often more expensive. They require a complex hydraulic system to supply high pressures. The presence of leakage is also an issue for haptic interfaces. For instance, in [78], the authors show that oil can quickly soil a large area. It may cause damage to human operators due to the high pressurization. It is also risky as oil is flammable.

In [62], the authors proposed hydraulic force feedback systems in hazardous/hostile environments in an effort to improve tracking performance and perception of operators.

2.4 Novel Actuators for Haptic Devices

All the previous actuators require a quite high power density. For some applications, it is necessary to have a portable haptic device that may be not attainable with standard technology. Some researches are therefore currently developing new actuation technologies. Piezoelectric actuators, shape memory alloys, electro- and magneto-rheological fluids, and electro active polymers can be used to design new haptic interfaces. These actuators are developed based on smart materials. Their properties can be significantly changed according to external stimuli, such as stress, strain, temperature, moisture, pH, electric or magnetic fields [76]. Recently, haptic interfaces based on these actuators are mostly available in laboratories and not yet commercially available. Below, a subset of the latest actuators and their potential for impact as mediums of haptic feedback are introduced and discussed.

2.4.1 Piezoelectric Actuators

The most known technology that can be used in a haptic interface is piezoelectric actuators. By applying a high voltage, crystals change their shape and produce a force. This property is very interesting for feedback of tactile information. The displacement induced by this property is, however, very tiny (around a few μm). This may be a significant obstacle for a haptic interface based on this technology.

However, as vibration around 1 kHz can be achieved they could be used to render tactile feedback and thus simulate different textures [12].

2.4.2 Shape Memory Alloys

Another technology that could be used is shape memory alloys (SMA). Their main properties are transformation to their original shape when the appropriate thermal constraint is applied. If resistance is applied during its transformation, SMA will produce forces. This property could be used to design a force feedback device. Constrained and free recovery, elastic interfaces, and proportional control are a few examples of the many possible applications of such a technology [5]. SMA actuators can produce much more useful work per unit volume than most actuating mechanisms. However, environmental conditions have to be taken into account such as forces, displacements, and temperature conditions. Their main drawback relies on their long relaxation time, which could be unacceptable for a haptic interface.

2.4.3 Electro-Rheological Fluids

Electro-rheological fluids (ERF) are materials that modify their viscosity behavior depending on the applied electric field [83]. The most common cause for this behavioral change is yield stress, which is amplified by the applied field. Simply by removing the electric field, ERF comes back to its original state. High frequencies (around 1 kHz) between states can be achieved that allows them to be used as a haptic interface. In [31], the authors proposed a haptic interface based on electro-rheological actuators. The main drawback lies in the high voltage needed (around 2–4 kV per millimeter gap between the electrodes). This may lead to safety consequences for users. Another drawback is due to the inhomogeneity of the used fluid which can change its properties resulting in a degraded electric-mechanic power converter of the actuators.

2.4.4 Magneto-Rheological Fluids

Magneto-rheological fluids (MRF) are quite similar to ERF but they are based on magnetic fields instead of electric fields. The principle is the same: viscosity changes according to the applied field. MRF have much higher shear stress than ERF. In [94], the authors proposed a haptic interface based on MRF. To our knowledge, these kinds of actuators are only used in laboratories for haptic interface design. This is mainly due to the costs of fluids, their cumbersomeness, and their heaviness (presence of iron particles).

2.4.5 Electro-Active Polymers

Electro-active polymers (EAP) exhibit a significant shape or size change in response to electrical stimulation. Features including power efficiency, flexibility, high dynamics, resilience to damage, and lightweight make them interesting for micro electromechanical systems [47] and bionic robots [77] for instance. Their drawbacks mainly include required high voltages and non-ideal polymers. This characteristic of the polymers is not yet completely understood. Moreover, the manufacturing process is still very complicated and not suitable for mass production. They are therefore quite difficult to find on the market.

2.5   Comparison of the Different Technologies

Table 7.1 summarizes the advantages and drawbacks of the technologies introduced in the previous subsection.

Standard actuators and, in particular, electrical DC motors are currently equipped in most available haptic interfaces. This is because they are well known and provide high bandwidth. They are quite cheap and their maintenance is reduced. The main drawback lies in their poor force-to-weight ratio and thus tend to be heavy. To overcome this drawback, haptic interfaces include transmission mechanisms which lead to the introduction of some friction and backlash that degrade performance.

The other standard actuators, such as hydraulic and pneumatic actuators, offer a higher force-to-weight ratio than DC motors. They thus offer another way to deal with the design and control of mechanical compliance.

The compactness, mass, and power consumption of the novel actuators introduced in the previous section offer new solutions to design haptic interfaces. Whereas they offer new possibilities, most of their developments stay within the laboratory and only a few of them are commercially available. Promising applications do exist but will grow slowly until several problems such as poor knowledge of material characteristics, complicated manufacturing process, and cost are overcome.

3   Haptics During Surgical Procedures

The first sensory modality that develops in humans is the sense of touch. Haptic perception encompasses tactile perception via skin stimulation and kinesthetic perception via muscle stimulation, which informs a human about an object's material, stiffness, and shape properties [52]. Haptic interaction is the human's most basic way to understand an environment and effect change in it. Haptic feedback provides humans who operate machines with a sense of touching objects they are not actually touching but are manipulating through machines. Haptic feedback allows

Table 7.1 Comparison of various technologies that could be used in a haptic interface

Technologies	Advantages	Disadvantages
Electric	Simple energy transmission from power supply to the device (flexible wires)	Low force-to-weight ratio
	Easy to control and high bandwidth	High inertia
	Clean and quiet	Backlash (distortion force/torque)
	Low cost	Over-heating with heavy load
Pneumatic	Lightweight, small size, high mass-to-weight ratio	Low bandwidth
	Compliance	High stiction friction
	No return lines required for exhausted fluid	Highly nonlinear properties (compressibility of air, friction effect, dynamics of valve)
	Clean system, non-flammable fluid	Leakage
Hydraulic	Very high power-to-weigh ratio	Viscosity of oil changes with temperature
	Compliance	High-pressure oil leaks are unsafe
	High bandwidth	Flammable oil
	Self-lubricating (low friction)	Costly to produce and maintain
		Leakage
Piezoelectric	Small size	Limited strains and low tensile strength
	Low actuation power	High driving voltage
	High bandwidth, high frequencies	Fragile materials
	Wide range of operating temperature	
SMA	Large active stresses	Low bandwidth
	High energy density	Low efficiency
	High material strength and high elasticity	High hysteresis (long relaxation time)
		Problem in fatigue failure
ERF	High dynamics	Limited shear stress
	Wide range of operating temperature	Sedimentation problem
MRF	High shear stress	Costly (high-quality fluids)
	Low voltage power	Heavy (high loading of dense iron)
EAP	Large active strains	Moderate active stresses
	High energy density	Limited temperature range
	Small size and intrinsic softness	Complicated manufacturing process
	High bandwidth	High driving voltage

the human operators of machines to handle objects more gently, safely, reliably, and precisely.

To recreate haptic feedback about an environment that is accessed indirectly rather than touched directly by a user, a haptic interface displays forces received from a virtual or a robotic proxy (slave) probing the environment. Depending on whether the slave is virtual or robotic, a haptic virtual environment or a haptic teleoperation system is formed. Regardless, as the user utilizes the haptic interface to operate the slave that interacts with the environment, haptic feedback about slave-environment interaction displayed by the haptic interface engages the user's sense of touch and gives transparency (i.e., realism and fidelity) to the interaction. Haptic systems have applications in various domains including medicine, e.g., for surgical training with a virtual slave, (tele-)surgery with a robotic slave, and post-disability (tele-)rehabilitation with a virtual or robotic slave.

Surgical procedures are being transformed by robots entering operating rooms. Robots are enhancing surgical techniques and expanding surgeons' capabilities. Surgical robots present the patient to the surgeon through a haptic interface and replicate the surgeon's forces and motions on the patient. Haptics enhances robot-assisted surgery by providing a surgeon the sensation of operating on a patient via direct touch even though it is done through a haptic interface and then a surgical robot. Haptics has the potential to increase the surgeon's task performance ability and empower him/her to complete more complex surgical tasks. In a teleoperated robotic surgical system, the patient side surgical manipulators are controlled via a human interface, operated by the surgeon. The haptic interface connecting the surgeon to the surgical manipulator and environment is an integral part of any robot-assisted surgical system and should be able to intuitively transfer surgical maneuvers to the surgical robot [30, 85, 89]. The haptic interface should also provide sufficient sensory feedback such as haptic sense to the surgeon to more intuitively control the robotic manipulators. The haptic force feedback can reduce unintentional injuries [84], surgeon's fatigue [56], and assist tissue characterization [95].

3.1 Desired Characteristics of Haptic Interfaces

High-fidelity haptic feedback, which is critical to the safety and success of any interaction, requires appropriate haptic design and control. To this end, the haptic interfaces are required to meet the following criteria.

1. Criteria for high kinematic performance:

    • Large workspace: The workspace of the haptic interface needs to be large enough to match the workspace of the human arm and the slave. Otherwise, the user has to employ clutching to repeatedly move the haptic interface to another position and orientation (pose) without moving the slave, slowing down the operation.
    • High manipulability: Manipulability or dexterity of the haptic interface measures its ability to take any arbitrary pose and apply any arbitrary force and torque (wrench) across the workspace [98]. Also of importance is the haptic interface's isotropy, which measures if the haptic interface can move and apply force equally well in all directions (i.e., directional uniformity). Singularities conspire to reduce the haptic interface's manipulability and isotropy.

- Large stiffness and maximum force feedback: The haptic interface should be able to recreate highly stiff environments by providing large forces against the user's hand despite a limited amount of joint torque. Otherwise, stiff environments would be perceived as soft.

2. Criteria for high dynamic performance:

- Lightweight, low-friction, and back-driveable: The haptic interface should not exert any wrench on the user's hand if the slave is moving in free space. Thus, the haptic interface should have a low apparent impedance (mainly inertia) and low friction, especially if the motions involve high accelerations. The haptic interface's mechanical structure, actuators, and configuration determine the inertia and friction. Measuring the ratio of force output to motion input, an impedance has inertia, damping and stiffness components.
- Fast response (large bandwidth): Bandwidth, which defines the maximum speed or frequency at which the haptic interface can operate, needs to be high to enable rendering of transitions between free space and stiff contact and of different textures. Lightweight haptic interface design leads to a larger bandwidth.
- Large Z-width: A haptic interface's Z-width is the range of impedances that it can stably display [17]. The larger the Z-width, the richer the haptic information presented to the user. A small Z-width can make it hard to distinguish between different environments because they are presented as similar impedances.

There are trade-offs between the desirable characteristics of haptic interfaces such as maximum force feedback capability vs. minimum inertia or maximum stiffness vs. workspace size. Indeed, a large force feedback capability requires large actuators, increasing the haptic interface's inertia. A large workspace requires long links, decreasing the haptic interface's stiffness (and increasing its inertia). Therefore, the design of the haptic interface has to be optimized for a specific surgical application.

3.2 Critical Review of Available Haptic Interfaces vis-a-vis the Desired Characteristics

Several haptic interfaces have been designed by different researchers. One of the most widespread haptic interfaces is the PHANToM haptic device (3D Systems Inc., Morrisville, NC, USA) [55]. There are various versions of the PHANToM haptic device which are designed for different applications. 3D Systems Touch is probably the most popular one due to its relatively low price. It has six degrees of freedom (DoFs) in position and orientation sensing with the workspace size of $1.3E-3\,m^3$ and can provide maximum force feedback of 3.3 N in three translational DoFs. The specifications of the 3D Systems touch are enhanced for its modified version, 3D Systems Touch X. This interface has a bigger workspace, a better position resolution, and a better force feedback capability (7.9 N). 3D Systems Inc. has designed and commercialized PHANToM Premium models to fulfill the requirements of the applications which require a larger workspace, better position resolution, and better range of force feedback rendering than the Touch and Touch X models. There are three main variants for the PHANToM Premium family; PHANToM Premium 1.0, PHANToM Premium 1.5, and PHANToM Premium 3.0. Also, A version of the PHANToM Premium 1.5 and PHANToM Premium 3.0 is available that provides 3 DoFs force feedback and 3 DoFs torque feedback. For instance, the six-DoF version of Premium 3.0 has the workspace size of $198E-3\,m^3$ and can provide 22 N force feedback and 515 mN.m torque feedback. The 3D Systems haptic interfaces have a large workspace thanks to their serial kinematics design, but they can provide lower force and torque feedback than the haptic interfaces with the parallel kinematics design.

A family of haptic hand-controllers with a parallel kinematic design is manufactured by Force Dimension (Nyon, Switzerland) [29, 87]. The Force Dimension haptic interfaces can be categorized into three groups as Omega, Delta, and Sigma. As the kinematic design of these haptic interfaces is parallel, they can provide larger force feedback but with a relatively smaller workspace than the haptic interfaces with the serial kinematics design. Among the Force Dimension haptic interfaces, the Sigma 7 provides the largest force and torque feedback (20.0 N and 400 mN.m, respectively) and the best position resolution in all six DoFs. It has the workspace size of $4E-3\,m^3$. Omega.6 has six DoF position sensing and can provide three DoF force feedback. The economic version of the Omega.3 is Novint Falcon, which is designed and manufactured by Novint Technologies (Albuquerque, New Mexico, USA). The Falcon haptic interface has lower force feedback capability (9 N), position resolution, and workspace size ($1.1E-3\,m^3$) than the Omega 3 haptic interface.

Virtuose haptic interfaces family (Haption, Soulg-sur-Ouette, France) are designed with serial kinematics with 6 DoF of position sensing. The 3D models provide three DoF active translational force feedback and the 6D models provide three DoF force feedback and three DoF torque feedback. The workspace and force feedback capability is various among the Virtuose haptic interfaces. For example, Virtuose 6D provides a maximum force up to 35 N with the workspace size of $92E-3\,m^3$. The Freedom 7 (MPB Technologies Inc., Pointe-Claire, Quebec, Canada) is a haptic interface which is designed specifically for medical simulations [32]. This haptic interface has a serial kinematic design with the workspace size of $12E-3\,m^3$ and is capable of providing high position resolution, low joint friction, and low apparent inertia. However, it can only offer 0.6 N force feedback, which is small relative to other haptic interfaces.

The haptic interfaces based on pantograph kinematics are introduced by Quanser Inc. (Markham, Ontario, Canada). One of the Quanser haptic interfaces is High-Definition Haptic Device (HD^2) which provides 6 DoF haptic feedback with 6 DoF of position sensing [79]. The dual-pantograph kinematics allows the HD^2 to have a relatively large workspace ($70E-3\,m^3$) while generating relatively large force and torque feedback of 11 N and 0.95 Nm, respectively. The Maglev 200 (Butterfly Haptics, LLC, Pittsburgh, PA, USA) is a 6 DoF magnetic levitation haptic interface [9]. In this device, the mechanical mechanism is replaced with a single moving part levitated by magnetic fields. Therefore, it has zero static friction, zero mechanical backlash, high position resolution, and wide stiffness range. This haptic interface can also provide 3-DoF force feedback with a maximum of 40 N along the z-axis and 3-DoF torque feedback with a maximum of 3.6 Nm. The main disadvantage of the Maglev 200 is its very small workspace which is $0.01E-3 \, m^3$.

So far, the discussed haptic interfaces were all back-drivable with either low force and torque feedback capability or small workspace. There is another category of haptic interfaces which can provide large force and torque feedback while having a large workspace. These haptic interfaces are not back-drivable and are admittance controlled. The Haptic Master (MOOG Inc., New York, USA) is one of the admittance controlled haptic interfaces [46], which has 3 DoFs of force feedback and can generate up to 100 N of force feedback with the workspace size of $80E-3 \, m^3$. Another example of admittance controlled haptic interfaces is VISHARD10 [92], which has a workspace size $173E-3 \, m^3$ and can provide 170 N force feedback and up to 13 Nm torque feedback. The drawback of admittance controlled haptic interfaces is that small force feedback can be obscured by the mechanical properties (including apparent inertia) and joint frictions of the haptic interface [90], which degrade the haptic feedback resolution and sensitivity for the user. Therefore, the admittance controlled haptic interfaces are generally designed for rehabilitation applications in which the high force feedback capability is more important than the force feedback resolution.

Haptic interfaces with exoskeleton construction are designed and fabricated by various researchers. This type of haptic interface provides a large workspace with high force feedback capability. L-EXOS [24] is an example of portable exoskeleton haptic interfaces and SARCOS dexterous master (Sarcos Robotics Co., Salt Lake City, UT, USA) [36] is an example of grounded exoskeleton haptic interfaces. Portable exoskeleton haptic interfaces have a large workspace, but their weight must be supported by the user. Another disadvantage of this type of haptic interface is that force feedback cannot be adequately displayed as they are not grounded to a base. The grounded exoskeleton haptic interfaces do not have this disadvantage; however, their workspace is relatively smaller than the portable interfaces. The major drawback of exoskeleton haptic interfaces is that the design needs to be user-specific to be compatible with the user's hand kinematics; otherwise, the exoskeleton haptic interface may not be safe or ergonomic for the user.

In addition to the haptic interfaces discussed above, there are some haptic interfaces designed for a specific application. The da Vinci Surgical System (Intuitive Surgical, Inc. CA, USA) has a sophisticated haptic interface, but no design-related information is available for proprietary reasons [69]. The Xitact CHP (Mentice AB, Gothenburg, Sweden) is a haptic interface which is designed for the simulation of medical procedures such as cardiology, peripheral interventions, and interventional radiology. Simodont dental trainer (MOOG Inc., New York, USA) is another example of the haptic interfaces which is designed for a specific medical procedure [6].

3.3   Nature of Medical Intervention Dictates the Haptic Interface Selection

It is noticeable that there are many available haptic interfaces, each of which has its advantages and disadvantages. This, in part, relates to an unavoidable trade-off to optimize the overall design specifications to match a specific application. In teleoperated surgeries, the haptic interface's workspace, maneuverability, degrees of freedom, and sensory feedback should ideally match the intuitive movements of the surgeon's hand to induce the experience and sensation of conventional surgery. As mentioned in Sect. 3.2, the haptic interface should simultaneously satisfy requirements of back-drivability, low apparent inertia and low friction for the best perception of small reflected forces, and large intrinsic stiffness and force feedback capability for the best perception of large reflected forces. Then, the haptic interface can recreate soft and stiff contact experiences for the user with high fidelity. Also, uniformity of haptic feedback and sufficient sensitivity over the practical workspace is of paramount importance to ensure surgical safety and accuracy. Additionally, a haptic interface needs to have a practical workspace that will allow efficient and smooth navigation of the surgical site.

The design of a new haptic interface or the selection of a commercially available haptic interface is application-driven. The specific surgical application regulates the constraints of haptic interface design or selection. A detailed surgical application analysis is needed to determine the following required specifications of the haptic interface. These specifications are determined by evaluating and analyzing conventional freehand surgery.

- The required number of DoF of the haptic interface's end effector: The required number of DoF is the minimum number of DoF that is required to perform a task in the Cartesian space. Most surgical applications are a 7-DoF operation; 3 DoFs for translational movement, 3 DoFs for orientational movement, and one DoF for grasping motion. However, some surgical applications demand lower DoF for the haptic interface's end effector. For instance, needle insertion is a 5-DoF action; 3 DoFs for translational movement and 2 DoFs for orientational movement.
- The required number of DoF of haptic feedback: Ideally, the number of DoF of the force and torque feedback should be matched with the number of DoF of the end effector of the haptic interface. However, in practice, the number of DoF of haptic feedback is a trade-off between the interface complexity, cost, and the degree of benefit of the feedback. In addition to the application requirements, the force and torque feedback DoFs are dictated from the existing measured force and torque at the end effector of the slave. Furthermore, some DoFs of the haptic feedback can be substituted with the graphical cues. Although a number

of studies showed the benefits of haptic feedback [91], some studies showed that the task accuracy and performance is comparable for both haptic feedback and graphical cues modalities, which means that in cases where the haptic feedback is not available, graphical cues can adequately and cost-effectively substitute for haptic feedback [84].

- The required resolution of position and orientation sensing: The resolution of both position and orientation sensing must satisfy the requirements of the surgical application. For instance, a brain tumor ablation in neurosurgery requires sub-millimeter and sub-radian precision; therefore, the haptic interface should meet this requirement by sensing the position and orientation of the surgeon's hand with sufficient resolution and transferring this information to the slave.
- The required force and torque feedback capability and resolution: To determine the force and torque feedback capability and resolution, the requirements of the medical intervention need to be inspected. The haptic interface must provide the range of forces and torques that are presented in the particular medical intervention with a sufficient resolution. For example, a common minimally invasive surgery demands a force range of $\pm 10\,\text{N}$ with a resolution of $0.2\,\text{N}$ [42]. For the applications in which solid objects such as bones need to be rendered, the haptic interface must provide enough force feedback to render an immovable object. A virtual object with the stiffness of $2\,\text{N/mm}$ is stiff enough to represent a solid object [55]. The human's hand ability to distinguish a difference in force is 5 to 15 percent of the reference force [82]. High force and torque output resolution are needed to provide the small changes in the environment distinguishable for the user.
- The required workspace size: The workspace of the haptic interface must be large enough to guarantee that the desired task is executable. Also, the workspace should be free of singularities. In the process of workspace selection for the haptic interface, careful attention must be paid to the motion scaling between the interface and the slave. Also, the interface footprint to the workspace ration must be small enough for mobility and ease of integration in the operation theatre.

It is obvious that a haptic interface with the characteristics higher than the required one can provide superior performance; however, such a device will be expensive, excessively complex, computationally expensive for real-time control, and over-designed. Using a commercially available haptic interface and performing some modifications on it, is more straightforward and cost-effective than the design and fabrication of a new haptic interface. However, all of the desired characteristics of the haptic interface cannot be achieved, and the fidelity of the haptic interaction might be compromised as the modifications of the commercially available haptic interface are limited to some minor alterations.

After determining the application-driven specifications for the haptic interface, the device-based and the user-based performance measures need to be evaluated to further assess the haptic interface. The device-based measures quantify the performance of the haptic interface for its specifications that are not application-related and are required for the practical and effective operation of the haptic interface.

The back-drivability, friction forces, minimum controllable force, force bandwidth, dynamic range of impedance, manipulability, backlash, Jacobian conditioning, and apparent inertia of the interface can be evaluated with device-based measures. In other words, the device-based measures determine how the kinematics and dynamics of the haptic interface interfere with the interaction between the operator and the environment. The user-based measures quantify the usability of a haptic interface with regard to perceiving and manipulating the environment by the user. The user-based measures are application dependent and determine the perceptual effectiveness of the haptic interface. For the user-based measures, psychophysical experiments need to be performed to evaluate how a haptic interface affects the perception of the remote or virtual environment for the user. The user-based evaluation of the haptic interfaces can be performed based on five different haptic modes [39]. These modes are target acquisition, object manipulation, geometric perception, material perception, and environmental monitoring. Other user-related factors such as user ergonomics, fatigue, and discomfort during the use of a haptic interface also have to be assessed. A detailed discussion of the device-based and the user-based measures is beyond the scope of this chapter.

4 Surgical Skill Acquisition Using Haptic Guidance

Imagine that you are a surgeon guiding your tools between a patient's organs and suturing a ruptured spleen using Minimally Invasive Surgical (MIS) techniques. Your vision to the surgical site is limited as it is performed using an endoscope, and you must rely on your tactile feedback from long-shafted instruments, which cross the patient's skin, to guide their motion and perform your tasks. You must get used to the distorted touch perception these tools provide you. This imaginary but realistic scenario highlights the importance of the quality of tactile or haptic feedback during a surgical performance. Surgeons do know how important vision and haptic feedback are in guiding performance and building skills. This is even more valid when vision is faulty. However, what we know about haptic feedback and its effects on surgical performance is quite limited. We know even less about the functions of haptics (i.e., those which rely on sensory information) in building surgical skills. The study of haptic feedback is important to understand the interaction of a surgeon's brain and body with surgical tools, and how better haptics perception improves dexterity.

4.1 How Do Surgeons Learn?

At first, surgery was an art exercised by barbers without any formal training, surveillance, or regulation. Early in the 1800s, a high mortality rate from this type of practice led decision-makers to regulate this profession and to require that its activities should fall under the control of certified doctors [53]. Among the most renowned of the modern surgical education system, William Halsted from the John Hopkins hospital developed in 1904 a model of surgical training (based on a German model of that era), creating the first paradigm of residency in the hospital system [59]. Thirty years later, various organizations, including the American College of Surgeons (ACS), American Board of Surgery (ABS) and the American College of Medicine (AMA), set forth a committee on graduate training and surgery. They published a more structured program in the Minimum Standard for Graduate training in surgery [54]. According to this program and their year of training, residents were proposed annual goals to attain, through gradual and systematic training in specific skills, and skill development in surgery [68].

Entry in most postgraduate surgical training programs requires not only an excellent universal medical knowledge, but also a certain dexterity (in contrast to clinical medical training), to acquire both technical and non-technical skills. Surgeons must acquire personal skills and good visuo-spatial perception for tissue identification [7]. Previous studies have shown that the most important and desired features in a surgical resident are the right attitude and surgical dexterity, both of which are not generally assessed in medical knowledge examinations [7].

Compared to traditional open surgery, situations where visual feedback is not reliable, such as teleoperations or bleeding in the surgical field during MIS for instance, require teaching new surgical techniques in order for surgeons to gain a higher level of dexterity [3]. The additional information that haptic feedback provides could prove advantageous in recognizing blood vessels, abnormal tissue, and gauging appropriate force to apply during tissue manipulation [8, 20, 58].

In the current training model, surgeons obtain skills and learn the perception of the environment in order to implement appropriate motion coordination (which may differ in various anatomical contexts) in three stages [66]. First is the cognition stage where the novice breaks down in simple steps the task to execute. They observe and memorize these steps and how to deal with complications. Next, in the integration stage, they learn these tasks and eliminate ineffective movements by improving their motor abilities. At this stage of the training, the use of wet labs helps. An acceleration movement automation and an increase in the integration of cerebral visual learning can be obtained thanks to simple surgical assistantships. Finally, the repetition of automated movements creates automatized neural loops which create confidence, speed, and accuracy through engagement. An efficient evolution allows for more complex tasks to be acquired, accompanied by a flattening of the learning curve [66]. Each trainee presents a different motor skill learning curve for each type of invasive procedure. They thus need to be addressed differently [67]. Experience to deal with complications is perhaps the most vital aspect for assessing learning and patient safety [67]. Therefore, the ultimate strategy for efficient surgical education is to flatten the learning curve and train surgeons to keep surgical complications at a minimal level from the beginning of their training [66]. The use of wet labs helps training in order to attain movement automatization before facing the patient. They enhance dexterity and have allowed the reinforcement of certain cognitive aspects, especially in the area of laparoscopic surgery allowing to decrease microvascular complications [86]. Peyton's educational model is ideal for teaching surgical skills in these laboratory [41, 64] conditions. It is divided into four stages following this order: demonstration, fragmentation, development, and implementation. Moreover, in surgical procedures, visual and haptic senses are both the most frequently used [57], particularly when testing organ viability and controlling a hemorrhage. As an example, a surgeon dedicates less than 40% of its time doing dissection, palpation, and tissue manipulation tasks [93]. However, individuals use the three varieties of sensory inputs (visual, auditory, and kinesthetic) to learn motor coordination based on their innate sensory preferences [23]. Research on perception has allowed an understanding of how humans acquire skillful movement using these three sensory inputs, either separately or in combination over time. Visual learners are people who gain the most from being shown a skill, while kinesthetic learners benefit the most from imitation and obtaining information through movement and touch. Hence, training can be more effective when combining simultaneous feedback from visual and haptic sensory systems, even if visual feedback is more beneficial [19]. Moreover, a previous study showed that when visual information is not reliable in procedures which require replay of specific gestures, haptic feedback enhances motor learning more than vision alone [37].

The tactile sensation that the surgeon both actively and passively perceives defines haptics in surgery [58]. The use of gloves involved only an indirect reduction of haptic feedback in open surgery [25], but in MIS the lack of haptic feedback is much more predominant [11]. Nevertheless, modern bench-top simulators utilizing actual surgical instruments provide realistic haptic feedback close to the one used in real surgeries [26].

Our knowledge of haptics and its role in learning a motor skill remains largely understudied in surgery and skill training. In the following sections, we provide an overview of the latest knowledge of haptics in various fields, some challenges and issues associated with haptic feedback, and explore some possibilities for improving skill training via haptic guidance.

4.2   Haptics as an Underexplored Teaching Tool

The term Haptics (derived from the Greek word *haptikos* meaning "to perceive"), refers to the combination of kinesthetic and tactile sensations humans use to perceive complementary information about their environment [43]. Humans get kinesthetic information (proprioception) from the cutaneous system (touch) and the motor system, which is perceived by the brain [43]. Created from an organic stimulus, the information is propagated from the skin and muscle receptors up to the central nervous system. It continues through the autonomous system to the thalamus and then to the cerebral cortex. Each stimulus applied to the body is transcribed by the receptors that transmit the signal to the spinal medulla which divides the stimuli into two groups. The first one perceives pain and temperature and forms the spinothalamic tract. The second one forms the medial lemniscus and carries on the information coming from articulations, muscle, and tendons. Kinesthesia corresponds to the information received by posture and body movement. It provides the perception of oneself and the ability of planning and controlling motions already in effect, as well as the awareness of the relative location of neighboring body parts [26].

Through kinesthesia, one knows where each joint is, even without visual feedback. However, it is altered in surgery by the distance between the tip of the instrument and the hand of the surgeon. This alteration can be compensated with training that develops extended proprioception through tool interaction [26]. Motor skill learning is a continuum of innate mechanisms which allow building movement coordination thanks to haptic perception. There are different phases in the process of skill acquisition and is heavily related to motor skill practice [1]. Several phases and cycles, cognitive, associative, and autonomous, are involved in surgical skill acquisition [22, 66]. During the cognitive phase, a trainee should recognize the individual components of the procedural skill. In the associative phase, the trainee links these components into a continuous action using feedback to optimize movement. During the autonomous phase, movements become automatic, requiring little cognitive input. Gallagher et al. introduce an attention model describing a part of the process of learning a complex motor task in surgery [27]. In this model, complex tasks are split into simpler movements, which are more easily mastered by trainees before contextualizing them as parts of a more complex motor activity. This is validated through research into the methods by which individuals acquire skills when learning to play musical instruments. Humans tend to dedicate more attention to the cognitive phase of training since, as it is currently understood, more efficient identification of task components results in a faster associative phase of learning [96]. Hence, the better trainees are in the identification of the step, the more quickly and efficiently they will learn.

Most people rely on visual feedback when learning (i.e., they learn based on what they see). However, as haptics speeds up the process, haptic feedback is more effective for learning. Indeed, haptics offers information on body position and how to replicate it, and this why it is a critical element for creating muscle memory [37]. Also, according to Fitts' law which states that there is a direct relationship between the difficulty of a task and its execution time [21], one can consider that a task has become instinctive as soon as the trainees obtain a faster task completion with haptic feedback. Nevertheless, despite its benefits in motor skill learning and motor skill acquisition, solely using haptics has some challenges. One of the limitations of haptic training alone is that it is less effective than visual training to distinguish shapes and position. Moreover, when vision is reintroduced after haptic training, visual information overrides the acquired haptic information [20]. Surprisingly, during surgical training, relying on visual feedback alone increases the trainees' workload and the time they require to train themselves [93]. Note that skill transfer between people is often performed via visual and verbal channels. Also, haptics can help demonstrate movements which can hardly be described through the verbal channel. This can be explained as follows: haptic perception is the only human sense that allows direct manipulation of the environment. It can also help in complex situations when visual information is not reliable, such as during a video signal delay in teleoperation, wherein preliminary training with haptic feedback can help complete a task.

In summary, haptic training requires less mental workload since it occurs in the motor loop and visual training requires complex sensorimotor transformations. Yet, haptic training has been displaced in favor of vision for surgical training [20]. Emphasizing haptic feedback during the associative phase (when trainees acquire novel surgical skills) can reduce the time needed for attaining proficiency in motor coordination [96]. Kinesthetic learning takes advantage of kinesthetic memory, or one's ability to remember and recall motor patterns. Kinesthetic memory, or muscle memory as it is sometimes called, is employed within kinesthetic training and is formed using the kinesthetic sensory and motor information obtained from movement [20]. An example of kinesthetic training is when professional golfers guide the golf swing of trainees to teach them to hit with greater accuracy. Two types of feedback are delivered when developing a motor skill: during (or after) and inherent to the current task or extrinsically related [70]. They may be originated from within the body (proprioception) or they may arise from the external environment: exteroception or augmented feedback [80]. Exteroception is heavily assisted by audition and vision, while proprioception uses the skin receptors, muscles, joints and vestibular system [70]. The two types of augmented feedback are knowledge of performance (KP) and knowledge of results (KR) [71]. KP is the kinematic information obtained by regarding the trajectory characteristics after task completion, while KR is the feedback related to the undertaking and completion of the task [71]. Even though these forms of feedback are associated with different mechanisms, they may still be considered to operate under the same underlying principles [71]. Various roles have been attributed to augmented feedback in motor skill acquisition. Kinematic feedback not only provides information about the task performed either by identifying errors made and possible corrections but also works as a motivator and as an assessment tool [2]. Schmidt et al. describe the benefits of kinematic feedback in task completion through various forms. They notify the trainees of their motion patterns, or they provide them information not otherwise given by intrinsic sources [70]. However, there are two main challenges in the implementation of kinematic feedback: that task simplicity at the bench-top level makes enhanced feedback redundant, and that it is challenging to extrapolate results to complex tasks [71].

4.3   Surgical Skill Acquisition Using Haptic Guidance

Kinesthetic information can be stored and contained within memory, and once it is retrieved, it can then be utilized to guide one's movement with high accuracy. In this section, we recall how kinesthetic guidance, when provided by an expert to a novice, can assist the novice in learning a laparoscopic skill more efficiently. During these surgical procedures, kinesthetic sensation makes the surgeons aware of the position of their joints and limbs as they perform surgical maneuvers. It requires many repetitions so that these maneuvers are committed to memory. This repetition improves muscle memory and performance, which are a necessity and are characteristics of an expert surgeon.

Novices who attempt to attain this level of dexterity often spend countless hours doing unstructured training. Their training could be accelerated if they were guided in the way they execute a movement, fostering kinesthetic information for muscle memory such that in the next attempt, the movement itself can become instinctive [20].

For many years, it has been quite a challenge to implement computer-simulation-based haptic training emphasizing kinesthetic feedback as the technology had not yet reached the maturity necessary to allow for recording and playing back kinesthetic information from a human performer. New haptic interfaces are now capable of that, as well as quantitatively characterizing the performance through several metrics [74]. Several studies on kinesthetic feedback can be found in the literature. Within these studies, engineers have discovered and proposed innovative applications for kinesthetic interfaces, including the SensAble™ PHANToM Omni (see Sect. 3.2) [35, 40, 75]. One instance of training using haptic guidance is the What-You-Feel-Is-What-I-Feel (WIFIWIF) model. In this model, during the training of a complex motor skill, novices are provided kinesthetic guidance from expert surgeons via a computer master-slave. With the haptic devices, kinesthetic information can be captured from the expert surgeons and delivered instantly via force feedback to novice surgeons. It permits them to track the experts' movements and to learn a surgical task [15]. It is analogous to golf swing training mentioned previously. However, haptic devices such as the PHANToM can provide guidance through laparoscopic instruments attached to the haptic device. Moreover, the PHANToM can provide force and position feedback during the training that can be customized to different tasks and levels of skill.

5 Supervised Haptic Training

Haptic simulators usually provide solutions to autonomously train oneself without any danger, typically to perform repetitive attempts to get familiar or to improve oneself on a given gesture. It can correspond to usual gestures or to rare and difficult cases which could be encountered in real life but which need to be mastered, particularly under stressful conditions. Nevertheless, for difficult cases, it remains useful for a trainer to guide the trainees' motions, while keeping the advantages of haptic simulators, for more accurate and efficient training. Classically, a trainer can directly guide the hands of a trainee to perform a correct motion. Yet, this "four-hand fellowship" does not permit the trainee to feel and dose the correct level of force to apply on their device in case of interaction of their tool with its environment as the efforts are shared between both persons.

Dual-user systems have been designed in the first approach as a means for two operators to cooperate remotely on a shared task. To do so, these systems extend the master-slave classical teleoperation architecture by adding a second master manipulated by a second operator, as shown in Fig. 7.3. In fact, a dual-user system can be seen as a particular case of a more generic multiple master/single slave (MMSS) teleoperation system. The main concept with dual-user systems is that the users share the slave control according to a dominance factor ($\alpha \in [0, 1]$). When $\alpha = 1$ (respectively 0), the trainer (respectively trainee) has full control over the trainee's (respectively trainer's) device and on the slave. When $0 < \alpha < 1$, both users share the slave control with dominance (over the other user) which is a function of α. According to the architectures found in the literature, the effect of α on the force feedback provided to the users differs. These differences are highlighted in Sect. 5.2.

Their usage has then been enlarged from cooperation to training purposes by Chebbi et al. [14]. Indeed, dual-user systems are a practical solution to the problem of rendering the four-hand fellowship in the haptic training: it can reproduce this important force information simultaneously to both users, each one interacting with their own haptic interface.

The next section exposes a typical use-case highlighting why these systems fit with this need.

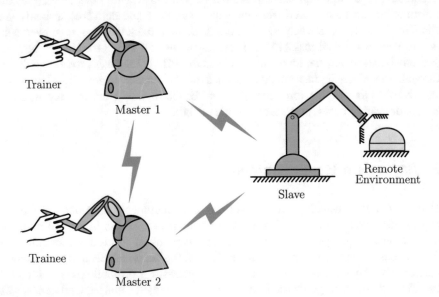

Fig. 7.3 Architecture of a dual-user system

5.1 Typical Use-Case and Requirements

In the context of hands-on training, users are typically a trainer and a trainee. Each one manipulates his/her own haptic interface. They share the same "slave" which can be either a real surgical robot dived into a real environment, located in a near-black box for instance, or a virtual robot in a virtual environment, provided by software simulation. Such an application, providing **haptic mentoring** with a virtual tool in a virtual world, has been introduced by Chebbi et al. in [13].

The following typical use case helps determine the main requirements of the system. Suppose at first that the trainer (an experienced surgeon) aims at demonstrating the right trajectories of his or her surgical tool to perform a task featuring some tool–environment contact. This implies that he/she requires realistic force feedback to dose the forces, as in a bilateral teleoperation context. The trainer manually sets $\alpha = 1$ to become the **leader** (the trainee becomes the **follower**): it is a **mentoring mode**. He/she then gets full force feedback from the slave in order to perform his/her task as if he/she was handling the real instruments. Meanwhile, the trainee's device follows the trajectory of the leader's. If the trainee deviates from this reference trajectory when in free motion, the compliance of the device brings him/her back to the right position. In case of interaction between the tool and its environment, the trainee can also feel in his/her hands the right level of effort to provide to the tool, by means of a display which guides him/her to set the device at the right position with the right applied force. Afterward, the functioning can be inverted by reversing α so that the trainee manipulates and the trainer follows and evaluates trainee's motions and applied forces; it is an **evaluation mode**.

5.2 Existing Dual-User Architectures for Haptic Training Purpose

Various architectures have been proposed for the control of dual-user haptic systems. The shared-control concept has been introduced by Nudehi et al. in [61] as a form of minimally invasive telesurgical training for experienced surgeons to transmit their knowledge and guidance to trainees under a shared robotic control environment. The aforementioned principle of α is introduced in this work. In this case, the leading user controls the slave position, but the only feedback he/she receives is transmitted visually by the remote camera image displayed on a local monitor. It corresponds to a unilateral teleoperation scheme. The control of the slave is shared between both users, and each one experiences feedback forces proportional to the difference of position of his/her own device and to α. In any case, an interaction between the robot and its environment would not feed interaction forces back to any user. Therefore, this kind of training system can only be used to train users on motions, not on efforts. The same limitation occurs with the proposal of Chebbi et al. with a virtual tool [13]: both users can guide each other according

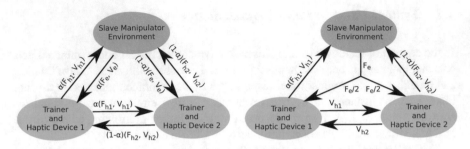

Fig. 7.4 CLC (left) and MCET (right) dual-user system architectures from [38]

to the selected mode, but the following user cannot feel interaction forces generated by the leader during tool-environment interactions.

Khademian and Hashtrudi-Zaad overcame this limitation by providing a three-port multilateral control architecture [38] (Fig. 7.4). Two distinct architectures are proposed by the authors. The first one is the "Complementary Linear Combination" (CLC) architecture which provides feedback forces combining the environment and the other user forces (respectively F_e and F_{h2}, for user 1). The desired position and force commands for each device are a complementary weighted sum of positions and forces of the other two devices. When $\alpha = 1$ (respectively 0), master 1 (respectively 2) and the slave form a four-channel bilateral teleoperation system. The position of the user 1 device and the forces he/she applied on it, also serve as input for the master 2 controller (the behavior is symmetric when $\alpha = 0$). When $0 < \alpha < 1$ both users can perform a task collaboratively in a shared environment. Unfortunately, when trying to perform a mentoring (respectively evaluation) session, the trainee (respectively trainer) has no direct force feedback from the environment, only from the other user. This is why the authors advise using this system with values of $\alpha = 0.2, 0.5, 0.8$ to get the best transparency. Yet, in these cases, both users influence the slave trajectory, which may lead from inaccurate to trembling motions, not compliant with mentoring and evaluation modes. The second architecture Khademian and Hashtrudi-Zaad introduced is called "Masters Correspondence with Environment Transfer" (MCET) [38], where both user devices follow the motion of each other and the effort fed back to both users is $F_e/2$. The command to the slave is weighted accordingly by α and $(1 - \alpha)$. Kinesthetic performance analysis and experimental user perception studies, performed by the same authors, showed that the MCET architecture provides better kinesthetic feedback to both users with the least sensitivity to the dominance factor, compared with CLC. Yet, still according to the authors, this architecture transmits a distorted environment force to both users (which was equal to $F_e/2$ in addition with a force expressed as a function of the difference of the positions of the devices of both users).

Ghorbanian et al. refined the shared control concept by splitting the dominance into two factors α and β [28]. α determines the balance of authority of the trainer over the trainee (and reciprocally), while β indicates the supremacy of both users over the slave robot. This provides an additional degree of freedom in the authority mechanism which is used to set a nonlinear relation between α and β in order to adjust the authority of the leader (the user for which $\alpha > 0.5$) over the slave with respect to the authority over the follower. It also supports uncertain communication channels in the presence of delays, packet loss, data duplication and packet swapping, which corresponds to practical issues to overcome as soon as the masters and the slave are away. However, when $\alpha = 1$, the desired master 2 force (F_{h2_d}) is only linked to the position of Master 1, not to the tool-environment interaction force. It should not help this user evaluate the right force to apply in a training situation. As this work has not ever been compared with CLC or MCET, it cannot be determined whether the transparency of this architecture is better.

Notice that the architecture introduced by Razi et al. [65] transmits to the follower a force $F_{h2_d} = (F_e - \alpha F_{h_1})$, where F_{h_1} is the force applied by user 1 on their device, which also does not correspond to the aforementioned hands-on training requirements. An (also) interesting work has been introduced by Shamaei et al. [73] where two dominance factors are used. Several architectures are provided. The one which tracks the positions of all three devices has been evaluated experimentally. This architecture's performance is analyzed in terms of impedance matching between trainer and trainee during soft and hard interactions. This provides interesting but partial information during interaction: no information about position nor force absolute tracking error is provided, only a comparison versus the two dominance factors. Also, the authors specify that this approach is limited in terms of control freedom as derivative operators may destabilize it.

In [72], the users' haptic feedback is a weighted (with α) sum of the virtual tool-environment interaction force F_e and the other user interaction (with his master device) force F_h. More precisely, the desired haptic feedback applied to user 1 $F_{h_1d} = \alpha.F_e + (1 - \alpha)F_{h_2}$. Thus, when $\alpha = 1$, user 1 only feels F_e and user 2 only feels F_{h_1}. Experiments show that when $\alpha = 0.5$ and both users follow very close trajectories, they feel very close force feedback, in the presence of time delays between each master and the slave. This is an interesting result. Yet, both users must partially cooperate to guide the slave tool with the same dominance. In our point of view, it may disturb the mentoring/evaluation as each user is implied in the motions and the interactions without distinction. Also, the experiments have been performed with constant environment interaction characteristics. More realistic experiments should mix free motion and interaction phases in the same session.

Furthermore, in a dual-user hands-on training system, the authority α should be switched anytime by the trainer. The architectures should guarantee that fast changes of α may not destabilize them. Yet CLC, MCET, and Razi et al. architectures have been modeled as LTI systems. Therefore, the robustness of these architectures should be evaluated according to the varying parameter α, which has not been ensured yet as far as we know. The only architectures in which robustness versus α variations is proven are the one from Ghorbanian et al. [28], through a Lyapunov function, and the one from Shamaei et al. [73], through an unconditional stability approach based on Llewellyn's criterion. Also, Zakerimanesh et al. proposed to consider the actuator saturation and boundedness of inputs [99] (but only validated this proposal in simulation so far).

In 2011, Nuno et al. published a survey [60] concerning passivity-based controllers for non-linear bilateral teleoperation. In our point of view, an energy approach seems promising as it already provided interesting results in more classical teleoperation studies (with two examples [44, 45]). In [49], this kind of approach has been used in the "Energy-based Shared Control (ESC)" architecture. It was developed only for single-degree-of-freedom (DOF) mechanisms, with experiments conducted with only two real haptic devices and a virtual one. ESC has been compared with two other dual-user architectures from Khademian and Hashtrudi-Zaad, introduced in [38] (the "Complementary Linear Combination" (CLC) and the "Masters Correspondence with Environment Transfer" (MCET)), by means of simulations [50]. This comparative study concludes, from simulation results, that the ESC has position tracking performance equivalent to that of CLC and MCET architectures. In [48], we proposed to embed an Adaptive Authority Adjustment (AAA) function to automatically select, in evaluation mode, the best balance between the authority of the trainer and the trainee according to the trainee's motion tracking quality. However, this solution required tuning two task-dependent parameters which were hard to be optimized by any trainer. In between, Lu et al. proposed the use of three dominance factors to provide more flexibility to the users [51]. These factors are dynamically adapted, in the presence of time delays, by extending the AAA mechanism introduced in [48]. They use linear modeling to prove the stability and study the transparency of the system.

6 Conclusion

As in many other disciplines where dexterous manipulation is necessary, this chapter illustrates how haptics is relevant for hands-on training and per-operation enhancement. Technological evolution brings new possibilities with haptic interfaces more and more precise, which permits one to obtain more realistic simulators and more precise interfaces for operations at a distance or through medical robots.

This chapter also presents the issues and different ways of providing this haptic guidance on medical procedures. Control of haptic devices allows one to mimic the characteristics of the real world but also allows adding information to "guide" trainees. Several strategies were presented, and they suggest a variety of consequences for users' manipulations. These strategies can be implemented through multiple methods of control and could use different types of information.

The main objectives of hands-on training on simulators, mentoring and evaluation without any risk for the users and the tools (the latter is only true with virtual tools), seem reachable with modern haptic technologies. Yet, the typical use case provided in Sect. 5.1 shows that training a gesture in terms of motions and interactions requires the setup of situations where the active guidance of the following user and the tool-environment force feedback should not be mixed, in order to provide the trainee (in mentoring mode) and the trainer (in evaluation mode) with clear information about the force applied by the other user. This opportunity to be able to feel the same interaction force as the other user was not feasible in traditional four-hands training. This is a promising innovation. Nevertheless, as far as we could observe, no real user study proved that this characteristic really improves hands-on training. This work is still to be done. Also, this state of the art shows that technically speaking, the realization of such behavior is complex. Several approaches have been proposed with mixed results. As still today, progress is brought to teleoperation architectures when the first ones were proposed approximately thirty years ago [4], we can imagine that automatic control theory progress should help improve these younger dual-user architectures in order to obtain in the near future a simulator with several degrees of freedom able to perform supervised hands-on training even with a remote trainer/trainee.

Acknowledgements The authors acknowledge the financial support of the China Scholarship Council (CSC) and the Consejo Nacional de Ciencia y Tecnologia (CONACyT) in Mexico.

References

1. Adams, J.A.: A closed-loop theory of motor learning. Journal of Motor Behavior **3**(2), 111–150 (1971). https://doi.org/10.1080/00222895.1971.10734898. PMID: 15155169
2. Adams, J.A.: Historical review and appraisal of research on the learning, retention, and transfer of human motor skills. Psychological Bulletin **101**(1), 41–74 (1987). https://doi.org/10.1037/0033-2909.101.1.41
3. Aggarwal, R., Moorthy, K., Darzi, A.: Laparoscopic skills training and assessment. British Journal of Surgery **91**(12), 1549–1558 (2004)
4. Anderson, R.J., Spong, M.W.: Bilateral control of teleoperators with time delay. In: Proceedings of the 27th IEEE Conference on Decision and Control, pp. 167–173 (1988). https://doi.org/10.1109/CDC.1988.194290
5. Auricchio, F., Taylor, R.L., Lubliner, J.: Shape-memory alloys: macromodelling and numerical simulations of the superelastic behavior. Computer Methods in Applied Mechanics and Engineering **146**(3), 281–312 (1997). https://doi.org/10.1016/S0045-7825(96)01232-7. URL http://www.sciencedirect.com/science/article/pii/S0045782596012327
6. Bakr, M.M., Massey, W., Alexander, H.: Evaluation of simodont® haptic 3d virtual reality dental training simulator. International journal of dental clinics **5**(4) (2013)
7. Bann, S., Darzi, A.: Selection of individuals for training in surgery. The American Journal of Surgery **190**(1), 98–102 (2005)
8. Basdogan, C., De, S., Kim, J., Muniyandi, M., Kim, H., Srinivasan, M.A.: Haptics in minimally invasive surgical simulation and training. IEEE Computer Graphics and Applications **24**(2), 56–64 (2004)
9. Berkelman, P.J., Hollis, R.L.: Lorentz magnetic levitation for haptic interaction: Device design, performance, and integration with physical simulations. The International Journal of Robotics Research **19**(7), 644–667 (2000). https://doi.org/10.1177/027836490001900703
10. Beyl, P., Van Damme, M., Van Ham, R., Vanderborght, B., Lefeber, D.: Pleated pneumatic artificial muscle-based actuator system as a torque source for compliant lower limb exoskeletons. IEEE/ASME Transactions on Mechatronics **19**(3), 1046–1056 (2014). https://doi.org/10.1109/TMECH.2013.2268942
11. Bholat, O.S., Haluck, R.S., Murray, W.B., Gorman, P.J., Krummel, T.M.: Tactile feedback is present during minimally invasive surgery. Journal of the American College of Surgeons **189**(4), 349–355 (1999). https://doi.org/10.1016/S1072-7515(99)00184-2. URL http://www.sciencedirect.com/science/article/pii/S1072751599001842

12. Caldwell, D.G., Gosney, C.: Enhanced tactile feedback (tele-taction) using a multi-functional sensory system. In: [1993] Proceedings IEEE International Conference on Robotics and Automation, pp. 955–960 vol.1 (1993). https://doi.org/10.1109/ROBOT.1993.292099

13. Chebbi, B., Lazaroff, D., Bogsany, F., Liu, P.X., Ni, L., Rossi, M.: Design and implementation of a collaborative virtual haptic surgical training system. In: IEEE International Conference Mechatronics and Automation, 2005, vol. 1, pp. 315–320 Vol. 1 (2005). https://doi.org/10.1109/ICMA.2005.1626566

14. Chebbi, B., Lazaroff, D., Liu, P.X.: A collaborative virtual haptic environment for surgical training and tele-mentoring. International Journal of Robotics and Automation 22(1), 69–78 (2007). https://doi.org/10.2316/Journal.206.2007.1.206-1007

15. Chellali, A., Dumas, C., Milleville-Pennel, I.: Haptic communication to support biopsy procedures learning in virtual environments. Presence: Teleoperators and Virtual Environments 21(4), 470–489 (2012). https://doi.org/10.1162/PRES_a_00128

16. Chou, C.P., Hannaford, B.: Measurement and modeling of mckibben pneumatic artificial muscles. IEEE Transactions on Robotics and Automation 12(1), 90–102 (1996). https://doi.org/10.1109/70.481753

17. Colgate, J.E., Brown, J.M.: Factors affecting the z-width of a haptic display. In: Proceedings of the 1994 IEEE International Conference on Robotics and Automation, pp. 3205–3210 vol.4 (1994). https://doi.org/10.1109/ROBOT.1994.351077

18. Delorme, S., Laroche, D., DiRaddo, R., F. Del Maestro, R.: Neurotouch: A physics-based virtual simulator for cranial microneurosurgery training. Neurosurgery 71 (suppl 1), 32–42 (2012)

19. Ernst, M., Banks, M.: Humans integrate visual and haptic information in a statistically optimal fashion. Nature 415(6870), 429–433 (2002)

20. Feygin, D., Keehner, M., Tendick, R.: Haptic guidance: Experimental evaluation of a haptic training method for a perceptual motor skill. In: Proceedings of the 10th Symposium on Haptic Interfaces for Virtual Environment and Teleoperator Systems. HAPTICS 2002, pp. 40–47. IEEE (2002)

21. Fitts, P., Peterson, J.: Information capacity of discrete motor responses. Journal of Experimental Psychology 67, 103–112 (1964)

22. Fitts, P.M., Posner, M.I.: Human Performance, Belmont, CA edn. Brooks/Cole Pub. Co. (1967)

23. Fleming, N.: I'm different; not dumb. modes of presentation (vark) in the tertiary classroom. In: Proceedings of the 1995 Annual Conference of the Higher Education and Research Development Society of Australasia (HERDSA), Research and Development in Higher Education, vol. 18, pp. 308–313. HERDSA (1995). Zelmer,A., (ed.)

24. Frisoli, A., Rocchi, F., Marcheschi, S., Dettori, A., Salsedo, F., Bergamasco, M.: A new force-feedback arm exoskeleton for haptic interaction in virtual environments. In: First Joint Eurohaptics Conference and Symposium on Haptic Interfaces for Virtual Environment and Teleoperator Systems. World Haptics Conference, pp. 195–201 (2005). https://doi.org/10.1109/WHC.2005.15

25. Fry, D.E., Harris, W.E., Kohnke, E.N., Twomey, C.L.: Influence of double-gloving on manual dexterity and tactile sensation of surgeons. Journal of the American College of Surgeons 210(3), 325–330 (2010). https://doi.org/10.1016/j.jamcollsurg.2009.11.001. URL http://www.sciencedirect.com/science/article/pii/S1072751509015555

26. Gallager, A., O'Sullivan, G.: Fundamentals of surgical simulation. Principles and practice. Dorchert Heidelrgerg, Springer, London (2012)

27. Gallagher, A.G., Ritter, E.M., Champion, H., Higgins, G., Fried, M.P., Moses, G., Smith, C.D., Satava, R.M.: Virtual reality simulation for the operating room: proficiency-based training as a paradigm shift in surgical skills training. Annals of surgery 241(2), 364–372 (2005). URL https://www.ncbi.nlm.nih.gov/pubmed/15650649

28. Ghorbanian, A., Rezaei, S., Khoogar, A., Zareinejad, M., Baghestan, K.: A novel control framework for nonlinear time-delayed dual-master/single-slave teleoperation. ISA Transactions 52(2), 268–277 (2013)

29. Grange, S., Conti, F., Rouiller, P., Helmer, P., Baur, C.: The delta haptic device. Mecatronics (2001)
30. Greer, A.D., Newhook, P.M., Sutherland, G.R.: Human–machine interface for robotic surgery and stereotaxy. IEEE/ASME Transactions on Mechatronics 13(3), 355–361 (2008)
31. Han, Y.M., Choi, S.B.: Force-feedback control of a spherical haptic device featuring an electrorheological fluid. Smart Materials and Structures 15(5), 1438–1446 (2006). https://doi.org/10.1088/0964-1726/15/5/033
32. Hayward, V., Gregorio, P., Astley, O., Greenish, S., Doyon, M., Lessard, L., McDougall, J., Sinclair, I., Boelen, S., Chen, X., Demers, J.G., Poulin, J., Benguigui, I., Almey, N., Makuc, B., Zhang, X.: Freedom-7: A high fidelity seven axis haptic device with application to surgical training. In: A. Casals, A.T. de Almeida (eds.) Experimental Robotics V, pp. 443–456. Springer Berlin Heidelberg, Berlin, Heidelberg (1998)
33. Herzig, N., Moreau, R., Leleve, A., Pham, M.T.: Stiffness Control of Pneumatic Actuators to Simulate Human Behavior on Medical Haptic Simulators. In: IEEE (ed.) 2016 IEEE AIM. IEEE, IEEE, Banff, Canada (2016). https://doi.org/10.1109/AIM.2016.7576997. URL https://hal.archives-ouvertes.fr/hal-01333383. A paraître. Nominé parmi les 5 meilleurs articles pour le prix du "Best Paper Award" de la conférence AIM 2016
34. Herzig, N., Moreau, R., Redarce, T., Abry, F., Brun, X.: Nonlinear position and stiffness Backstepping controller for a two Degrees of Freedom pneumatic robot. Control Engineering Practice 73, 26–39 (2018). https://doi.org/10.1016/j.conengprac.2017.12.007. URL https://hal.archives-ouvertes.fr/hal-01682127
35. Hu, T., Castellanos, A.E., Tholey, G., Desai, J.P.: Real-time haptic feedback in laparoscopic tools for use in gastro-intestinal surgery*. In: T. Dohi, R. Kikinis (eds.) Medical Image Computing and Computer-Assisted Intervention – MICCAI 2002, pp. 66–74. Springer Berlin Heidelberg, Berlin, Heidelberg (2002)
36. Jacobsen, S., Smith, F., Backman, D., Iversen, E.: High performance, high dexterity, force reflective teleoperator. In: ANS topical meeting on robotics and remote systems, pp. 24–27 (1991)
37. Kerr, R.: Intersensory integration: A kinesthetic bias. Perceptual and Motor Skills 79(3), 1068–1070 (1994). https://doi.org/10.2466/pms.1994.79.3.1068
38. Khademian, B., Hashtrudi-Zaad, K.: Shared control architectures for haptic training: performance and coupled stability analysis. The International Journal of Robotics Research 30(13), 1627–1642 (2011)
39. Kirkpatrick, A.E., Douglas, S.A.: Application-based evaluation of haptic interfaces. In: Proceedings 10th Symposium on Haptic Interfaces for Virtual Environment and Teleoperator Systems. HAPTICS 2002, pp. 32–39 (2002). https://doi.org/10.1109/HAPTIC.2002.998938
40. Konstantinova, J., Li, M., Aminzadeh, V., Althoefer, K., Dasgupta, P.: Evaluating manual palpation trajectory patterns in tele-manipulation for soft tissue examination. In: 2013 IEEE International Conference on Systems, Man, and Cybernetics, pp. 4190–4195 (2013). https://doi.org/10.1109/SMC.2013.714
41. Krautter, M., Weyrich, P., Schultz, J.H., Buss, S.J., Maatouk, I., Jünger, J., Nikendei, C.: Effects of peyton's four-step approach on objective performance measures in technical skills training: A controlled trial. Teaching and Learning in Medicine 23(3), 244–250 (2011). https://doi.org/10.1080/10401334.2011.586917. URL https://doi.org/10.1080/10401334.2011.586917. PMID: 21745059
42. Lazeroms, M., Villavicencio, G., Jongkind, W., Honderd, G.: Optical fibre force sensor for minimal-invasive-surgery grasping instruments. In: Proceedings of 18th Annual International Conference of the IEEE Engineering in Medicine and Biology Society, vol. 1, pp. 234–235 vol.1 (1996). https://doi.org/10.1109/IEMBS.1996.656931
43. Lederman, S.J., Klatzky, R.L.: Haptic perception: A tutorial. Attention, Perception, & Psychophysics 71(7), 1439–1459 (2009). https://doi.org/10.3758/APP.71.7.1439. URL https://doi.org/10.3758/APP.71.7.1439
44. Lee, D., Li, P.: Passive bilateral feedforward control of linear dynamically similar teleoperated manipulators. Robotics and Automation, IEEE Transactions on 19(3), 443–456 (2003)

45. Lee, D., Spong, M.: Passive bilateral teleoperation with constant time delay. Robotics, IEEE Transactions on **22**(2), 269–281 (2006)
46. van der Linde, R., Lammertse, P.: Hapticmaster – a generic force controlled robot for human interaction. Industrial Robot: the international journal of robotics research and application **30**(6), 515–524 (2003)
47. Liu, C., Bar-Cohen, Y.: Scaling laws of microactuators and potential applications of electroactive polymers in mems. Proceedings of SPIE – The International Society for Optical Engineering **3669** (1999). https://doi.org/10.1117/12.349692
48. Liu, F., Leleve, A., Eberard, D., Redarce, T.: A dual-user teleoperation system with adaptive authority adjustement for haptic training. In: Medical and Service Robots, Proceedings of 4th International Workshop on (2015)
49. Liu, F., Leleve, A., Eberard, D., Redarce, T.: A dual-user teleoperation system with online authority adjustment for haptic training. In: IEEE Engineering in Medicine and Biology Society (EMBC), Proceedings of 37th Annual International Conference of (2015)
50. Liu, F., Leleve, A., Eberard, D., Redarce, T.: An energy based approach for passive dual-user haptic training systems. In: Proceedings of the 2016 IEEE International Conference on Intelligent Robots and Systems (IROS 2016). Daejeon, South Corea (2016)
51. Lu, Z., Huang, P., Dai, P., Liu, Z., Meng, Z.: Enhanced transparency dual-user shared control teleoperation architecture with multiple adaptive dominance factors. International Journal of Control, Automation and Systems **15**(5), 2301–2312 (2017). https://doi.org/10.1007/s12555-016-0467-y
52. MacLean, K.E.: Haptic interaction design for everyday interfaces. Reviews of Human Factors and Ergonomics **4**(1), 149–194 (2008). https://doi.org/10.1518/155723408X342826
53. Magee, R.: Medical practice and medical education 1500–2001: an overview. ANZ Journal of Surgery **74**(4), 272–276 (2004). https://doi.org/10.1111/j.1445-2197.2004.02960.x. URL https://onlinelibrary.wiley.com/doi/abs/10.1111/j.1445-2197.2004.02960.x
54. Mason, M.L.: Significance of the american college of surgeons to progress of surgery in america. The American Journal of Surgery **51**(1), 267–286 (1941). https://doi.org/10.1016/S0002-9610(41)90056-9. URL http://www.sciencedirect.com/science/article/pii/S0002961041900569
55. Massie, T.H., Salisbury, J.K.: The phantom haptic interface: A device for probing virtual objects. In: Proceedings of the ASME Dynamic Systems and Control Division, pp. 295–301 (1994)
56. Mayer, H., Nagy, I., Knoll, A., Braun, E.U., Bauernschmitt, R., Lange, R.: Haptic feedback in a telepresence system for endoscopic heart surgery. Presence: Teleoperators and Virtual Environments **16**(5), 459–470 (2007). https://doi.org/10.1162/pres.16.5.459
57. Meier, A.H., Rawn, C.L., Krummel, T.M.: Virtual reality: surgical application challenge for the new millennium11no competing interests declared. Journal of the American College of Surgeons **192**(3), 372–384 (2001). https://doi.org/10.1016/S1072-7515(01)00769-4. URL http://www.sciencedirect.com/science/article/pii/S1072751501007694
58. Van der Meijden, O.A., Schijven, M.P.: The value of haptic feedback in conventional and robot-assisted minimal invasive surgery and virtual reality training: a current review. Surgical endoscopy **23**(6), 1180–1190 (2009)
59. Nguyen, L., Brunicardi, F.C., DiBardino, D.J., Scott, B.G., Awad, S.S., Bush, R.L., Brandt, M.L.: Education of the modern surgical resident: Novel approaches to learning in the era of the 80-hour workweek. World Journal of Surgery **30**(6), 1120–1127 (2006). https://doi.org/10.1007/s00268-005-0038-5. URL https://doi.org/10.1007/s00268-005-0038-5
60. Nuño, E., Basáñez, L., R., O.: Passivity-based control for bilateral teleoperation: A tutorial. Automatica **47**(3), 485–495 (2011)
61. Nudehi, S., Mukherjee, R., Ghodoussi, M.: A shared-control approach to haptic interface design for minimally invasive telesurgical training. Control Systems Technology, IEEE Transactions on **13**(4), 588–592 (2005)
62. Ostoja-Starzewski, M., Skibniewski, M.: A master-slave manipulator for excavation and construction tasks. Robotics and Autonomous Systems **4**, 333–337 (1989). https://doi.org/10.1016/0921-8890(89)90032-8

63. Panait, L., Akkary, E., Bell, R., Roberts, K., Dudrick, S., Duffy, A.: The role of haptic feedback in laparoscopic simulation training. Journal of Surgical Research **156**(2), 312–316 (2009)
64. Peyton, J.: Teaching in the theatre., chap. Teaching and learning in medical practice. Manticore Europe (1998)
65. Razi, K., Hashtrudi-Zaad, K.: Analysis of coupled stability in multilateral dual-user teleoperation systems. Robotics, IEEE Transactions on **30**(3), 631–641 (2014)
66. Reznick, R.K., MacRae, H.: Teaching surgical skills–changes in the wind. New England Journal of Medicine **355**(25), 2664–2669 (2006)
67. Rogers, D.A., Elstein, A.S., Bordage, G.: Improving continuing medical education for surgical techniques: Applying the lessons learned in the first decade of minimal access surgery. Annals of Surgery **233**(2), 159–166 (2001)
68. Sachdeva, A.K.: The changing paradigm of residency education in surgery: a perspective from the american college of surgeons. The American surgeon **73**(2), 120 (2007)
69. Salisbury Jr, J.K., Madhani, A.J., Guthart, G.S., Niemeyer, G.D., Duval, E.F.: Master having redundant degrees of freedom (2004). US Patent 6,684,129
70. Schmidt, R.A., Lee, T.D., Winstein, C., Wulf, G., Zelaznik, H.N.: Motor Control and Learning: A Behavioral Emphasis. Human Kinetics (2018)
71. Schmidt, R.A., Young, D.E.: Methodology for Motor Learning: A Paradigm for Kinematic Feedback, vol. 23:1. Routledge (1991). https://doi.org/10.1080/00222895.1991.9941590
72. Shahbazi, M., Talebi, H., Atashzar, S., Towhidkhah, F., Patel, R., Shojaei, S.: A new set of desired objectives for dual-user systems in the presence of unknown communication delay. In: Advanced Intelligent Mechatronics (AIM), 2011 IEEE/ASME International Conference on, pp. 146–151 (2011). https://doi.org/10.1109/AIM.2011.6027064
73. Shamaei, K., Kim, L., Okamura, A.: Design and evaluation of a trilateral shared-control architecture for teleoperated training robots. In: Engineering in Medicine and Biology Society (EMBC), 2015 37th Annual International Conference of the IEEE, pp. 4887–4893 (2015). https://doi.org/10.1109/EMBC.2015.7319488
74. Singapogu, R.B., Long, L.O., Smith, D.E., Burg, T.C., Pagano, C.C., Prabhu, V.V., Burg, K.J.L.: Simulator-based assessment of haptic surgical skill: A comparative study. Surgical Innovation **22**(2), 183–188 (2015). https://doi.org/10.1177/1553350614537119. PMID: 25053621
75. Singapogu, R.B., Long, L.O., Smith, D.E., Burg, T.C., Pagano, C.C., Prabhu, V.V., Burg, K.J.L.: Simulator-based assessment of haptic surgical skill: A comparative study. Surgical Innovation **22**(2), 183–188 (2015). https://doi.org/10.1177/1553350614537119. PMID: 25053621
76. Smith, R.: Smart Material Systems: Model Development. No. 32 in Frontiers in Applied Mathematics. SIAM (2005)
77. Staab, H., Sonnenburg, A., Hieger, C.: The dohelix-muscle: A novel technical muscle for bionic robots and actuating drive applications. In: 2007 IEEE International Conference on Automation Science and Engineering, pp. 306–311 (2007). https://doi.org/10.1109/COASE.2007.4341812
78. Stadler, W.: Analytical robotics and mechatronics, international ed edn. McGraw-Hill Education, New York, USA (1995)
79. Stocco, L.J., Salcudean, S.E., Sassani, F.: Optimal kinematic design of a haptic pen. IEEE/ASME Transactions on Mechatronics **6**(3), 210–220 (2001). https://doi.org/10.1109/3516.951359
80. Swinnen, S.: Advances in Motor learning and control, chap. Information feedback for motor skill learning. Human Kinetics, Champaign (1996)
81. Talasaz, A., Trejos, A.L., Patel, R.V.: The role of direct and visual force feedback in suturing using a 7-dof dual-arm teleoperated system. IEEE Transactions on Haptics **10**(2), 276–287 (2017). https://doi.org/10.1109/TOH.2016.2616874
82. Tan, H.Z., Pang, X.D., Durlach, N.I., et al.: Manual resolution of length, force, and compliance. Advances in Robotics **42**, 13–18 (1992)
83. Tao, R., Roy, G.D.: Electrorheological Fluids. WORLD SCIENTIFIC (1994). https://doi.org/10.1142/2245. URL https://www.worldscientific.com/doi/abs/10.1142/2245

84. Tavakoli, M., Aziminejad, A., Patel, R., Moallem, M.: High-fidelity bilateral teleoperation systems and the effect of multimodal haptics. IEEE Transactions on Systems, Man and Cybernetics – Part B **37**(6), 1512–1528 (2007)
85. Tavakoli, M., Patel, R.V., Moallem, M.: A haptic interface for computer-integrated endoscopic surgery and training. Virtual Reality **9**(2-3), 160–176 (2006). https://doi.org/10.1007/s10055-005-0017-z
86. Tedman, S., Thornton, E., Baker, G.: Development of a scale to measure core beliefs and perceived self efficacy in adults with epilepsy. Seizure: the journal of the British Epilepsy Association **4**, 221–31 (1995). https://doi.org/10.1016/S1059-1311(05)80065-2
87. Tobergte, A., Helmer, P., Hagn, U., Rouiller, P., Thielmann, S., Grange, S., Albu-Schäffer, A., Conti, F., Hirzinger, G.: The sigma. 7 haptic interface for mirosurge: A new bi-manual surgical console. In: 2011 IEEE/RSJ International Conference on Intelligent Robots and Systems, pp. 3023–3030. IEEE (2011)
88. Tondu, B., Ippolito, S., Guiochet, J., Daidié, A.: A Seven-degrees-of-freedom Robot-arm Driven by Pneumatic Artificial Muscles for Humanoid Robots. International Journal of Robotics Research **24**(4), p.257–274 (2005). https://doi.org/10.1177/0278364905052437. URL https://hal.archives-ouvertes.fr/hal-01292939
89. Torabi, A., Khadem, M., Zareinia, K., Sutherland, G.R., Tavakoli, M.: Manipulability of teleoperated surgical robots with application in design of master/slave manipulators. In: 2018 International Symposium on Medical Robotics (ISMR), pp. 1–6 (2018). https://doi.org/10.1109/ISMR.2018.8333307
90. Torabi, A., Khadem, M., Zareinia, K., Sutherland, G.R., Tavakoli, M.: Application of a redundant haptic interface in enhancing soft-tissue stiffness discrimination. IEEE Robotics and Automation Letters **4**(2), 1037–1044 (2019). https://doi.org/10.1109/LRA.2019.2893606
91. Trejos, A.L., Patel, R.V., Naish, M.D.: Force sensing and its application in minimally invasive surgery and therapy: A survey. Proceedings of the Institution of Mechanical Engineers, Part C: Journal of Mechanical Engineering Science **224**(7), 1435–1454 (2010). https://doi.org/10.1243/09544062JMES1917
92. Ueberle, M., Mock, N., Buss, M.: VISHARD10, a novel hyper-redundant haptic interface. In: Proceedings – 12th International Symposium on Haptic Interfaces for Virtual Environment and Teleoperator Systems, HAPTICS, pp. 58–65 (2004). https://doi.org/10.1109/HAPTIC.2004.1287178
93. Wagner, C.R., Stylopoulos, N., Jackson, P.G., Howe, R.D.: The benefit of force feedback in surgery: Examination of blunt dissection. Presence: Teleoperators and Virtual Environments **16**(3), 252–262 (2007). https://doi.org/10.1162/pres.16.3.252
94. Winter, S.H., Bouzit, M.: Use of magnetorheological fluid in a force feedback glove. IEEE Transactions on Neural Systems and Rehabilitation Engineering **15**(1), 2–8 (2007). https://doi.org/10.1109/TNSRE.2007.891401
95. Yamamoto, T., Abolhassani, N., Jung, S., Okamura, A.M., Judkins, T.N.: Augmented reality and haptic interfaces for robot-assisted surgery. The International Journal of Medical Robotics and Computer Assisted Surgery **8**(1), 45–56 (2012)
96. Yang, X., Bischof, W.F., Boulanger, P.: Validating the performance of haptic motor skill training. In: 2008 Symposium on Haptic Interfaces for Virtual Environment and Teleoperator Systems, pp. 129–135 (2008). https://doi.org/10.1109/HAPTICS.2008.4479929
97. Yiannakopoulou, E., Nikiteas, N., Perrea, D., Tsigris, C.: Virtual reality simulators and training in laparoscopic surgery. International Journal of Surgery **13**(9), 60–64 (2014)
98. Yoshikawa, T.: Manipulability of Robotic Mechanisms. The International Journal of Robotics Research **4**(2), 3–9 (1985)
99. Zakerimanesh, A., Hashemzadeh, F., Ghiasi, A.R.: Dual-user nonlinear teleoperation subjected to varying time delay and bounded inputs. ISA Transactions **68**, 33–47 (2017). https://doi.org/10.1016/j.isatra.2017.02.010

Part III
Haptics for Physical Impairments

Chapter 8
Assistive Soft Exoskeletons
with Pneumatic Artificial Muscles

Yuichi Kurita, Chetan Thakur, and Swagata Das

Abstract Active aging improves health, life satisfaction, and sense of living. This chapter describes state-of-the-art assistive devices and introduces a variety of exoskeletons including hard and soft types. Newly developed pneumatic artificial muscles with low air pressure driven capability are presented, and their application to motion assistance for daily activity and to enhance sports experience are demonstrated. The developed pneumatic gel actuator (PGM) can be driven under very low air pressure and volume. The PGMs are applied to a walk assist suit that does not use any electric devices, named the Unplugged Powered Suit. Control of the assistance timing is attempted by adding foot sensors, electric valves, and micro-computers to the suit. The application of PGMs to assist wrist motion is explored. The developed ForceHand glove can support the user in performing basic movements with the help of pneumatic gel muscles attached to the glove. Applications described include an attempt to take advantage of lightweight and flexible characteristics of PGMs in the development of an assistive suit for sports players to enhance their physical abilities and to create a more exciting experience. Tennis and laser tag games were augmented by this soft-exoskeleton technology and its performance was evaluated in this context.

1 Introduction

Arthur C. Clarke, who is a famous British science fiction writer, mentioned in his literature that the role of technology in the advancement of humankind should not be underestimated. Many people already use technology that augments their abilities in day-to-day life. Smartphones, for example, are like an external memory. Can we improve our physical abilities with wearable assistive technology?

Y. Kurita (✉) · C. Thakur · S. Das
Graduate School of Engineering, Hiroshima University, Hiroshima, Japan
e-mail: ykurita@hiroshima-u.ac.jp; chetanthakur@hiroshima-u.ac.jp;
swagatadas@hiroshima-u.ac.jp

© Springer Nature Switzerland AG 2020
T. McDaniel, S. Panchanathan (eds.), *Haptic Interfaces for Accessibility, Health, and Enhanced Quality of Life*, https://doi.org/10.1007/978-3-030-34230-2_8

An active, healthy life span is the period during which one can spend daily life in a healthy state with high quality of life. A reduction in healthy life span results in a lower quality of life and increases national care and medical expenses. The rate of increase of healthy life expectancy is less than that of average life expectancy; thus the period for which nursing care is required is extending. To address this problem, it is necessary to help the elderly age while being active and independent. Active aging improves health, life satisfaction, and sense of living. However, limitations on motion and the possibility of injuries and accidents are greater for the elderly. There is thus a vicious cycle in which limited motion prevents one from leaving the home easily, which further limits mobility and reduces quality of life. It is important to break this cycle by developing assistive devices that can be comfortably worn and enable someone to be mobile and extend their healthy life span.

In this chapter, we introduce the survey of assistive devices and exoskeletons including hard and soft types. We also describe the soft exoskeleton with pneumatic artificial muscles we developed in detail and the applications of the soft exoskeleton to motion assistance for daily activity and to enhance sports experience.

2 Assistive Exoskeletons

Medical rehabilitation has benefited from the growth of human-computer interfaces. Recent years have witnessed many excellent interfaces or interaction systems that can help in training for motor rehabilitation. Holden et al. discuss the state of the art for virtual reality (VR) applications in the motor rehabilitation field during the early 2000s [1], when virtual reality was in its growing stage. PHANToM [2] was one of the earliest devices for providing haptic feedback to the user during their interaction with the virtual environment. This is a grounded type haptic interface since it is fixed to the table while the user holds the hand-held stylus to feel the forces rendered. The set-up was widely used for surgical training. Cybergrasp [3] is another widely discussed interface that can be used to apply forces on individual fingers to resist flexion. VR has been reported to be a powerful tool to provide participants with repetitive practice, feedback about performance, and motivation to endure practice, all of which happen to be important aspects in the motor rehabilitation process. Several researchers have been working on the development of VR systems for the upper extremity (UE) to rehabilitate stroke patients. Holden et al. were the first group to report that VR could be used for reinstating movements in stroke patients [4]. They used the concept of "learning by imitation" from a virtual teacher to encourage the subjects to perform certain tasks that could restore UE movements. The Rutgers group [5] developed another VR rehabilitation system for the hand. They used Cyberglove to get kinematics and positions of the hand during training and their self-developed glove, Rutgers Master II [6] to provide haptic monitoring and feedback combined with position sensing. The training system developed by this group included the aspects of range-of-motion (ROM), speed, fractionation, and strength.

Another recent systematic review on advances in upper limb stroke rehabilitation [7] discusses the recent state of the art on UE robot-enabled therapy leading to facilitation of fine-motor control, whole arm movements, reach-to-touch and promotion of rehabilitation outside the hospital. This review gives a decent discussion on the home rehabilitation robotic systems that have gained popularity over the last few years. Home rehabilitation has become quite a necessity because of the increasing number of elderly individuals across the world and limited physical therapists. However, a major concern in providing telerehabilitation or self-rehabilitation is the inability of augmenting the therapist's skills into the system. The primary target group for robotic rehabilitation of the UE has been stroke survivors. The first telerehabilitation application utilizing inexpensive robotics is the Java Therapy system [8]. It focused on wrist flexion-extension and forearm pronation-supination therapy following brain injury. The system was based on a web site with a library of evaluation and therapy activities. A joystick helped in providing assistive or resistive force while performing therapy. The rehabilitation progress could be tracked on the web. Rutger Master II was once again used by Popescu et al. [9] to design a telerehabilitation system that enabled the therapy parameters to be remotely monitored and modified by therapists. Therajoy [10] is a home-based rehabilitation system using low-cost joysticks and wheels such as TheraDrive [11] and a unified custom-designed software called UniTherapy [12]. It was designed for stroke patients. ReJoyce [13] is a combination of active and passive robotic systems with functional electrical stimulation (FES) meant for the neurorehabilitation of the stroke or spinal cord injury impaired hand. The 2-channel FES aids in opening and closing of the hand and an end-effector allows the user to interact with games and perform activities requiring grasping and reaching.

Lower limb exoskeletons (LLE) for human augmentation were reported in 1967 by R. S. Mosher at General Electric Company in article Handiman to Hardiman [14]. The design and device could not generate real-time and synchronous motion due to a complex controller and computation limitations. Over many years researchers developed and improved the design and performances of LLE. Rehabilitation and augmentation have been two major application areas for LLE [15–17]. LLE are broadly categorized based on their application into assistive devices, robotics exoskeletons and new-generation soft exosuits. Assistive devices were developed for rehabilitation and training. These devices focus on advancing rehabilitation techniques and reducing the burden of physiotherapists. LOKOMAT [18–20] is a commercial LLE designed for rehabilitation training for patients with paraplegic conditions or other disorders. Hanyang exoskeleton assistive robot (HEXAR)-CR50 [21] was developed for augmentation with a load carrying capacity designed for use by factory workers. Driven gait orthosis (DGO) [22] was designed to assist therapists during gait training to overcome physical limits; it does so by helping to mimic the physiological gait of patients during training. Exoskeleton power assist device (EPAD) [23] is an untethered tendon-driven walking assistance device with a caster walker carrying important control units. Power from the caster to exoskeleton is transferred using Bowden cables for simplicity and flexibility. Discussions on post-stroke soft exosuit training include the preliminary investigation of clinical trials of tendon-driven soft walking assistance exosuits [24]. An article on wearable rehabilitative power assistance equipment using PAMs emphasizes assisting and training elderly subjects for walking, standing up, and sitting down on chairs [25]. OpenSim [26], Anybody [27] and Dhaibaworks [28] are 3D musculoskeletal simulation tools used for evaluation of assistive devices based on effects-simulated human musculoskeletal models. Studies and analysis can be generalized using a standard model or customized evaluation for finding more accurate rehabilitation techniques or parameters.

Hard exoskeletons have been extensively used in rehabilitation as assistive devices, human ability enhancers, haptic interfaces and systems with position sensing. Electric motors are generally used for actuation in hard exoskeletons. Pneumatic actuation is also becoming popular to reduce weight. CAREX [29] is a cable-driven 5DOF exoskeleton designated for neural rehabilitation. Cable-driven systems help reduce weight and contribute to cost-effectiveness. RiceWrist [30] is another work developed by Rice University with a motive of rehabilitation. It supports 4-DOF of the upper limb (elbow rotation, forearm rotation, wrist flexion-extension, and abduction-adduction) and is reported to be low-friction and backlash-free, with high manipulability in the workspace of interest. It was also integrated with motor image motion enabler (MIME) [31] to provide post-stroke physical therapy for the shoulder and elbow using assisted reaching movements. The three unilateral modes of MIME that were extended by RiceWrist are passive, active-assisted and constrained modes. The passive and active-assisted modes are implemented using a joint-level PD controller. An impedance controller is used to implement the constrained mode. HEXORR or hand exoskeleton rehabilitation robot [32] is designed to support all hand digits with full range of motion. Stroke patients with mild and moderate motor function impairment tried this robot. It was reported that the subjects' range of motion (ROM) of the fingers were successfully increased through the trials with the robot. SUEFUL-7 [33] is a 7DOF exoskeleton robot with muscle-model oriented EMG-based control. It assists the motions of shoulder vertical and horizontal flexion-extension, shoulder internal-external rotation, elbow flexion-extension, forearm pronation-supination, wrist flexion-extension, and wrist abduction-adduction. The EMG-based control of the exoskeleton is implemented by using a neuro-fuzzy modifier to modify the muscle-model matrix which in turn is driven by the user's EMG. The CADEN-7 [34] is another 7-DOF exoskeleton that enables full glenohumeral, elbow, and wrist joint functionality. This prototype eliminates the difficulty of inserting the arm from the end and sliding axially down the length of the arm, which can be quite troublesome for impaired people. It uses an open mHMI (mechanical human-machine interface) for both upper and lower arm segments to eliminate this difficulty. Another example of a 3-DOF wrist exoskeleton is the Wrist Gimbal [35]. This was developed for stroke rehabilitation primarily focusing on developing a distal device that could benefit both distal and proximal joints of the arm.

Researchers have also introduced pneumatic actuators for realizing exoskeletons. Noda et al. showed a novel approach of combining pneumatic-electric hybrid actuation with Bowden cables [36]. Bowden cables are capable of transmitting large torques that are generated by pneumatic actuators. This approach successfully reduces a large amount of weight of the exoskeleton arm. SUE [37] is designed to introduce the supination-extension degrees of freedom to robotic therapy. Electric motors used in these types of applications must be highly geared in order to be able to provide effective forces to the human wrist. To overcome the disadvantages of highly-geared electric motors such as difficulty to perform back-driving and high impedance of gear trains that limit the bandwidth of the robot, pneumatic cylinders were introduced in this research that were lightweight and thus low-impedance. EXOWRIST [38] is one of the initial approaches using pneumatic actuators only in configuring a wrist exoskeleton. It was a 2-DOF (flexion-extension and abduction-adduction) wrist exoskeleton using pneumatic actuators. A closed loop control was implemented through a PID-based control algorithm.

Robotics exoskeleton and augmentation devices are mainly non-stationary amputee or human augmentation devices. Hybrid assistive limb (HAL) [39–41] is a commercial exoskeleton developed to support various use cases including support for caregivers in hospitals and nursing homes and support for factory workers in stressful work environments which require carrying heavy loads. A variety of LLEs have been designed for rehabilitation and training for paraplegic patients. eLEGs [42] from Ekso Bionics uses hydraulic drives to assist paraplegic patients in regaining walking ability. The exoskeleton mimics a normal gait for the wearer. REX [43] is designed to support robot assisted physiotherapy and indoor walking assistance for paraplegic individuals. The ReWalk [44] exoskeleton from Argo Technology focuses on gaining the ability to walk for individuals with spinal cord injuries (SCI). This commercial exoskeleton can be used for rehabilitation or walking assistance. Indego [45] supports walking for people with SCI and can assist therapists in performing individualized gait training. BLEEX [46], a hydraulic actuator-driven exoskeleton, facilitates the ability to carry heavy loads with minimal effort over any terrain. Wearable-agri-robot [47] is another such device specifically designed for elderly people working in agricultural environments. Standalone power assist suit [48] was developed to assist nursing care workers to avoid back injuries with consideration of hospital work environments.

Robotic exoskeletons can be challenging to use freely in outside environments, and are often difficult to wear and difficult to use. To overcome these challenges, researchers have developed wearable exosuits. These include fabric-based or easily-worn robotic structures embedded with wearable sensors for tracking body segments. The choice of actuators for this type of device varies from certain types of PAM to tendons controlled by portable electric actuator units. A walking assistance device with a body weight support system was developed by Honda [49]. This device aims to support the life of elderly subjects or to assist with the work involved in production operation. It reduces the floor reaction force of the device's user and reduces muscle effort and metabolic energy required while walking. Passive walking assistance [50, 51] was designed to reduce the metabolic cost of walking by using specially-designed clutches to release a passive spring element to assist plantarflexion during the swing phase of the gait cycle. It reduces metabolic cost by approximately 7%. A plantarflexion assistance exosuit [52] is known to reduce the metabolic cost of walking. Myosuit [53] is a wearable exosuit designed for assisting walking for activities of daily living. The device uses bi-articular tendon actuators for assisted walking to reduce hip extensor activities. Bio-inspired soft exosuits [54] use fabrics and tendons for generating assistive torques, and experimental evaluation of these suits indicates a 6.4% reduction in the metabolic cost of walking. A soft knee rehabilitation device [55] has also been developed with thermoplastic polyurethane-based actuators designed to assist walking and exercising. Curara [56] is a wearable exoskeleton structure that assists in walking as well as rehabilitation. The exosuit reduces the feeling of restriction upon use.

Researchers have been trying to improve the wearability aspect in exoskeletons over the years. However, this often results in a trade-off between comfort and precision. Hard exoskeletons are quite excellent in precision and can generate higher forces. To realize wearability and comfort, some of these benefits must be forfeited. Therefore, this topic remains a good area of research and development. This section discusses some of the state of the art that have targeted the development of exoskeletons using soft materials and actuators. The recent advances in upper limb soft assistive systems can be broadly classified according to their target motion of the upper limb. There are prototypes for fingers (multi-finger and individual finger), hand, and elbow. Polygerinos et al. [57] focus on the grasping action of the hand. They utilize soft actuators comprised of molded elastomeric chambers with fiber reinforcements. These actuators can induce bending, twisting and extending trajectories under fluid pressurization. An open palm design has been introduced in this work. However, the fluid reservoir adds weight to the system which is placed along the waist. HANDEXOS [58] was designed for post-stroke patients to train the opening movement of the impaired hand. Iqbal et al. [59] and Conti et al. [60] also consider the multi-finger flexion and extension for hand rehabilitation and assistance, respectively. An index finger exoskeleton has been shown in [61] that allows bidirectional torque control using a Bowden cable-based series elastic actuation mechanism. A thumb exoskeleton has been designed in [62] to assist thumb opposition movements in the context of activities of daily living. It weighs as low as 150 g and generates a force of up to 10 N at the tip of the thumb.

Soft wearables targeted for the elbow and wrist happen to be quite limited in number. A soft elbow exoskeleton torque estimation control strategy has been discussed in [63] to provide effective power assistance. The sEMG from biceps of the user is used for estimating the motion intention. Bartlett et al. [64] and Sasaki et al. [65] use pneumatic actuators for supporting the wrist joint. The wrist orthosis in [64] supports flexion-extension and pronation-supination. However, a control mechanism has not been discussed for this device. The ASSIST glove [65] uses rotary-type soft actuators to develop an active support splint for the wrist. The E-glove [66] is another approach using vibrotactile feedback to motivate post-stroke patients to regain their mobility. The glove was also integrated with a gaming system to introduce a computer interaction element to the user.

3 Assistive Soft Exoskeletons with Pneumatic Gel Muscles

3.1 Characteristics of Pneumatic Gel Muscle

A pneumatic artificial muscle (PAM) is a lightweight flexible actuator driven by pneumatic pressure. With behavior similar to that of biological muscles, the McKibben artificial muscle is one of the most popular PAMs. As shown in Fig. 8.1, it is comprised of an inner tube surrounded by braided mesh. The inner tube is made of a stretchable rubber tube and the braided mesh provides protection and controls contraction of the artificial muscle. The pantograph structure of the braided mesh allows the muscle to extend and contract easily. The contraction of the actuator depends on the air pressure and volume, and the non-extensibility of the braided mesh shortens and produces tension when the endpoint is attached to a load. Such actuators are favorable for motion assistance because of their similarity to human muscle functions. However, the conventional McKibben actuator requires high pneumatic pressure and volume depending on the size of the muscle. Therefore, even when the McKibben actuator is light and flexible, it is not suitable for soft and wearable motion assistance devices owing to the requirement for a compressor or an air tank and a controller for the air pressure and volume flow to drive the pneumatic actuator. To overcome these problems, we developed an improved artificial muscle, a pneumatic gel actuator (PGM) that can be driven under very low air pressure and volume. The structural design of the PGM follows a principle similar to that of the conventional McKibben artificial muscle. The PGM is capable of contraction when no pressure is applied to the artificial muscle. This is achieved by foaming the inner tube with flexible gel made of a styrene-based thermoplastic elastomer to improve the viscosity and flexibility. Thanks to its thick and honeycomb structure, the surface of the inner tube is easily stretched in multiple directions. The flexible gel reduces the inside volume of the rubber tube and increases its flexibility, which facilitates smooth contraction and extension of the actuator. The detail of the PGM was reported in [67]. We compared the performance of the PGM with the commercially

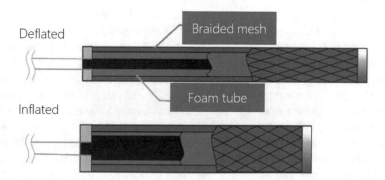

Fig. 8.1 Structure of a pneumatic gel muscle

Fig. 8.2 Contraction ratio of
developed PGM and
commercially available PAM

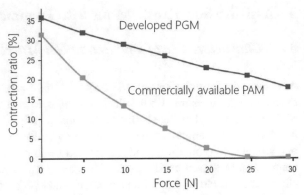

available PAM (PM-10 RF, Squse Co., Ltd.) at a pressure of 0.2 MPa. Figure 8.2 shows the results. It was confirmed that the PGM used in this study could generate a higher force and higher shrinkage ratio.

3.2 Unplugged Powered Suit

Walking motions involve muscle coordination of the lower limbs, body balance during the stance phase, flexion and extension at hip and knee joints, and ankle plantarflexion and dorsiflexion. The gait cycle involves eight stages, namely the initial contact, loading response, mid-stance, terminal stance, pre-swing, initial swing, mid-swing, and late swing. Walking improves subjective health, life satisfaction, and a sense of living. However, aging affects the performance of walking [68]. For example, degrading in the swing phase of the gait cycle results in stumbling. We thus decided to develop a walking assistance suit using a PGM for the swing phase as the first prototype by attaching the PGM along the rectus femoris. The purpose of this setup is to unload the effort of the rectus femoris in the swing phase of the gait cycle.

The traditional PAM requires compressors, air tanks, and controllers that supply air pressure. However, carrying a compressor is impractical because the compressor is noisy and needs a power supply. Because the PGM can actuate with low pressure, we tested an actuating PGM with a commercially-available rubber bulb pump. We used pumps made of silicon rubber with dimensions of 80×53 mm and mass of 46 g. The pump has a volume of 300 cubic units and can generate air pressure of 0.05 MPa with a person of 65 kg. We used these pumps to actuate the PGM in the assistive suit as it solves the problem of actuating a PGM with powered sources. Because we use a PGM that can actuate with low air pressure and generate a much stronger force than the traditional McKibben PAM, we should be able to use an unplugged air supply for the PGM. We thus take advantage of the dual-support phase of the gait cycle to design the mechanism of the assistive suit in providing

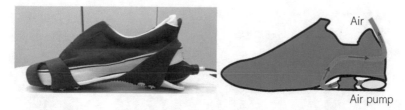

Fig. 8.3 Shoe design

Fig. 8.4 Developed
prototype of Unplugged
Powered Suit

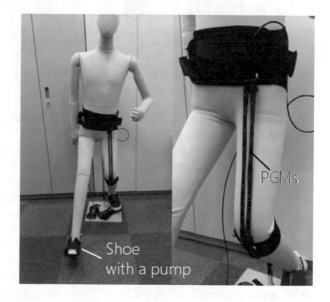

air pressure to drive the PGM during the swing phase of the gait cycle. We attached
pumps to the heel of the shoe as shown in Fig. 8.3 on the contralateral limb where
the PGM is connected. One leg enters the stance phase while the other enters the
swing phase in the dual-support phase. When the heel strikes the ground, the pumps
provide the driving pressure and actuates the PGM. This assists the initial swing
and mid-swing involving hamstring, soleus, and rectus femoris muscle activity. The
pumps and PGM are connected by pipework. The prototype design of the assistive
suit is shown in Fig. 8.4.

We conducted surface electromyography (EMG) measurements to verify the
unloading effect. We made EMG measurements of four muscles of the lower limb,
namely the rectus femoris, hamstring, soleus, and lateral gastrocnemius muscles. We
collected the measurements of nine healthy young participants. In the experiment,
surface EMG measurement was conducted for three full gait cycles in three trials
of each of assisted and unassisted walking. During the experiment, the maximum
voluntary contraction (MVC) was recorded for all four muscles under observation.
We compared the percentage MVC of assisted and unassisted walking for each

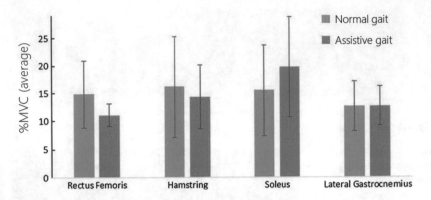

Fig. 8.5 Muscle activity with and without the suit during walking

muscle and took the average over all subjects for the selected muscle. The average for all subjects is shown in Fig. 8.5.

A t-test revealed a statistically significant difference in the muscle activity of rectus femoris but did not find this in other muscles. This suggests that the UPS can unload the muscle activity of the rectus femoris during walking thanks to the proposed mechanism to generate assistive force as needed. The rectus femoris contributes to flexion of the hip joint in the swing phase by generating a flexion torque. Since the PGM was attached along the rectus femoris of the assisted leg in the proposed mechanism of the UPS, the PGM creates flexion torque by the contraction of the PGM due to the pressure generated when the non-assisted leg of the user steps on the pump. This unloads the muscle activity of the rectus femoris. In the stance phase of the assisted leg, the hip joint extends. If the PGM still creates the flexion torque, the user needs to generate more extension torque to swing the assisted leg. Because hamstring is one of the major muscles that create the extension torque at the hip joint, wearing the UPS carries the risk of increasing the muscle activity of the hamstring. However, when the foot of the non-assisted leg leaves from the pump in the terminal stance phase of the non-assisted leg, the pressure supply is stopped and the contraction force of the PGM decreases as a result. This helps to avoid increasing the muscle activity of the hamstring. Experimental results show that there are no significant differences in the muscle activities of the soleus and lateral gastrocnemius with and without wearing the UPS. This fact also supports that the proposed mechanism avoids increasing the muscle activity of the assisted leg in the stance phase.

3.3 Assistive Wearable Gait Augment Suit

Most of the lower limb augmentation devices or LLE have limited use outside of a controlled environment due to the weight of the device, wearability and portability.

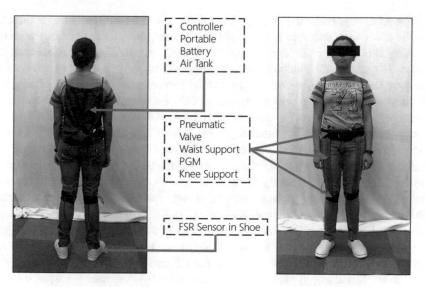

Fig. 8.6 Design of augmented walking suit as worn by subjects during experiment

It is possible to overcome these by developing an augmentation device which is wearable and uses soft and flexible actuators for providing assistive forces or torques. Augmented walking suit (AWS) [69] is a lightweight, portable, and easy-to-use example that augments normal walking. It can be used by abled elderly subjects to reduce muscle fatigue during their work hours. Unlike traditional exoskeletons where the actuators are aligned along joints to provide assistive torques, in exosuits the actuators are placed along the body segments aligned with muscle structure which supports motion assistance. The assistive control strategy of such devices relies on information about the change or abnormality in the gait cycle which is received from sensors embedded in the devices. The AWS uses synchronized gait motion and contralateral orientation of lower limbs during the gait cycle to decide the assistive phase. Based on clinical gait analysis [70] hip flexors and extensors are active during the full gait cycle and knee extensors are responsible for forward movement during the swing phase of the gait cycle.

Figure 8.6 shows the developed prototype of AWS for augmented walking. The developed prototype uses PGMs along hip flexors i.e. rectus femoris (RF), which are fixed with the help of velcro tapes to the lumbar support belt and knee support belt. Force-sensitive resistors are placed in shoes which detect the phases in the gait cycle while walking. The actuation control and gait phase detection systems are deployed on the controller in the backpack of the user. The backpack also contains a portable air tank, battery and pneumatic solenoid valves which are used as power sources for actuation of PGMs and for controlling actuation based on gait detection phases. The device weighs 1.2 Kg which is lighter than most of the state-of-the-art wearable and portable walking assistance suits.

The AWS was evaluated by recording surface EMG (sEMG) of 8 major muscles across the lower limb that are superficially accessible. The experiment involves walking straight on a flat surface for 15 m while maintaining a normal walking speed. The effects of assistive walking were measured for two levels of assistive air pressure i.e. wearing AWS but with no assist control and with 80 kPa of assistive air pressure and walking without wearing AWS for comparison. Each experiment was repeated three times for standardization and statistical analysis. sEMG of the rectus femoris (RF), vastus medialis (VM), vastus lateralis (VL), biceps femoris (BF), tibialis anterior (TA), lateral gastrocnemius (LG), medial gastrocnemius (MG) and soleus (SOL) were recorded for studying the effect of AWS on changes in muscle activity during walking. The experiments were conducted on 8 subjects, and sEMG of each subject was normalized with maximum voluntary contraction (MVC) and then averaged for comparison and statistical analysis. Figure 8.7 shows normalized averaged sEMG signals of hip extensors and flexors muscles. The graphs show data of a full gait cycle from swing to stance phase. Based on the outcome of the results, we observe that the RF and BF show reduced activity in both swing and stance phases of the gait cycle. To validate the significance of the reduction, a statistical analysis on normalized average sEMG for all eight muscles was conducted. Figure 8.8 shows the result of the statistical analysis comparing average sEMG of each muscle for assisted walking and unassisted walking. It shows significant difference in the reduction for most of the muscle groups for assisted walking.

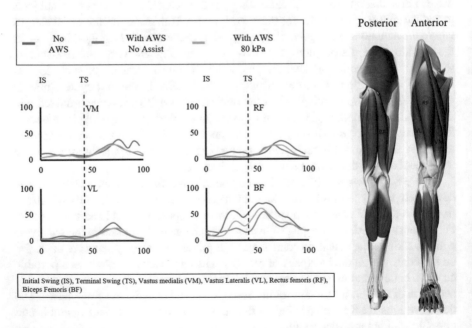

Fig. 8.7 Reduction in muscle activity of hip flexors and extensors measured during experimental evaluation. IS Initial Swing, TS Terminal Swing, VM vastus medialis, VL vastus lateralis, RF rectus femoris, BF biceps femoris

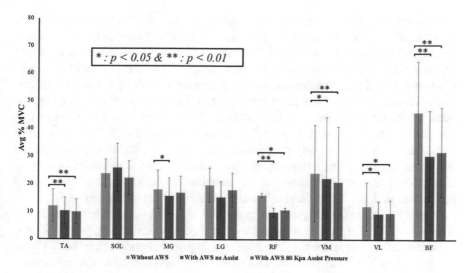

Fig. 8.8 Significance of reduction in muscle activity based on statistical evaluation of reduction in muscle activities for all the subjects. $*p < 0.005$; $**p < 0.01$

3.4 ForceHand Glove

The wrist is an integral part of the human body. Without proper functioning of the wrist, it becomes hard to perform simple tasks such as opening a doorknob, operating a light switch, or using a screwdriver. In the average human's lifestyle, wrist disorders are very likely to occur, primarily due to sedentary behavior. Two of the most common complications of the wrist and the hand include carpal tunnel syndrome and arthritis. Studies have shown that carpal tunnel syndrome is observed in a higher proportion in the obese population [71]. It can affect people at an age as low as 20 and as high as 80 [72]. On the other hand, arthritis is more likely to occur in elderly people but is sometimes found in younger people too. It can cause severe wrist pain, sometimes accompanied by swelling that makes it difficult to use the hand for ordinary purposes such as gripping or pinching. Arthritis can also be a cause for carpal tunnel syndrome. In addition, accidental injuries may sometimes create painful situations in the post-recovery stages if not healed well.

The ForceHand (FH) glove has been extensively discussed in [73, 74]. It was initially designed with a motive to assist people with a weak hand who have difficulties performing daily activities. The glove can support the user in performing basic movements with the help of the pneumatic gel muscles attached to it. The FH glove uses six pneumatic gel muscles in total. Two muscles are each meant to support flexion and extension of the wrist and one each to support pronation and supination of the forearm. Another major component used in operating the control of the glove is the stretch sensor. The stretch sensors change their resistance value with increases or decreases in the stretch applied to the sensor. These sensors are elastic to a definite extent. Four sensors are used and associated with each motion in the FH glove. In addition to assistance, the glove has potential to be used as an interface that generates force-feedback sensation through the pneumatic gel muscles. In recent years, the introduction of virtual reality has created a demand for better interfaces in terms of wearability to enhance the quality of human-computer interaction. Multiple force-feedback devices have been developed in the field; however, a majority of them are either grounded or non-wearable. There is hardly a full package that contains all the qualities of wearability, softness, lightness of weight, portability and an active capacity for responding to human motions or intentions. Through the FH glove, we attempt to achieve these qualities in a single interface that can have multiple prospects of application in fields including rehabilitation, assistance, sports augmentation, force-feedback provision, and human ability enhancement.

Figure 8.9 shows the FH glove prototype and its application as a source of force-feedback. The FH glove was designed to support flexion, extension, pronation, and supination of the forearm which complete two degrees of freedom. The third degree of freedom of the hand is radial and ulnar deviation (abduction and adduction, respectively). This was not considered in this work because of the limited range of motion involved. Figure 8.10 shows the two degrees of freedom of the human forearm that were considered for the FH glove's design. The PGMs that support flexion and extension are respectively attached to the ventral and dorsal sides of the hand. The PGMs for pronation and supination are attached spirally to the forearm with the starting point on the ventral side of the hand. Both the PGMs are spiraled in the opposite direction. The working principle of the glove is such that, whenever the user initiates any motion, the stretch sensors can detect the intent and the respective set of artificial muscles are actuated to help the user perform the specified motion. The FH glove is also integrated with Leap Motion to realize a force-feedback generating system. The Leap Motion sensor is used to get the real-time skeletal data of the hands of the user. A Unity environment is designed wherein there is a virtual object that can be contacted or touched through the skeletal hands. The PGM

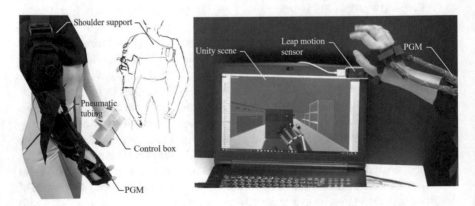

Fig. 8.9 ForceHand glove prototype and its application as a source of force-feedback

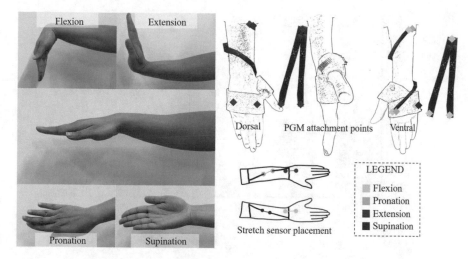

Fig. 8.10 Degrees of freedom of the wrist taken into consideration for the FH glove design

actuation was enabled in such a way that the force rendered at the user's wrist was proportional to the weight of the virtual object being pushed. The PGMs relevant to the specific motions were actuated to induce the force on the user's forearm.

An experiment was conducted to evaluate if there is a reduction in the muscle activity of the user when the FH glove assisted in performing the motions. The users performed some hand motions with and without the FH glove so that we could study the differences in the associated muscle activity. sEMG in small muscles is directly in proportion to the force generated by the user [75]. Therefore, sEMG can be used as a unit to measure the force generated by the user during the forearm motions supported by the FH glove. No visual information was provided to the subjects so that they could concentrate only on the forearm motion task. The experiment was conducted on eight healthy subjects with an age range of 20–25 years. The muscles used for recording sEMG data were flexor carpi ulnaris (FCU), extensor carpi radialis (ECR), and brachioradialis (supinator longus or SL). A paired t-test was used on the normalized data to check the statistical difference between with- and without FH glove supported activities. The null hypothesis of having no significant differences between with- and without-glove cases was considered. Figure 8.11 shows the results of average %MVC of all subjects for all motions when the statistical difference was compared for with- and without-glove cases. The FCU and ECR muscles were respectively observed during the flexion and extension motions. FCU and SL muscles were observed for both pronation and supination motions. There was a statistically significant reduction in the muscle activities for all four motions in at least one associated muscle. This statement can be used to draw an inference that the FH glove can substantially reduce the muscle activity associated with wrist motions.

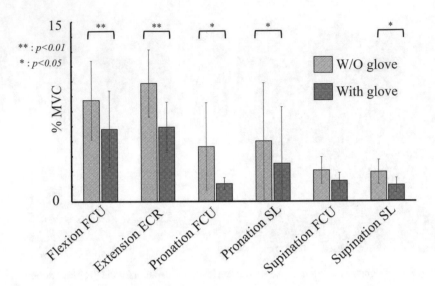

Fig. 8.11 Average %MVC of 8 subjects compared during with- and without-FH-glove cases.
$*p < 0.005$; $**p < 0.01$

3.5 Sports Augmentation Suit

3.5.1 Tennis Swing Augmentation

Interests in enabling the disabled and the elderly to participate in sports has increased. Universal sports or superhuman sports that enable everyone to play under the same competition regardless of the presence or absence of disability, age difference, or gender has also attracted increased attention. Taking advantage of the lightweight and flexible characteristics of PGMs, we address the challenge of developing an assistive suit for sports players to enhance their physical abilities and to create a more exciting experience.

The developed assistive suit for tennis players has the following features: (1) the air pressure for driving the artificial muscles is created by motion during sporting activities; (2) the timing of the assistive force for the swing motion is manually determined by the wearer by pushing a mechanical valve. We utilize the weight of the player to supply pressure from shoes equipped with a pump. This enables the player to accumulate the air pressure to drive the PGM without an air compressor. The player engages the assistive force by switching on an electric valve located at the grip of the tennis racket. Figure 8.12 shows a schematic diagram and a photo of the developed device. The apparatus comprises a pump for generating compressed air, a tank for accumulating the generated compressed air, an artificial muscle (PGM) for generating contraction forces by pressure supply, and a mechanical valve for switching the supply of compressed air from the tank to the artificial muscle. The PGM was attached to a sports shoulder supporter by using a surface fastener. To

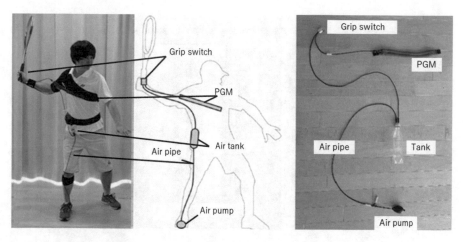

Fig. 8.12 Tennis augmentation suit with PGMs

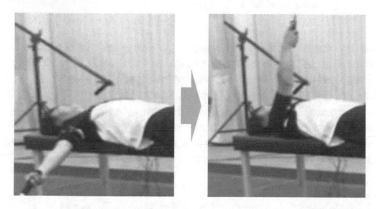

Fig. 8.13 Evaluation task of racket swing speed

assist the forehand swing motion (horizontal bending movement of the shoulder joint), it was attached along the major pectoral muscle, which is mainly responsible for the forehand swing. A rubber blower with an internal volume of 44 ml was used as the air pump. It was placed on the inner heel of the shoe. In addition, a 500 ml bottle was used as an air tank to store the air pressure. It was connected to the pump and the valve by piping via a check valve attached to the exhaust port of the blower. The internal pressure increased to approximately 0.16 kPa over 50 steps [76].

We evaluated the performance of the developed suit. Eight adult males (aged 22–38 years old, height 168–185 cm, body weight 62–75 kg) were studied for motion analysis and surface myoelectric potential measurement. The subjects were asked to bend their shoulder joints 10 times with and without the assistive force, and we measured the swing speed of the racket tip and the surface EMG of the large breast muscle. In the experiment illustrated in Fig. 8.13, we instructed the subjects to move only the shoulder joint in the horizontal bending direction, without using

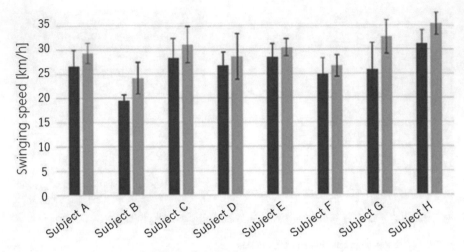

Fig. 8.14 Average swing speed for each subject

the elbow and the wrist, while in the supine position such that the influence on the large pectoral muscle by the assist force could be evaluated. In the case of assistance, the subjects themselves pushed the switch of the mechanical valve to generate the assist force of the PGM. The supplied pressure was 0.15 MPa. Figure 8.14 shows the results of swing speed. We confirmed that the swing speed was improved with a statistically-significant difference between with-suit and without-suit conditions by a paired t-test ($p < 0.01$). The measured surface EMG was also significantly decreased ($p < 0.05$).

3.5.2 Muscleblazer: A Wearable Laser Tag Module Powered by PGM-Induced Force-Feedback

Exergaming and exertion interfaces provides a firm connection between players and the game mechanics, resulting in higher motivation and involvement [77, 78]. Recently, similar efforts have been introduced in the field of sports. Nojima et al. presented "Augmented Dodgeball" as an interactive sport, supplemented with entertainment [79, 80]. By adding technology to a traditional dodgeball game, they tried to increase the level of amusement and balance the skills of the widely-varying types of players. Another similar implementation is the HoverBall, which is a flight-path-controllable ball using quadcopter technology. This technology enables the re-programming of the physical dynamics of a ball to expand its interactive features with users through the addition of characteristics such as hovering, anti-gravity and proximity [81]. To augment physical activity and technology in the game, we propose a novel laser tag game by integrating multiple ideas of human augmentation. In the proposed game, the general rule is that each player uses an IR-gun to shoot the sensors attached to the players on the opposite team. The

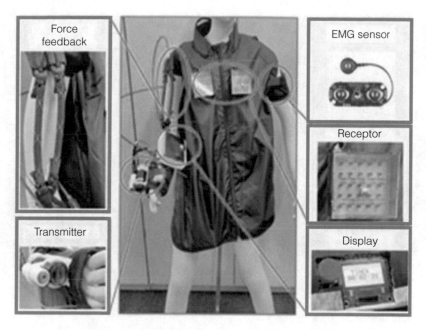

Fig. 8.15 Overview of the system for Muscleblazer

players who are hit a sufficient number of times by an IR ray transmitted from the guns of opposite team players are defeated and must leave the game. The team that successfully defeats all players of the opposite side wins the game.

We developed a wearable transmitter that typically keeps users hands-free. This transmitter is augmented with a set of pneumatic gel muscles (PGMs) [67] that are used to provide the user with a force-feedback sensation during every shot. Since PGMs are very soft and lightweight, they are extremely suitable for use in sports. This augmentation gives the players a realistic feeling of shooting the Infrared (IR) gun with their own arm. An overview of the system is shown in Fig. 8.15. Major components of this system include a transmitter and a force-feedback system. The transmitter consists of an IR LED, an OLED display, a Bluetooth module, a micro-controller, a set of PGMs (Pneumatic Gel Muscles), push buttons, and LEDs. All elements are connected to the microcontroller that can track the statistics of all users. The force-feedback system is comprised of two PGMs, an air tank, a solenoid valve, a pressure sensor and an Arduino MEGA board. This system is driven by the transmitter module simultaneously with the IR. One end of the PGMs is attached to the ventral side of the forearm. The other end is connected to the shoulder joint. The actuation of PGMs is driven by the solenoid valves controlled by the Arduino MEGA. Whenever an IR-gun shot is detected, the solenoid valve is switched on for 100 ms. These valves are operated in a normally-closed configuration. The pressure sensor monitors the air pressure available in the air tank. Whenever the air pressure is lower than 0.25 MPa, a warning LED glows, and the player has to replace his or her air tank with a new one to be able to play again.

Fig. 8.16 Muscleblazer game

We conducted an experiment to gather user feedback based on multiple aspects. We introduced the game to our subjects just before they participated in this experiment. Four participants (20–25 years old, all male) joined this experiment. They were given the full suit and were introduced to the game rules. They played one round of the game and then were asked to complete a questionnaire with quantitative as well as qualitative responses (Fig. 8.16). First, the participants were asked to rate the game based on the following 7 postulates: Fun (0: High fun; 5: Low fun), Difficulty (0: Difficult; 5: Easy), Engagement (0: Not engaged; 5: Highly engaged), Physical liveliness (0: Indifferent; 5: Feel the urge), Augmentation (0: Indifferent; 5: Felt powerful), Hardware design (0: Bad; 5: Good), and Innovation (0: Not innovative; 5: Highly innovative). Through this experiment, we acquired information which will be helpful in making improvements to the game. Figure 8.17 shows the results of the questionnaire. We observed players enjoying and taking this sport to a new level while having fun.

The proposed game can be played by all individuals irrespective of their age and physical fitness. Therefore, through this game, we want to encourage more people to reinstate exercise into their lives. Some of the advantages of this game are that while playing this game, players have a short time to decide the strategy and game planning. This improves the level of engagement of players towards the game. In addition, the shooting device can be attached anywhere on the body, which in turn allows players to bring out their own shooting style based on their imagination. This game doesn't use any guns like paintball, and therefore, there is no element

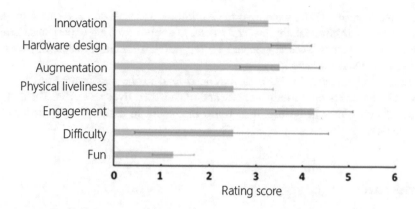

Fig. 8.17 Force-feedback system of Muscleblazer

of violence or war. Also, while playing the game, bodily damage is minimized or eliminated entirely due to the wearable system. Thus, the game is completely safe for both the players and audience.

4   Conclusion

This chapter described state-of-the-art assistive devices and introduced a variety of exoskeletons including hard and soft types. In this chapter, newly developed pneumatic artificial muscles with low air pressure driven capability was presented in detail, and its applications to motion assistance for daily activity and to enhance sports experience were demonstrated. The developed pneumatic gel actuator (PGM) can be driven under very low air pressure and volume. The structural design of the PGM follows a principle similar to that of the conventional McKibben artificial muscle. Thanks to the specially designed structure of the inner tube, the PGM has a capability of increasing its flexibility, which facilitates smooth contraction and extension of the actuator. We applied the PGMs to a walk assist suit that does not use any electric devices, named Unplugged Powered Suit (UPS). Small air pumps attached at the bottom of the shoes were used to actuate the PGMs. EMG measurements confirmed statistically significant differences in the muscle activity of the rectus femoris but did not find this in other muscles. This suggests that the UPS can unload the muscle activity of the rectus femoris during walking thanks to the proposed mechanism to generate assist force as needed. We also attempted to control the assistance timing by adding foot sensors, electric valves, and micro-computers to the suit. We explored the application of PGMs to assist wrist motion. The developed ForceHand glove can support the user in performing basic movements with the help of pneumatic gel muscles attached to the glove. We suggested that the FH glove can be used as a source of force-feedback. Taking advantage of lightweight and flexible characteristics of PGMs, we strove to develop an assistive suit for sports players to enhance their physical abilities and to create a more exciting experience. Tennis and laser tag games were augmented by our soft-exoskeleton technology and its performance was evaluated.

Continued research in this area may also lead to the development of a novel device that helps people's daily activity and makes their lives more fun. The authors hope this chapter will be useful to those interested in assistive soft exoskeleton technology.

References

1. Holden, M. K.: Virtual environments for motor rehabilitation. Cyberpsychology & behavior 8(3), 187–211 (2005)
2. Massie, T. H., Salisbury, J. K.: The phantom haptic interface: A device for probing virtual objects. In: Proceedings of the ASME winter annual meeting, symposium on haptic interfaces for virtual environment and teleoperator systems 55(1), 295–300 (1994)
3. CyberGlove Systems Inc. http://www.cyberglovesystems.com/
4. Holden, M., Todorov, E., Callahan, J., Bizzi, E.: Virtual environment training improves motor performance in two patients with stroke: case report. Neurology Report 23(2), 57–67 (1999)
5. Jack, D., Boian, R., Merians, A.S., Tremaine, M., Burdea, G. C., Adamovich, S. V., Recce, M., Poizner, H.: Virtual reality–enhanced stroke rehabilitation. IEEE Transactions on Neural Systems and Rehabilitation Engineering 9(3), 308–318 (2001)
6. Bouzit, M., Burdea, G., Popescu, G., Boian, R.: The Rutgers Master II—new design force-feedback glove. IEEE/ASME Transactions on Mechatronics 7(2), 256–263 (2002)
7. Loureiro, R. C., Harwin, W. S., Nagai, K., Johnson, M.: Advances in upper limb stroke rehabilitation: a technology push. Medical & biological engineering & computing 49(10), 1103–1118 (2011)
8. Reinkensmeyer, D. J., Pang, C. T., Nessler, J. A., Painter, C. C.: Reinkensmeyer, D. J., Pang, C. T., Nessler, J. A., Painter, C. C.: web-based robotic rehabilitation. In: Mounir Mokhtari (ed) Integration of assistive technology in the information age, pp. 9:66–71. IOS Press (2001)
9. Popescu, V. G., Burdea, G. C., Bouzit, M., Hentz, V. R.: A virtual-reality-based telerehabilitation system with force feedback. IEEE Transactions on Information Technology in Biomedicine 4(1), 45–51 (2000)
10. Johnson, M. J., Feng, X., Johnson, L. M., Winters, J.: Potential of a suite of robot/computer-assisted motivating systems for personalized, home-based, stroke rehabilitation. Journal of NeuroEngineering and Rehabilitation 4(1), 6 (2007)
11. Feng, X., Winters, J. M.: An interactive framework for personalized computer-assisted neurorehabilitation. IEEE Transactions on Information Technology in Biomedicine 11(5), 518–526 (2007)
12. Feng, X., Ellsworth, C., Johnson, L., Winters, J. M.: UniTherapy: software design and hardware tools of teletherapy. In: Rehabilitation Engineering and Assistive Technology Society North America. Orlando, FL, USA (2004)
13. Kowalczewski, J., Chong, S. L., Galea, M., Prochazka, A.: In-home tele-rehabilitation improves tetraplegic hand function. Neurorehabilitation and Neural Repair 25(5), 412–422 (2011)
14. Mosher R. S.: Handiman to Hardiman. Society of Automotive Engineers Transactions 76, 588–597 (1967)
15. Yan T., Cempini, M., Oddo, C. M., Vitiello, N.: Review of assistive strategies in powered lower-limb orthoses and exoskeletons. Robotics and Autonomous Systems 64, 120–136 (2015)

16. Chen, B., Ma, H., Qin, L. Y. Gao, F., Chan, K. M., Law, S. W., Liao, W. H.: Recent developments and challenges of lower extremity exoskeletons. Journal of Orthopaedic Translation 5, 26–37 (2016)
17. Aliman, N., Ramli, R., Haris, S. M. M.: Design and development of lower limb exoskeletons: A survey. Robotics and Autonomous Systems 95, 102–116 (2017)
18. Low, K. H., Liu, X., Goh, C. H., Yu, H.: Locomotive control of a wearable lower exoskeleton for walking enhancement. Journal of Vibration and Control 12(12), 1311–1336 (2006)
19. Jezernik, S, Colombo, G., Morari, M.: Automatic gait-pattern adaptation algorithms for rehabilitation with a 4-DOF robotic orthosis. IEEE Transactions on Robotics and Automation 20(3), 574–582 (2004)
20. Riener, R., Lünenburger, L., Maier, I., Colombo, G., Dietz, V.: Locomotor Training in Subjects with Sensori-Motor Deficits: An Overview of the Robotic Gait Orthosis Lokomat. Journal of Healthcare Engineering 1(2), 197–216 (2010)
21. Lim, D. Kim, W., Lee, H., Kim, H., Shin, K., Park, T., Lee, J., Han, C.: Development of a lower extremity exoskeleton robot with a quasi-anthropomorphic design approach for load carriage. In: 2015 IEEE/RSJ International Conference on Intelligent Robots and Systems (IROS), pp. 5345–5350 (2015)
22. Colombo, G., Joerg, M., Schreier, R., Dietz, V.: Treadmill training of paraplegic patients using a robotic orthosis. Journal of Rehabilitation Research and Development 37(6), 693–700 (2000)
23. Kong, K., Jeon, D.: Fuzzy Control of a New Tendon-Driven Exoskeletal Power Assistive Device. In: IEEE/ASME International Conference on Advanced Intelligent Mechatronics, AIM, pp. 146–151 (2005)
24. Awad, L. N., Bae, J., O'Donnell, K., De Rossi, S., Hendron, K., Sloot, L. H., Kudzia, P., Allen, S., Holt, K. G., Ellis, T. D., Walsh, C. J.: A soft robotic exosuit improves walking in patients after stroke. Science Translational Medicine 9(400), eaai9084 (2017)
25. Noritsugu, T., Sasaki, D., Kameda, M., Fukunaga, A., Takaiwa, M.: Wearable power assist device for standing up motion using pneumatic rubber artificial muscles. Journal of Robotics and Mechatronics 19(6), 619–628 (2007)
26. Sherman, M. A., Seth, A., Delp, S. L.: Simbody: Multibody dynamics for biomedical research, Procedia IUTAM 2, 241–261 (2011)
27. Rasmussen, J., Damsgaard, M., Surma, E., Christensen, S. T., de Zee, M., Vondrak, V.: AnyBody – a software system for ergonomic optimization. In: Fifth World Congress on Structural and Multidisciplinary Optimization, Vol. 4, p. 6 (2003)
28. DhaibaWorks. https://www.dhaibaworks.com/
29. Mao, Y., Agrawal, S. K.: Design of a cable-driven arm exoskeleton (CAREX) for neural rehabilitation. IEEE Transactions on Robotics 28(4), 922–931 (2012)
30. Gupta, A., O'Malley, M. K., Patoglu, V., Burgar, C.: Design, control and performance of RiceWrist: a force feedback wrist exoskeleton for rehabilitation and training. International Journal of Robotics Research 27(2), 233–251 (2008)
31. Burgar, C. G., Lum, P., Shor, P. C., Van der Loos, H. F. M.: Development of Robots for Rehabilitation Therapy: The Palo Alto VA/Stanford Experience. Journal of Rehabilitation Research and Development 37(6), 663–673 (2000)
32. Schabowsky, C. N., Godfrey, S. B., Holley, R. J., Lum, P. S.: Development and pilot testing of HEXORR: hand EXOskeleton rehabilitation robot. Journal of neuroengineering and rehabilitation 7(1), 36 (2010)
33. Gopura, R. A. R. C., Kiguchi, K., Li, Y.: SUEFUL-7: A 7DOF upper-limb exoskeleton robot with muscle-model-oriented EMG-based control. In: IEEE/RSJ International Conference on Intelligent Robots and Systems, pp. 1126–1131. St. Louis, MO, USA (2009)
34. Perry, J. C., Rosen, J., Burns, S.: Upper-limb powered exoskeleton design. IEEE/ASME transactions on mechatronics 12(4), 408–417 (2007)
35. Martinez, J. A., Ng, P., Lu, S., Campagna, M. S., Celik, O.: Design of Wrist Gimbal: a forearm and wrist exoskeleton for stroke rehabilitation. In: 2013 IEEE 13th International Conference on Rehabilitation Robotics (ICORR), pp. 1–6. Seattle, WA, USA (2013)

36. Noda, T., Teramae, T., Ugurlu, B., Morimoto, J.: Development of an Upper Limb Exoskeleton Powered via Pneumatic Electric Hybrid Actuators with Bowden Cable. In: 2014 IEEE/RSJ International Conference on Intelligent Robots and Systems (IROS 2014), pp. 3573–3578. Chicago, IL, USA (2014)

37. Allington, J., Spencer, S. J., Klein, J., Buell, M., Reinkensmeyer, D. J., Bobrow, J.: Supinator Extender (SUE): A Pneumatically Actuated Robot for Forearm/Wrist Rehabilitation after Stroke. In: 2011 33rd Annual International Conference of the IEEE Engineering in Medicine and Biology Society, pp. 1579–1582. Boston, MA, USA (2011)

38. Andrikopoulos, G., Nikolakopoulos, G., Manesis, S.: Design and development of an exoskeletal wrist prototype via pneumatic artificial muscles. Meccanica 50(11), 2709–2730 (2015)

39. Sankai, Y.: HAL: Hybrid Assistive Limb Based on Cybernics. In: Robotics Research, pp. 25–34. Springer, Berlin, Heidelberg (2010)

40. Kawamoto, H., Kandone, H., Sakurai, T., Ariyasu, R., Ueno, Y., Eguchi, K., Sankai, Y.: Development of an assist controller with robot suit HAL for hemiplegic patients using motion data on the unaffected side. In: 36th Annual International Conference of the IEEE Engineering in Medicine and Biology Society, pp. 3077–3080. Chicago, IL, USA (2014)

41. Kawamoto H., Kanbe S., Sankai, Y.: Power assist method for HAL-3 estimating operator's intention based on motion information. In: The 12th IEEE International Workshop on Robot and Human Interactive Communication, Proc: ROMAN 2003, pp. 67–72. Millbrae, CA, USA (2003)

42. Ekso Bionics. https://eksobionics.com/

43. Rex Bionics. https://www.rexbionics.com/

44. ReWalk. https://rewalk.com/

45. Quintero, H., Farris, R. J., Members, M. G.: Preliminary Evaluation of a Powered Lower Limb Othosis to Aid Walking in Paraplegic Individuals. IEEE Transactions on Neural Systems and Rehabilitation Engineering 19(6), 652–659 (2011)

46. Zoss, A., Kazerooni, H.: Design of an electrically actuated lower extremity exoskeleton. Advanced Robotics 20(9), 967–988 (2006)

47. Toyama, S., Yamamoto, G.: Development of wearable-agri-robot – Mechanism for agricultural work. In: 2009 IEEE/RSJ International Conference on Intelligent Robots and Systems, pp. 5801–5806. St. Louis, MO, USA (2009)

48. Ishii, M., Yamamoto, K., Hyodo, K.: Stand-Alone Wearable Power Assist Suit – Development and Availability. Journal of Robotics and Mechatronics 17(5), 575–583 (2005)

49. Ikeuchi, Y., Ashihara, J., Hiki, Y., Kudoh, H., Noda, T.: Walking assist device with bodyweight support system. In: IEEE/RSJ International Conference on Intelligent Robots and Systems, pp. 4073–4079, St. Louis, MO, USA (2009)

50. Pratt, J. E., Krupp, B. T., Morse, C. J., Collins, S. H.: The RoboKnee: an exoskeleton for enhancing strength and endurance during walking. In: IEEE International Conference on Robotics and Automation, Proc. ICRA'04, pp. 3:2430–2435. New Orleans, LA, USA (2004)

51. Collins, S. H., Wiggin, M. B., Sawicki, G. S.: Reducing the energy cost of human walking using an unpowered exoskeleton. Nature 522(7555), 212–215 (2015)

52. Malcolm, P., Derave, W., Galle, S., De Clercq, D.: A Simple Exoskeleton That Assists Plantarflexion Can Reduce the Metabolic Cost of Human Walking. PLoS One 8(2), e56137 (2013)

53. Schmidt, K., Duarte, J. E. Grimmer, M., Sancho-Puchades, A., Wei, H., Easthope, C. S., Riener, R.: The myosuit: Bi-articular anti-gravity exosuit that reduces hip extensor activity in sitting transfers. Frontiers in Neurorobotics 11, 57 (2017)

54. Asbeck, A. T., De Rossi, S. M. M., Holt, K. G., Walsh, C. J.: A biologically inspired soft exosuit for walking assistance. The International Journal of Robotics Research 34(6), 744–762 (2015)

55. Sridar, S., Nguyen, P. H., Zhu, M., Lam, Q. P., Polygerinos, P.: Development of a soft-inflatable exosuit for knee rehabilitation. In: 2017 IEEE/RSJ International Conference on Intelligent Robots and Systems (IROS), pp. 3722–3727. Vancouver, BC, Canada (2017)

56. Tanaka H., Hashimoto, M.: Development of a non-exoskeletal structure for a robotic suit. International Journal of Automation Technology 8(2), 201–207 (2014)

57. Polygerinos, P., Wang, Z., Galloway, K. C., Wood, R. J., Walsh, C. J.: Soft robotic glove for combined assistance and at-home rehabilitation. Robotics and Autonomous Systems 73, 135–143 (2015)

58. Chiri, A., Giovacchini, F., Vitiello, N., Cattin, E., Roccella, S., Vecchi, F., Carrozza, M. C.: HANDEXOS: Towards an exoskeleton device for the rehabilitation of the hand. In: IEEE/RSJ International Conference on Intelligent Robots and Systems, pp. 1106–1111, St. Louis, MO, USA (2009)

59. Iqbal, J., Khan, H., Tsagarakis, N. G., Caldwell, D. G.: A novel exoskeleton robotic system for hand rehabilitation–conceptualization to prototyping. Biocybernetics and biomedical engineering 34(2), 79–89 (2014)

60. Conti, R., Allotta, B., Meli, E., Ridolfi, A.: Development, design and validation of an assistive device for hand disabilities based on an innovative mechanism. Robotica 35(4), 892–906 (2017)

61. Agarwal, P., Fox, J., Yun, Y., O'Malley, M. K., Deshpande, A. D.: An index finger exoskeleton with series elastic actuation for rehabilitation: Design, control and performance characterization. International Journal of Robotics Research 34(14), 1747–1772 (2015)

62. Lambercy, O., Schröder, D., Zwicker, S., Gassert, R.: Design of a thumb exoskeleton for hand rehabilitation. In: International Convention on Rehabilitation Engineering and Assistive Technology, Singapore Therapeutic, Assistive & Rehabilitative Technologies (START) Centre, Kaki Bukit TechPark II, Article 41. Singapore (2013)

63. Lu, L., Wu, Q., Chen, X., Shao, Z., Chen, B., Wu, H.: Development of a sEMG-based torque estimation control strategy for a soft elbow exoskeleton. Robotics and Autonomous Systems 111, 88–98 (2019)

64. Bartlett, N. W., Lyau, V., Raiford, W. A., Holland, D., Gafford, J. B., Ellis, T. D., Walsh, C. J.: A soft robotic orthosis for wrist rehabilitation. Journal of Medical Devices 9(3), 030918 (2015)

65. Sasaki, D., Noritsugu, T., Takaiwa, M.: Development of active support splint driven by pneumatic soft actuator. In: IEEE International Conference on Robotics and Automation, pp. 520–525. Barcelona, Spain (2005)

66. Karime, A., Al-Osman, H., Gueaieb, W., El Saddik, A.: E-Glove: An electronic glove with vibro-tactile feedback for wrist rehabilitation of post-stroke patients. In: IEEE International Conference on Multimedia and Expo, pp. 1–6. Barcelona, Spain (2011)

67. Ogawa, K., Thakur, C., Ikeda, T., Tsuji, T., Kurita, Y.: Development of a Pneumatic Artificial Muscle Driven by Low Pressure and Its Application to the Unplugged Powered Suit. Advanced Robotics 31(21), 1135–1143 (2017)

68. Kim, J., Kung, S., Soma, R.: Relationship between reduction of hip joint and thigh muscle and walking ability in elderly people. The Japanese Journal of Physical Fitness and Sports Medicine 49(5), 589–596 (2000)

69. Thakur, C., Ogawa, K., Tsuji, T., Kurita, Y.: Soft Wearable Augmented Walking Suit with Pneumatic Gel Muscles and Stance Phase Detection System to Assist Gait. IEEE Robotics and Automation Letters 3(4), 4257–4264 (2018)

70. Perry, J., Davids, J. R.: Gait Analysis: Normal and Pathological Function. Journal of Pediatric Orthopaedics 12(6), 815 (1992)

71. Atroshi, I., Gummesson, C., Johnsson, R., Ornstein, E., Ranstam, J., Rosén, I.: Prevalence of carpal tunnel syndrome in a general population. JAMA 282(2), 153–158 (1999)

72. Phalen, G. S.: The Carpal-Tunnel Syndrome: seventeen years' experience in diagnosis and treatment of six hundred fifty-four hands. Journal of Bone and Joint Surgery 48(2), 211–228 (1966)

73. Das, S., Kishishita, Y., Tsuji, T., Lowell, C., Ogawa, K., Kurita, Y.: ForceHand glove: a wearable force-feedback glove with pneumatic artificial muscles (PAMs). IEEE Robotics and Automation Letters 3(3), 2416–2423 (2018)

74. Das, S., Lowell, C., Kurita, Y.: Force Your Hand-PAM Enabled Wrist Support. In: International AsiaHaptics conference, pp. 239–245. Kashiwanoha, Chiba, Japan (2016)

75. Malanga, G. A., Jenp, Y. N., Growney, E. S., An, K. N.: EMG analysis of shoulder positioning in testing and strengthening the supraspinatus. Medicine & Science in Sports & Exercise 28(6), 661–664 (1996)

76. Ogawa, K., Ikeda, T., Kurita, Y.: Unplugged Powered Suit for Superhuman Tennis. In: 2018 12th France-Japan and 10th Europe-Asia Congress on Mechatronics, pp. 361–364. Tsu, Mie, Japan (2018)

77. Consolvo, S., Everitt, K., Smith, I., Landay, J. A.: Design requirements for technologies that encourage physical activity. In: Proceedings of the SIGCHI conference on Human Factors in computing systems – CHI '06, pp. 457–466. Montreal, Quebec, Canada (2006)

78. Sinclair, J., Hingston, P., Masek, M.: Considerations for the design of exergames. In: 5th international conference on Computer graphics and interactive techniques in Australia and Southeast Asia – GRAPHITE '07, pp. 289–295. Perth, Australia (2007)

79. Nojima, T., Phuong N., Kai, T., Sato, T., Koike, H.: Augmented Dodgeball: An Approach to Designing Augmented Sports. In: 6th Augmented Human International Conference, pp. 137–140. Singapore, Singapore (2015)

80. Rebane K., Kai, T., Endo, N., Imai, T., Nojima, T., Yanase, Y.: Insights of the augmented dodgeball game design and play test. In: 8th Augmented Human International Conference, Article 12. Silicon Valley, CA, USA (2017)

81. Nitta, K., Higuchi, K., Rekimoto, J.: HoverBall: augmented sports with a flying ball. In: 5th Augmented Human International Conference, Article 13. Kobe, Japan (2014)

Chapter 9
Haptics for Accessibility in Hardware for Rehabilitation

Ramin Tadayon

Abstract As haptic interfaces become more widely integrated into rehabilitative therapy, the issue of accessibility of haptic information delivery remains a significant challenge to pursue. This chapter explores the latest research related to the task of making motor task information accessible through haptics in hardware (in combination with software) for physical rehabilitation. The concept of accessibility is first defined within the scope of rehabilitative interfaces is discussed. A literature review is presented which features highlights of related work within several categories of haptic information communication, including environmental augmentation, motion progression, postural correction, and guidance of pacing. A person-centric approach featuring haptic information delivery in autonomous at-home training is detailed in this chapter, including findings related to the proposed system and their implications for accessibility. Finally, methods for evaluation of these interfaces and fading of haptic feedback as user skill improves are discussed with directions for future research.

1 Introduction

Perhaps one of the most significant areas of application in technology and research on human-computer interaction is that of rehabilitation. One common cause for the widespread importance of rehabilitation, as an example, is stroke, which is the third leading cause of death and the leading cause of long-term disability in the United States, costing an average of $34 billion each year [1]. Stroke often causes impairments in the neurological pathways that facilitate control of the limbs in the body. Guided physical activity focusing on the areas of impairment can help restore these damaged pathways, particularly when completed in a timely manner.

R. Tadayon (✉)
Center for Cognitive Ubiquitous Computing, Arizona State University, Tempe, AZ, USA
e-mail: rtadayon@asu.edu

© Springer Nature Switzerland AG 2020
T. McDaniel, S. Panchanathan (eds.), *Haptic Interfaces for Accessibility, Health, and Enhanced Quality of Life*, https://doi.org/10.1007/978-3-030-34230-2_9

Therefore, rehabilitation for stroke typically aims to provide a patient with focused motor tasks intended to promote active recovery of the impaired limbs.

Rehabilitation covers a wide range of interventions aimed at restoring function which is impaired due to various forms of injury or disability. Rehabilitation can be designed to restore physical or cognitive capabilities, or restore the broader function of an individual within society. This presents a rather broad scope, and the technology and techniques applied can vary greatly depending on the area of rehabilitation. Hence, the scope of this chapter is restricted to physical rehabilitation and physical therapy, or the restoration of an individual's ability to move. The role of haptics in this process, and in facilitating the interaction between the human and machine within guided physical activity, is explored through the lens of accessibility.

The process of physical rehabilitation relies entirely on the type of physical impairment, the level of impairment and affected region, the individual undergoing rehabilitation, and the protocol for therapy. In some cases, the most intensive therapy occurs in the first few months of the rehabilitation process [2], and in others, this recovery may take much longer. Furthermore, the goal of therapy may vary from full recovery of function to restoration of quality of life depending on the circumstances. However, in general, in the case of physical therapy, two types of rehabilitative exercise exist: clinical exercise, or exercise that is performed under the direct supervision and presence of a physician, trainer or therapist; and at-home exercise, performed at home without guided supervision. In this chapter, these two forms of physical rehabilitation will be referred to as "supervised" and "unsupervised" rehabilitation for the sake of simplicity.

The typical physical rehabilitation process entails the interaction between two key individuals: the subject and trainer. In this case, the subject is the individual receiving rehabilitative care, while the trainer can be any human individual who oversees the subject's rehabilitation, helps to set goals, assigns exercises and monitors the subject's improvement and functional recovery. This separation is made intentionally to frame the process of rehabilitation within the perspective of the technology, and to clearly define its role within the process. Every rehabilitation program begins with an initial meeting between the subject and trainer. During this session, the subject is given a baseline evaluation to determine the type of impairment, the limbs affected, the impact on the subject's quality of life and ability to live independently, and the functional ability of the subject in the impaired areas. This baseline evaluation is used to set short-term and long-term goals for recovery, and to assign an initial set of exercises for the subject to complete to work toward those goals.

Next, a schedule is determined wherein the subject regularly completes supervised rehabilitation with the trainer. This supervised training typically occurs within a clinic or rehabilitation center, and consequently the terms "clinical" and "supervised" rehabilitation are used interchangeably here. During these sessions, the subject completes the assigned motor tasks while the trainer observes the subject, provides real-time feedback on the subject's performance with the intention of motivating and assisting the subject, guides the subject by demonstrating the correct motion or physically assisting with the subject's motion, and, upon completion of the task, sets new goals and assigns new motor tasks for the subject. These supervised training sessions typically do not cover all exercises required for the subject to maintain steady recovery [3], so in most cases, unsupervised rehabilitation is also assigned to the subject between supervised sessions. During unsupervised rehabilitation, the subject completes the assigned exercises without the presence nor supervision of the trainer, typically at home. Hence, the terms "at-home" rehabilitation and "unsupervised" rehabilitation are used interchangeably, although not all unsupervised rehabilitation need occur within the home environment.

Technology has been applied in nearly every aspect of this process, to assist both the trainer and subject. Therefore, to determine where haptics can fit within this field, one can consider the general role of haptics in human-computer interaction. Two primary roles can be determined when observing the spectrum of haptics technology: information communication and human augmentation. This chapter focuses on haptics for information communication, and provides examples of its manifestation in rehabilitative devices for the purpose of accessibility and methods for evaluating effectiveness. The chapter includes a discussion of the definition of accessibility within the context of physical rehabilitation, as well as its significance and necessity in the design of rehabilitative devices and software. Next, a representative set of the latest work on accessibility in haptics for physical rehabilitation is discussed, with recent examples for techniques to achieve accessibility and a common strategy for evaluation. Finally, the chapter concludes with remaining challenges, limitations of current technology and directions for future work.

2 Accessibility within Physical Rehabilitation

A wide variety of rehabilitative technology has been designed to assist subjects at all levels of involvement in the rehabilitation process. One of the concentrations of this technology is to make the rehabilitation process more accessible for users. This includes a subject's access to the information necessary to learn about his or her performance as well as matching the subject's physical ability to the motion task. It is in this context that the term accessibility is used here, as it relates most closely with the role of haptics in rehabilitation. How can an individual "access" a rehabilitative exercise task? In supervised rehabilitation, therapists provide the solution to this challenge by providing real-time feedback (information communication) as a subject completes motor tasks.

Under this understanding, the challenge of accessibility in rehabilitative devices can then be framed as the challenge of either utilizing a trainer's protocol for interaction when the trainer cannot be present, or by enhancing or supplementing that protocol under the trainer's presence. It is important to note that this definition of accessibility does not necessarily require that the subject have a disability outside of the motor impairment directly related to the physical rehabilitation, such as blindness or visual impairment, for which haptics could be used as a form of sensory substitution; it is a more generalized form wherein information or capabilities that would otherwise not be available to an individual are made available through haptics technology. The 'disability' for which this type of accessibility is applied is the lack of information necessary for motor learning, and the lack of physical capability necessary to complete an assigned motor task. The necessity of accessibility can be viewed as almost self-evident. Without having access to feedback on performance and other aspects of a motor task, proper improvement and learning can be very difficult [4]. Furthermore, without having the proper ability to complete a motor task, a subject can lose the motivation to exercise [5].

3   Haptics for Accessibility Through Task Information Communication

Within research on technologies for rehabilitation, the accessibility of task information, including information about how to complete a motor task as well as how well the subject is performing, or information that augments the sensation of the task or environment, is of critical importance. While other roles may be identified for haptics in this space in current and future work, task information comprises a significant, growing and ongoing body of research and is described in this work as it relates haptic technology directly to principles of motor learning [6]. When applied extensively, well-designed haptic systems for information communication in rehabilitation can lead to significant lifestyle and health improvements through retained motivation for motor learning and physical activity [7].

In the course of making motor task information accessible, haptics research has a series of questions to answer:

- *What types of information about a motion task and the subject's performance should be made accessible?* Therapists, physicians, medical researchers and trainers have provided answers to this question by identifying gaps in information delivery in rehabilitative programs where technology can play a role. In general, information about a motor task can include the trajectory of movement, speed of movement, the required posture of the subject when completing that movement, the number of repetitions required, the degree of motion, and the frequency of exercise required over a given time interval. Information about the subject's performance relates to these properties as well: the subject's posture, progression through the motion and pacing as they relate to the goals and parameters set for that subject, the subject's successful completion of repetitions of the motion task, or the subject's frequency of exercise in a given time period, are all elements of information which are used by the subject to learn or relearn the assigned task and by the trainer to make decisions on dynamic adjustment of the task, its parameters and goals for the subject, and the assignment of new tasks as needed.
- *How can the aforementioned information be communicated through haptics, and how can this communication be evaluated for a subject?* This question

is explored throughout this chapter. In the section entitled *Related Work on Haptic Information Communication*, strategies in recent work for haptic feedback are reviewed. In *A Person-Centric Approach: Haptics in Autonomous Training Assistance*, the person-centric approach, in which the role of haptic feedback is related to an individual's bias toward the modality, is presented using an example of an autonomous training device and software. Finally, in *Evaluation: Information Transfer*, the evaluation of this interaction through information transfer is discussed.

- *Since the communication of information affects learning or relearning of the motion task, how should the protocol of communication change as the subject's skill improves or the task changes?* This question is also explored throughout this chapter. A discussion of considerations for fading feedback, and specific applications to haptic feedback, are given in *Fading of Haptic Interaction*.

4   Related Work on Haptic Information Communication

Rehabilitative technology includes a range of implementations, from virtual reality to assistive robotics to wearables and serious games, among others [8]. In each of these cases, haptic solutions have been designed to communicate real-time information during rehabilitative exercise for various purposes. Haptic information can be beneficial to individuals across the entire spectrum of ability, acting as a replacement for lost function in the case of total impairment; or an augmenting supportive mechanism in the case of partial impairment; or simply as a supplemental training mechanism in the case of no impairment [9]. This review covers some of the most common types and purposes of information communication through haptics and common findings related to each category.

4.1   Haptics for Environmental Augmentation and Immersion in Virtual Reality

Perhaps the most common applications of technology in the rehabilitation space, particularly in unsupervised rehabilitation, is virtual reality (VR). While this review focuses on VR applications implemented within the last decade, this platform has been an integral part of rehabilitative research for far longer. One of the primary motivators is the relationship between the way a motion task is presented to a subject and the ability of that subject to succeed in performing the task [10]. Researchers explored the challenge of making unsupervised rehabilitative exercise accessible through the design of environments in which the distinction between correct and incorrect motion is made more clear and easier to quickly understand. Furthermore, implementation of VR to enable this unsupervised learning is becoming more

Table 9.1 Haptic immersion methods for VR

Immersion method	Related work	Haptic interface	Notable finding
Virtual shape/object surface augmentation	Andaluz et al.	Force feedback	Subjects more involved with the virtual task and goal objects.
Movement augmentation	Sadihov et al.	Tactile feedback	Subjects acquire motion task optimal performance more quickly with higher engagement across various game implementations.
Latency reduction	Tokuyama et al.	Force feedback	Immersion increased when latency between visual and haptic feedback is optimized.
Phantom limb voluntary movement simulation	Wake et al.	Tactile feedback	Phantom pain amelioration in addition to increased immersion and perception of limb movement.
Finger dexterity	Lin et al.	Tactile feedback	Improved finger dexterity with the addition of a haptic glove to the virtual environment.

practical than ever before as this technology continues to become more affordable and easier to set up and maintain [11]. A representative subset of recent work is summarized in Table 9.1 and described below.

A primary advantage in utilizing VR is to immerse the user into a meaningful abstraction of a motion task, thus improving motivation [12]. One of the most common abstractions is the interaction with virtual objects, such as shapes and textures, for which haptic feedback can be provided to improve immersion of the experience by simulating these surfaces. Andaluz et al. [13] achieve this through the generation of interaction experiences with virtual objects with real-time force feedback in VR, and indicated in evaluation that the combination of this technology facilitated meaningful interactions between the subject and the virtual task abstraction. Fluet and Deutsch's [14] review of the state of the art confirms that in many cases of haptic feedback application, greater sense of immersion and, in a few cases, greater real-world transfer of outcomes is reported by the inclusion of haptic feedback elements compared to the traditional audio-visual VR environment, although it is stressed that not enough findings are present to-date to consider the transfer outcome clinically reliable, making it an active subject of continuing research.

Combined with serious game applications, haptics-enabled virtual environments have achieved high levels of sustained engagement and motivation of users, which has led to the widespread use of serious games with haptic interfaces in cases including neurorehabilitation of children [15]. Sadihov et al. [16] apply vibrotactile feedback as a method of movement augmentation as subjects perform motions within various rehabilitative games projected in VR environments, and achieve the desired immersion across a variety of game scenarios and applications. Tokuyama et al. [17] found that this immersion increases when the latency between visual information and its associated haptic feedback is optimized within the serious game environment.

The effects of immersion from tactile stimulation in rehabilitative environments need not restrict themselves to motivational purposes. This phenomenon can have direct physical impact on the health of subjects as well. For example, Wake et al. [18] applied tactile stimulation to the rehabilitation of phantom limb pain by simulation of voluntary phantom limb movements, and indicated significantly high levels of pain amelioration in subjects in the haptic feedback condition. Lin et al. [19] reported improved finger dexterity with the addition of a haptic glove to supplement audio-visual finger exercises in VR.

4.2   Haptics for Motion Progression: Guidance and Correction

One application for which haptic information communication seems to provide quite a great deal of benefit to the learning subject is the guidance and correction of a physical motion task. Without the presence of a physical trainer, how can a rehabilitative system utilize haptic signals, including electrotactile, vibrotactile, force-based, pressure-based and other formats of haptic feedback, to inform the subject on how to move the impaired limb and to provide information on how well the subject is moving and what corrections need to be made as the subject progresses through the movement? The two strategies of augmentation, including error augmentation wherein errors in the motion or deviations from the ideal trajectory of motion are augmented through haptic stimulation, and error avoidance (also referred to as error reduction or simply haptic guidance), where haptic actuation is used to represent a correct trajectory or continuously guide the subject along the correct path of motion, are commonly presented and compared in the latest work. The difference between these two strategies is illustrated in Fig. 9.1. Liu et al.'s [20] review of haptic error correction strategies found that compared to haptic guidance strategies aimed at augmenting correct motion (by sending haptic cues to confirm when the subject's motion follows a correct trajectory or form), strategies that augmented error in the motion (by sending haptic cues when the motion deviates from the correct form) were found to be more effective for motor learning among the studies involved, although more evidence with higher sample sizes would need to be conducted to validate this claim. Hybrid strategies that balance between these two methods also require further evaluation, but current review indicates little advantage in their combination over the use of one or the other [21].

Many key findings related to best practices in the implementation of progression information communication in haptics have been made within the last decade. Stanley and Kuchenbecker [22] observed in a review of a multitude of haptic guidance strategies that those which utilized tactile stimulus location to convey direction and an easily discernible alternative communication format for distance tended to outperform other strategies. Ben-Tzvi et al. [23] found that when utilizing force feedback on an exoskeletal glove to provide simulations of object grasping in

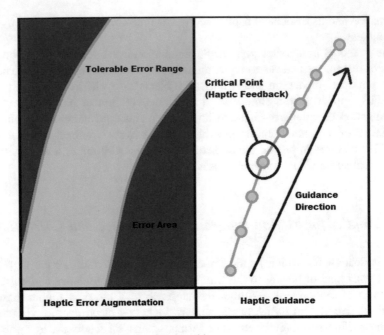

Fig. 9.1 Haptic error augmentation vs. haptic guidance

the assistance of hand rehabilitation, with trajectories designed to mimic a variety of grasping tasks, errors in grasping strategy could be immediately and significantly reduced. Chen et al. [24] provide force along the forearm to assist in pronation and supination movements using a wearable sleeve equipped with elastic bands that simulate muscle contractions to pull the forearm in the desired direction of the motion, making a case for rotational usage of the medium through cleverly designed motion-illusion patterns.

Akdogan et al.'s [25] rehabilitation system, utilizing EMG data comparisons between subject motion performance and desired therapist-provided patterns, corrected deviations between the two through tactile stimulation in the absence of visual feedback, indicating that this form of information accessibility can also be useful when the subject is blind or has limited to no access to visual information during exercise. In fact, tactile stimuli have even been shown to recalibrate an individual's auditory spatial localization sense, and can itself be a form of sensory rehabilitative training protocol for individuals who are blind [26]. Even when applied in conjunction to audiovisual feedback, rather than in replacement of this feedback, haptic information is shown to improve the reception of corrective information related to motor learning [27].

While it is possible to provide haptic guidance to help control the trajectory of a motion while simultaneously providing haptic guidance on the completion of that motion or the successful arrival at a targeted range of motion, it can be difficult for the subject to distinguish between the two, as Pezent et al. [28] have found, which leads to the need to clearly separate these forms of feedback with consideration to the design of the haptic protocol for each. It is also important in the discussion of haptic guidance to consider cases in which the environment changes, rather than the motion itself, as these environmental changes can directly impact the trajectory of motion in real-time. Baldi et al. [29] solve this challenge through the implementation of reciprocal collision avoidance in their feedback algorithm, allowing the guidance to dynamically adjust itself to the appearance of obstacles, presenting a more genuine reflection of real-world cases of motion.

An added benefit in the implementation of vibrotactile feedback for motion guidance and training is that it allows a trainer to provide personalized goals, movements and repetitions for each individual, thus obviating the need for group therapy sessions in which the individual goals and needs of subjects become lost [30]. In other words, it allows a single therapist's training protocol to become 'person-centric' in the presence of multiple simultaneous subjects of training, thus making the trainer, in addition to the information about the motor task, more accessible to the subject. This type of training customization is shown in Fig. 9.2, where the trainer programs a device to assist three subjects in an elbow flexion task with varying levels of repetition count, degree of motion, and duration of exercise for each individual. The theme of person-centricity is further discussed with an example below.

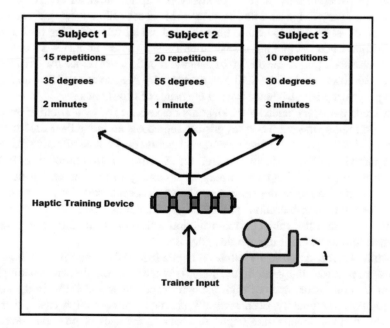

Fig. 9.2 Person-centric customization of haptic motion guidance

4.3 Haptics for Postural Correction

In rehabilitation exercise, not only is it necessary for a subject to complete a motion with as little error in trajectory as possible, but also, and perhaps even more importantly in many cases, to maintain correct body posture while exercising so that the appropriate muscle groups and neural regions are activated during the completion of the task and safety is maintained [31]. Therefore, haptic interfaces have also been designed to inform the subject about correct posture during exercise and to correct deviations in this posture. Unlike progression information, however, postural information can simultaneously involve several areas of the body, and the design of a haptic interface for this more complex interaction has merited recent discussion. Using the activity of cycling, for example, Peeters et al. [32] determined that vibrotactile signals across various parts of the body could be perceived to recognize differences between a variety of cycling positions, suggesting that they can be highly effective in improving an individual's postural awareness during physical activity.

Proprioception, or the awareness of the position and motion of one's body within space, is critically related to correct posture but is often impaired in many individuals as a result of disabilities due to stroke and other conditions. Tzorakoleftherakis et al. [33] examined the application of haptics in impairment of proprioception post-stroke, wherein loss of somatosensory perception affected control of limbs and posture. They found that through real-time vibrotactile cueing and correction, some of the lost information could be recovered in real-time allowing for improved performance when completing a basic motor task. These findings were confirmed by Cuppone et al. [34] in the design of a haptic robotic wrist device for wrist motion tasks in the absence of vision, and in a study by De Nunzio et al. [35] who applied tactile proprioceptive assistance toward the usage of prosthetics.

For some disorders, such as Parkinson's disease, or for rehabilitative exercises and motion tasks wherein incorrect posture represents an immediate risk to safety, solutions which provide reactive postural correction may not be effective due to the time-sensitivity of the feedback required for effective postural correction. In these cases, a more predictive approach may be necessary to provide earlier access to corrective information to the subject as shown in recent work [36]. Such devices also make careful considerations to the design of haptic feedback, as a haptic signal that distracts from the subject's main motion task may itself introduce a hazard, making it inaccessible for its intended purpose.

Postural correction typically relies on haptic feedback being given at the site of the desired postural change along the body. This can be a complex undertaking if the posture is complex or involves multiple active body sites and limbs in the correct configuration. Fortunately, even for multi-site postural correction cases, there are interfaces that can localize the haptic feedback at the sites of postural correction, such as the tactile shirt prototype designed by Kavathekar and Balkcom [37] for teaching motion tasks.

4.4 Haptics for Rhythmic Guidance of Pacing

While less popular than the other two categories of information, pacing or temporal information on motor performance has also seen a series of applications using haptics in recent work, to high degrees of success. Pacing information is typically most relevant in motor tasks that are cyclical in nature and sensitive to proper timing for successful completion. For these tasks, the goal of feedback is to encourage a subject to maintain a particular rhythm, and to guide the subject through the components of a rhythmic pattern through haptic cues. Alternatively, corrective haptic information may seek to speed up or slow down the rate of motion of a subject when he or she deviates from a desired pace by some threshold.

One rehabilitative exercise in which proper control over pacing is of utmost importance is breathing. Janidarmian et al. [38] utilize haptic biofeedback to pace the subject during therapeutic breathing exercises to help regulate breathing patterns for more optimal and healthier performance of the task, indicating how a simple interface can be critical in successful implementation of these temporally-sensitive tasks. While this device was worn on the umbilical region, other methods have implemented external objects for which hand contact is used to receive haptic information on breathing patterns [39].

Another rehabilitative motion task for which haptics is popularly assigned to provide rhythmic information is gait. Maintaining a steady gait pattern is of critical importance to avoid falls and injury during motion for individuals with disabilities due to Parkinson's or hemiparetic stroke, among others. Hence, these studies have found that haptic cueing to guide the pace of the gait cycle helps the subject maintain a steadier and safer rhythmic walking pattern [40–43]. A longitudinal study by Islam et al. [44] found that the improvements to symmetry of gait and walking pace maintained themselves even over longer periods of time using this type of interface.

5 A Person-Centric Approach: Haptics in Autonomous Training Assistance

Perhaps the most difficult challenge any technology must face when applied in the field of rehabilitation is that of interpersonal and intrapersonal variability within subjects. Individuals within the same rehabilitation program may each have unique types of impairment, disability, degree of impairment, physical activity goals and movement ability, preference toward certain types of exercises and feedback, differing levels of skill at different motor tasks, among others. These attributes combined can make every case of rehabilitation unique, often leading to low or partial adoption of rehabilitative technology and haptic interfaces by their intended targeted audience when the design does not account for this variability. For this reason, the design of haptics-enabled devices and interfaces in rehabilitation may benefit from focusing on the individual, under the guidance of a paradigm called

Person-Centered Multimedia Computing (PCMC) [45]. This section discusses the design of a rehabilitative exercise system for unsupervised rehabilitation of the upper extremity in the home environment that utilizes PCMC design principles to produce an effective platform for the distribution of audio, visual, and haptic feedback for a variety of subjects and trainers.

5.1 System Design

The system, called the Autonomous Training Assistant, was designed and developed through research interactions with a single subject and trainer. The subject was hemiparetic due to cerebral palsy resulting in one fully functional arm and one with impairment. The trainer designed bimanual motion tasks for the subject centered around the context of martial arts self-defense training using stick equipment. These exercises utilized the principle that the non-impaired limb can be used to guide and help control the impaired limb in a bimanual and harmonious fashion, with haptic feedback as the guiding medium between the two [40]. The subject and trainer would complete synchronized stick swing exercises where the trainer would stand opposite and facing the subject and lead by swinging the stick equipment using both arms, and the subject would be instructed to mirror the trainer's motion until their sticks made contact at a central point. This interaction is depicted in Fig. 9.3.

Fig. 9.3 Case study subject-trainer motion exercise

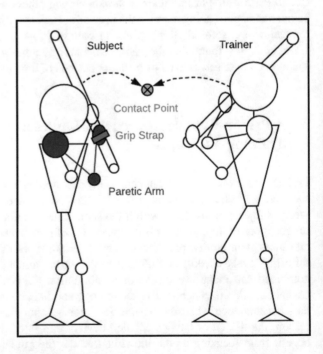

To make steady progress in rehabilitation, the subject was required to complete stick exercises assigned by the trainer in the home environment, without the trainer's presence. This unsupervised training environment was non-ideal due to the lack of guidance and feedback compared to supervised sessions with the trainer and subject, leading to the requirement of technology in the home environment to provide guidance and information communication as the subject exercised. The Autonomous Training Assistant was developed with this requirement in mind. The initial step in development was to carefully observe the training protocol between the subject and trainer, noting the type of feedback given during exercise. It was noted that the trainer provided all three categories of feedback listed above: posture (leg positioning, trunk rotation, and ensuring that the subject's paretic arm was in contact with the stick at all times), progression (maintaining a central point of contact between the two sticks, ensuring that the subject moved through the paretic arm's entire range of motion) and pacing (maintaining a steady rhythm of swings). Feedback was given verbally (audio) by describing the subject's error or giving correctional information in real-time; visually by demonstrating the correct form for the motion task and giving the subject a point of reference (mirror the trainer's motion); and through haptics by collision between the trainer's stick and subject's stick to indicate successful completion of one repetition of the task.

Using the observational data gathered from real exercise sessions between the subject and trainer, a training environment was designed to provide guidance to the subject while exercising at home. It is important to note that this environment was not designed to precisely mimic the modalities and details of feedback given by the trainer as the subject exercised, but rather to provide the same information in a multimodal manner that was most beneficial to the subject. The environment consisted of a Kinect for joint tracking, responsible for observing and tracking the subject's posture and motion trajectory; training software on-screen which included a virtual avatar that would mirror the subject's motion to provide audiovisual feedback; and a rod-shaped haptics-enabled intelligent training device entitled the Intelligent Stick, equipped with accelerometers and gyroscope to give more precise motion data and an array of vibrotactile actuators along the inner surface of the device to provide haptic feedback during exercise.

The Intelligent Stick device is depicted in Fig. 9.4. It was designed to be as accessible as possible for a variety of different users based on the observation that the case study subject had special requirements (like the inclusion of a strap to secure the grip of the paretic hand) that may completely differ from other subjects. Modules could easily be attached and detached to customize the form factor, features, width, and other aspects of the device. This modular customization design was a direct result of the case study format of this research and represented a key principle in PCMC: design for one, extend to many. The device communicates real-time accelerometer and gyroscope data to the training software in real-time through a wireless Bluetooth interface as the user completes an exercise.

When feedback is necessary, the software transmits the message back to the vibrotactile actuators on the stick, which can then actuate in a variety of patterns to convey information to the subject. In this case, a simple haptic guidance protocol

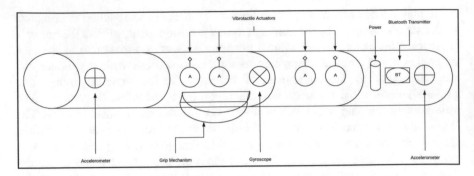

Fig. 9.4 Intelligent stick design

was used wherein a 3D motion trajectory was segmented into critical points along its path, and as the subject swings the stick, short vibrational cues would inform the subject as the stick passed through each critical point along the motion path in space. This included the final critical point, which was representative of the contact point between the subject and trainer's stick. Initial swing exercises consisted of only two critical points representing the two endpoints of the simple arc path, and haptic cues were provided when the Intelligent Stick reached those two endpoints. This protocol of haptic information was overseen and approved by the subject's trainer prior to implementation.

The haptic feedback loop of the Intelligent Stick device is described in5. The subject begins by activating the training software (which, in later design iteration, was implemented as a serious game with real-time adaptation [46]). The on-screen avatar acts as a virtual trainer and guides the subject through the completion of exercises designed by the subject's trainer, using motion parameters set and updated by the trainer for the subject. The subject attempts the motion tasks while the virtual avatar mirrors the subject's motion. At the two endpoints of the simple motion, which initially involved simple bimanual elbow flexion and extension, vibrotactile cues from the surface of the stick inform the subject of the endpoints of the motion, allowing the individual to move through repetitions of the motion task by changing the direction of motion whenever an endpoint is contacted. This continues for a set period prescribed by the trainer. The software then generates reports on the subject's progress including repetitions completed, maximum degree of motion, average rate of motion, and other information needed by the trainer to assess the subject's improvement bi-weekly and design new motion tasks or set new goals for the same task. The software can be remotely updated by the trainer with new tasks and goals. In later iterations, more complex motions involved more than two critical points; in this case, the subject simply follows the motion trajectory and receives confirmation along each critical point with a vibrotactile cue from the Intelligent Stick (Fig. 9.5).

Evaluation of the Autonomous Training Assistant is detailed in [47]. One primary outcome of this research is that in the presence of multiple modes of feedback including audio feedback on rhythm and pacing, visual feedback on posture, and

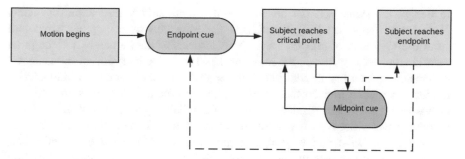

Fig. 9.5 Autonomous training assistant haptic feedback loop

haptic feedback on progression through the motion trajectory, the subject focused attention on the haptic feedback, indicating that in the design of a multimodal interface for any individual, bias toward modalities like haptics must be taken into account such that the most important information for the successful and safe completion of the motor task is encoded into the modality of feedback to which the individual exhibits the greatest bias of attention. In this particular case, haptic information was the most accessible format of concurrent feedback as the subject exercised, leading to the success of the critical point haptic guidance strategy described above. However, the Autonomous Training Assistant is capable of distributing feedback to a variety of modality mappings based on the needs and bias of different individuals. For example, in gait training, pacing information may be more important than progression through a range of motion, thereby changing the mapping of information categories to modalities and changing the role of haptics altogether.

6   Evaluation: Information Transfer

Evidence of the effectiveness of any rehabilitative system ultimately observes a subject's usage of the paretic or impaired region over a period of time, as this is perhaps the strongest indicator of regular activity and recovery [48]. However, to measure specifically how effective a haptic interface is at conveying information to the subject, a more direct measure is required. This is where the metric of Information Transfer (IT) is useful. IT studies in haptics research seek to determine the portion of an information set that is successfully understood and therefore transferred to a subject with the use of a specific protocol by measuring the number of correct distinct identifications of unique patterns or cues out of the entire presented set in what is called an 'identification task'. The details of calculating IT are outside the scope of this chapter, and are discussed in related work [49]. Here, several findings related to IT in rehabilitative haptics are discussed.

Typical studies on IT observe the range of distinguishable patterns and other sensitivities at various regions of the human body to determine not only how to design the signal set, but also where it is optimally placed along the body. Novich and Eagleman [50] investigated the capacities of perceptual bandwidth from the viewpoint of temporal and spatial aspects of the human skin through an identification experiment on the lower back, wherein the inclusion of the temporal dimension or intensity in addition to the spatial dimension achieved dramatically higher information transfer than when spatial information alone was utilized, and discovered a minimum distance of 6 cm for reliable perception of at least two distinct patterns between a pair of vibration motors. In comparing information transfer at the wrist and fingertip, Summers et al. [51] found that while their perception of differences in gaps between tactile signals and frequency variation were similar, the wrist provided greater accuracy of distinction between different amplitudes of tactile feedback than the fingertip.

Sonar and Paik [52] designed a soft actuating exo-skin prototype and indicated that high-resolution feedback could be achieved to present highly complex real-time information with high sensitivity of sensing, implying that an even greater level of bandwidth could be reached in the near future. However, Tan et al. [53] caution that there are a variety of factors that should be considered in the design of haptic signals when the goal is to improve IT rate, including the ease of training with the signal and information set, the relative compatibility between stimuli and response, and removal of external factors in the evaluation procedure for identification tasks.

Use of IT evaluation also enables comparative evaluation of various haptic mediums and the cross-interaction between those mediums. For example, D'Alonzo et al. [54] found that the combination of vibrotactile and electrotactile stimulation in a hybrid approach resulted in higher identification accuracy between varying intensities and improved the bandwidth of information encoded by the signal. Force feedback, as determined through IT studies, requires very little force amplitude to convey a fairly high resolution of information to the subject, but subject ability to distinguish characteristics such as surface quality increase with higher bandwidth, as indicated by Kilchenman and Goldfarb [55]. These findings may be validated in the future by determining their consistency when factors such as body region, individually-variant skin sensitivity, and exposure to training or daily exposure to haptic phenomena for a particular individual are considered.

7 Fading of Haptic Interaction

Of significant importance is the question of performance retainment. Once the guiding force of haptic feedback is removed, how well can subjects reproduce the motion with the same or similar quality of performance? Salazar et al. [56] demonstrate that for simple wrist motions, haptic trajectory guidance significantly reduced the performance error of subjects on a motion task, and immediately after this guidance was removed, subjects could still maintain similar performance by reproducing the motion represented by haptic feedback.

However, how long can subjects maintain this performance after the removal of haptic guidance? A review of the field by Jafari et al. [57] indicated that a primary concern for the use of haptic information communication is the likelihood that when this communication is removed, the subject's performance would degrade, rather than continuing to improve or maintaining itself. In a study by Bark et al. [58], the inclusion of haptic signals in feedback was not shown to have an effect on retention of motion knowledge and performance after a period as short as four days, even though inclusion of haptic augmentation helped lower the performance error during guided learning.

Sigrist et al. [59], in a review of strategies for various forms of concurrent and terminal feedback in the haptic, visual and auditory modalities, provided several suggestions that continue to warrant further investigation and validation within research, the first being a common conclusion from a large body of findings that highly unrestrictive haptic guidance, when adapted properly to subject skill level and applied as needed, can be highly effective and maintain this effectiveness as subject skill level increases with respect to the motor task; and the second suggestion [60] being that in highly-complex motion tasks, as user skill progresses, terminal feedback (delivered after completion of the motion task) may yield more effective learning than when that feedback is delivered concurrently (in-task).

8   Conclusions

While an extraordinary amount of progress within the last decade has opened the door to the use of haptics across a range of purposes within the rehabilitation domain, there remain many limitations to the technology which may be addressed in the near future. One potential challenge is ensuring haptic communication is easy to learn, easy to recall, grabs the user's attention, and is intuitive. Recent research in haptic languages seeks to avoid haptic communication misinterpretations by finding the most intuitive haptic patterns for conveying important spatiotemporal information [61]. Research toward optimizing the information transfer of haptic interfaces when encoding corrective or guiding motion information during exercise will lead society closer to widespread adoption of haptic interfaces in unsupervised rehabilitation settings.

Furthermore, the challenge of overdependence on haptic feedback after continued use remains a topic of current and future work. Many studies claim that subjects retain motor knowledge of a task after removal of haptic feedback fail to continue evaluation of subjects between periods of up to two weeks. Therefore, further evaluation is necessary to determine whether the learning imparted by haptic information communication during rehabilitative exercise is conserved over the full duration of rehabilitative treatment, particularly in cases where the subject relies heavily on unsupervised training to maintain functional improvement. Results shown by current implementations for the field are promising, however, and it is clear that haptics are an integral part of making unsupervised rehabilitation accessible across a broad spectrum of disabilities.

References

1. Benjamin, E.J., Blaha, M.J., Chiuve, S.E., Cushman, M., Das, S.R., Deo, R., Floyd, J., Fornage, M., Gillespie, C., Isasi, C.R.: Heart disease and stroke statistics-2017 update: a report from the American Heart Association. Circulation. 135(10), e146–e603 (2017).
2. Skilbeck, C.E., Wade, D.T., Hewer, R.L., Wood, V.A.: Recovery after stroke. Journal of Neurology, Neurosurgery & Psychiatry. 46(1), 5–8 (1983). https://doi.org/10.1136/jnnp.46.1.5.
3. Burdea, G.: The Role of Haptics in Physical Rehabilitation. In: Otaduy, M. (ed.) Haptic Rendering. pp. 517–529. A K Peters/CRC Press (2008). https://doi.org/10.1201/b10636-29.
4. Wulf, G., Shea, C., Lewthwaite, R.: Motor skill learning and performance: a review of influential factors. Medical Education. 44(1), 75–84 (2010). https://doi.org/10.1111/j.1365-2923.2009.03421.x.
5. Medalia, A., Saperstein, A.: The Role of Motivation for Treatment Success. Schizophrenia Bulletin. 37(S2), S122–S128 (2011). https://doi.org/10.1093/schbul/sbr063.
6. Williams, C.K., Carnahan, H.: Motor Learning Perspectives on Haptic Training for the Upper Extremities. IEEE Transactions on Haptics. 7(2), 240–250 (2014). https://doi.org/10.1109/TOH.2013.2297102.
7. Rajanna, V., Vo, P., Barth, J., Mjelde, M., Grey, T., Oduola, C., Hammond, T.: Kino-Haptics: An Automated, Wearable, Haptic Assisted, Physio-therapeutic System for Post-surgery Rehabilitation and Self-care. Journal of Medical Systems. 40(3), 60 (2015). https://doi.org/10.1007/s10916-015-0391-3.
8. Chen, Y., Abel, K.T., Janecek, J.T., Chen, Y., Zheng, K., Cramer, S.C.: Home-based technologies for stroke rehabilitation: A systematic review. International Journal of Medical Informatics. 123, 11–22 (2019). https://doi.org/10.1016/j.ijmedinf.2018.12.001.
9. Shull, P.B., Damian, D.D.: Haptic wearables as sensory replacement, sensory augmentation and trainer – a review. Journal of NeuroEngineering and Rehabilitation. 12(1), 59 (2015). https://doi.org/10.1186/s12984-015-0055-z.
10. Cooper, N., Milella, F., Pinto, C., Cant, I., White, M., Meyer, G.: The effects of substitute multisensory feedback on task performance and the sense of presence in a virtual reality environment. PLOS ONE. 13(2), e0191846 (2018). https://doi.org/10.1371/journal.pone.0191846.
11. Lange, B., Koenig, S., Chang, C.-Y., McConnell, E., Suma, E., Bolas, M., Rizzo, A.: Designing informed game-based rehabilitation tasks leveraging advances in virtual reality. Disability and Rehabilitation. 34(22), 1863–1870 (2012). https://doi.org/10.3109/09638288.2012.670029.
12. Pamungkas, D., Ward, K.: Electro-tactile feedback system to enhance virtual reality experience. International Journal of Computer Theory and Engineering. 8(6), 465–470 (2016). https://doi.org/10.7763/IJCTE.2016.V8.1090.
13. Andaluz, V.H., Salazar, P.J., Escudero V., M., Bustamante D., C., Silva S., M., Quevedo, W., Sánchez, J.S., Espinosa, E.G., Rivas, D.: Virtual Reality Integration with Force Feedback in Upper Limb Rehabilitation. In: Bebis, G., Boyle, R., Parvin, B., Koracin, D., Porikli, F., Skaff, S., Entezari, A., Min, J., Iwai, D., Sadagic, A., Scheidegger, C., and Isenberg, T. (eds.) Advances in Visual Computing. pp. 259–268. Springer International Publishing (2016).
14. Fluet, G.G., Deutsch, J.E.: Virtual Reality for Sensorimotor Rehabilitation Post-Stroke: The Promise and Current State of the Field. Current Physical Medicine and Rehabilitation Reports. 1(1), 9–20 (2013). https://doi.org/10.1007/s40141-013-0005-2.
15. Bortone, I., Leonardis, D., Solazzi, M., Procopio, C., Crecchi, A., Bonfiglio, L., Frisoli, A.: Integration of serious games and wearable haptic interfaces for Neuro Rehabilitation of children with movement disorders: A feasibility study. In: 2017 International Conference on Rehabilitation Robotics (ICORR). pp. 1094–1099 (2017). https://doi.org/10.1109/ICORR.2017.8009395.
16. Sadihov, D., Migge, B., Gassert, R., Kim, Y.: Prototype of a VR upper-limb rehabilitation system enhanced with motion-based tactile feedback. In: 2013 World Haptics Conference (WHC). pp. 449–454. IEEE (2013).

17. Tokuyama, Y., Rajapakse, R.P.C.J., Miya, S., Konno, K.: Development of a Whack-a-Mole Game with Haptic Feedback for Rehabilitation. In: 2016 Nicograph International (NicoInt). pp. 29–35 (2016). https://doi.org/10.1109/NicoInt.2016.6.
18. Wake, N., Sano, Y., Oya, R., Sumitani, M., Kumagaya, S., Kuniyoshi, Y.: Multimodal virtual reality platform for the rehabilitation of phantom limb pain. In: 2015 7th International IEEE/EMBS Conference on Neural Engineering (NER). pp. 787–790 (2015). https://doi.org/10.1109/NER.2015.7146741.
19. Lin, C.-Y., Tsai, C.-M., Shih, P.-C., Wu, H.-C.: Development of a novel haptic glove for improving finger dexterity in poststroke rehabilitation. Technology and Health Care. 24(S1), S97–S103 (2016). https://doi.org/10.3233/THC-151056.
20. Liu, L.Y., Li, Y., Lamontagne, A.: The effects of error-augmentation versus error-reduction paradigms in robotic therapy to enhance upper extremity performance and recovery post-stroke: a systematic review. Journal of NeuroEngineering and Rehabilitation. 15(1), 65 (2018). https://doi.org/10.1186/s12984-018-0408-5.
21. Lee, H., Choi, S.: Combining haptic guidance and haptic disturbance: an initial study of hybrid haptic assistance for virtual steering task. In: 2014 IEEE Haptics Symposium (HAPTICS). pp. 159–165 (2014). https://doi.org/10.1109/HAPTICS.2014.6775449.
22. Stanley, A.A., Kuchenbecker, K.J.: Evaluation of Tactile Feedback Methods for Wrist Rotation Guidance. IEEE Transactions on Haptics. 5(3), 240–251 (2012). https://doi.org/10.1109/TOH.2012.33.
23. Ben-Tzvi, P., Danoff, J., Ma, Z.: The Design Evolution of a Sensing and Force-Feedback Exoskeleton Robotic Glove for Hand Rehabilitation Application. Journal of Mechanisms and Robotics. 8(5), 051019-051019–9 (2016). https://doi.org/10.1115/1.4032270.
24. Chen, C.-Y., Chen, Y.-Y., Chung, Y.-J., Yu, N.-H.: Motion Guidance Sleeve: Guiding the Forearm Rotation Through External Artificial Muscles. In: Proceedings of the 2016 CHI Conference on Human Factors in Computing Systems. pp. 3272–3276. ACM, New York, NY, USA (2016). https://doi.org/10.1145/2858036.2858275.
25. Akdogan, E., Shima, K., Kataoka, H., Hasegawa, M., Otsuka, A., Tsuji, T.: The Cybernetic Rehabilitation Aid: Preliminary Results for Wrist and Elbow Motions in Healthy Subjects. IEEE Transactions on Neural Systems and Rehabilitation Engineering. 20(5), 697–707 (2012). https://doi.org/10.1109/TNSRE.2012.2198496.
26. Gori, M., Vercillo, T., Sandini, G., Burr, D.: Tactile feedback improves auditory spatial localization. Frontiers in Psychology. 5, 1121 (2014). https://doi.org/10.3389/fpsyg.2014.01121.
27. Sigrist, R., Rauter, G., Marchal-Crespo, L., Riener, R., Wolf, P.: Sonification and haptic feedback in addition to visual feedback enhances complex motor task learning. Experimental brain research. 233(3), 909–925 (2015).
28. Pezent, E., Fani, S., Bradley, J., Bianchi, M., O'Malley, M.K.: Separating haptic guidance from task dynamics: A practical solution via cutaneous devices. In: 2018 IEEE Haptics Symposium (HAPTICS). pp. 20–25 (2018). https://doi.org/10.1109/HAPTICS.2018.8357147.
29. Baldi, T.L., Scheggi, S., Aggravi, M., Prattichizzo, D.: Haptic Guidance in Dynamic Environments Using Optimal Reciprocal Collision Avoidance. IEEE Robotics and Automation Letters. 3(1), 265–272 (2018). https://doi.org/10.1109/LRA.2017.2738328.
30. Panchanathan, R., Rosenthal, J., McDaniel, T.: Rehabilitation and motor learning through vibrotactile feedback. In: Smart Biomedical and Physiological Sensor Technology XI (9107). p. 910717. International Society for Optics and Photonics (2014). https://doi.org/10.1117/12.2050204.
31. Borghese, N.A., Pirovano, M., Lanzi, P.L., Wüest, S., de Bruin, E.D.: Computational Intelligence and Game Design for Effective At-Home Stroke Rehabilitation. Games for Health Journal. 2(2), 81–88 (2013). https://doi.org/10.1089/g4h.2012.0073.
32. Peeters, T., Breda, E. van, Saeys, W., Schaerlaken, E., Vleugels, J., Truijen, S., Verwulgen, S.: Vibrotactile Feedback During Physical Exercise: Perception of Vibrotactile Cues in Cycling. International Journal of Sports Medicine. 40(6), 390–396 (2019). https://doi.org/10.1055/a-0854-2963.

33. Tzorakoleftherakis, E., Bengtson, M.C., Mussa-Ivaldi, F.A., Scheidt, R.A., Murphey, T.D.: Tactile proprioceptive input in robotic rehabilitation after stroke. In: 2015 IEEE International Conference on Robotics and Automation (ICRA). pp. 6475–6481 (2015). https://doi.org/10.1109/ICRA.2015.7140109.
34. Cuppone, A.V., Squeri, V., Semprini, M., Masia, L., Konczak, J.: Robot-Assisted Proprioceptive Training with Added Vibro-Tactile Feedback Enhances Somatosensory and Motor Performance. PLOS ONE. 11(10), e0164511 (2016). https://doi.org/10.1371/journal.pone.0164511.
35. De Nunzio, A.M., Dosen, S., Lemling, S., Markovic, M., Schweisfurth, M.A., Ge, N., Graimann, B., Falla, D., Farina, D.: Tactile feedback is an effective instrument for the training of grasping with a prosthesis at low- and medium-force levels. Experimental Brain Research. 235(8), 2547–2559 (2017). https://doi.org/10.1007/s00221-017-4991-7.
36. Tadayon, A., Zia, J., Anantuni, L., McDaniel, T., Krishnamurthi, N., Panchanathan, S.: A Shoe Mounted System for Parkinsonian Gait Detection and Real-Time Feedback. In: Stephanidis, C. (ed.) HCI International 2015 – Posters' Extended Abstracts. pp. 528–533. Springer International Publishing (2015).
37. Kavathekar, P.A., Balkcom, D.J.: A tactile shirt for teaching human motion tasks. In: 2017 IEEE/RSJ International Conference on Intelligent Robots and Systems (IROS). pp. 878–885 (2017). https://doi.org/10.1109/IROS.2017.8202249.
38. Janidarmian, M., Fekr, A.R., Radecka, K., Zilic, Z.: Haptic feedback and human performance in a wearable sensor system. In: 2016 IEEE-EMBS International Conference on Biomedical and Health Informatics (BHI). pp. 620–624 (2016). https://doi.org/10.1109/BHI.2016.7455975.
39. Yu, B., Feijs, L., Funk, M., Hu, J.: Breathe with Touch: A Tactile Interface for Breathing Assistance System. In: Abascal, J., Barbosa, S., Fetter, M., Gross, T., Palanque, P., and Winckler, M. (eds.) Human-Computer Interaction – INTERACT 2015. pp. 45–52. Springer International Publishing (2015).
40. Holland, S., Wright, R.L., Wing, A., Crevoisier, T., Hödl, O., Canelli, M.: A Gait Rehabilitation Pilot Study Using Tactile Cueing Following Hemiparetic Stroke. In: Proceedings of the 8th International Conference on Pervasive Computing Technologies for Healthcare. pp. 402–405. ICST (Institute for Computer Sciences, Social-Informatics and Telecommunications Engineering), ICST, Brussels, Belgium, Belgium (2014). https://doi.org/10.4108/icst.pervasivehealth.2014.255357.
41. Maculewicz, J., Erkut, C., Serafin, S.: An investigation on the impact of auditory and haptic feedback on rhythmic walking interactions. International Journal of Human-Computer Studies. 85, 40–46 (2016). https://doi.org/10.1016/j.ijhcs.2015.07.003.
42. Georgiou, T., Holland, S., Linden, J. van der, Tetley, J., Stockley, R.C., Donaldson, G., Garbutt, L., Pinzone, O., Grasselly, F., Deleaye, K.: A blended user centred design study for wearable haptic gait rehabilitation following hemiparetic stroke. In: 2015 9th International Conference on Pervasive Computing Technologies for Healthcare (PervasiveHealth). pp. 72–79 (2015). https://doi.org/10.4108/icst.pervasivehealth.2015.259073.
43. Georgiou, T., Holland, S., van der Linden, J.: Wearable Haptic Devices for Post-stroke Gait Rehabilitation. In: Proceedings of the 2016 ACM International Joint Conference on Pervasive and Ubiquitous Computing: Adjunct. pp. 1114–1119. ACM, New York, NY, USA (2016). https://doi.org/10.1145/2968219.2972718.
44. Islam, R., Holland, S., Georgiou, T., Price, B., Mulholland, P.: A longitudinal rehabilitation case study for hemiparetic gait using outdoor rhythmic haptic cueing via a wearable device. Presented at the 27th European Stroke Conference, Athens, Greece April 11 (2018).
45. Panchanathan, S., Chakraborty, S., McDaniel, T., Tadayon, R.: Person-Centered Multimedia Computing: A New Paradigm Inspired by Assistive and Rehabilitative Applications. IEEE MultiMedia. 23(3), 12–19 (2016). https://doi.org/10.1109/MMUL.2016.51.
46. Tadayon, R., Amresh, A., McDaniel, T., Panchanathan, S.: Real-time stealth intervention for motor learning using player flow-state. In: 2018 IEEE 6th International Conference on Serious Games and Applications for Health (SeGAH). pp. 1–8 (2018). https://doi.org/10.1109/SeGAH.2018.8401360.

47. Tadayon, R.: A Person-Centric Design Framework for At-Home Motor Learning in Serious Games. PhD Thesis, Arizona State University (2017).
48. Afzal, M.R., Pyo, S., Oh, M., Park, Y.S., Yoon, J.: Identifying the effects of using integrated haptic feedback for gait rehabilitation of stroke patients. In: 2017 International Conference on Rehabilitation Robotics (ICORR). pp. 1055–1060 (2017). https://doi.org/10.1109/ICORR.2017.8009389.
49. Tadayon, R., McDaniel, T., Panchanathan, S.: A survey of multimodal systems and techniques for motor learning. Journal of information processing systems. 13(1), 8–25 (2017).
50. Novich, S.D., Eagleman, D.M.: Using space and time to encode vibrotactile information: toward an estimate of the skin's achievable throughput. Experimental Brain Research. 233(10), 2777–2788 (2015). https://doi.org/10.1007/s00221-015-4346-1.
51. Summers, I.R., Whybrow, J.J., Gratton, D.A., Milnes, P., Brown, B.H., Stevens, J.C.: Tactile information transfer: A comparison of two stimulation sites. The Journal of the Acoustical Society of America. 118(4), 2527–2534 (2005). https://doi.org/10.1121/1.2031979.
52. Sonar, H.A., Paik, J.: Soft Pneumatic Actuator Skin with Piezoelectric Sensors for Vibrotactile Feedback. Frontiers in Robotics and AI. 2, 38 (2016). https://doi.org/10.3389/frobt.2015.00038.
53. Tan, H.Z., Reed, C.M., Durlach, N.I.: Optimum Information Transfer Rates for Communication through Haptic and Other Sensory Modalities. IEEE Transactions on Haptics. 3(2), 98–108 (2010). https://doi.org/10.1109/TOH.2009.46.
54. D'Alonzo, M., Dosen, S., Cipriani, C., Farina, D.: HyVE: Hybrid Vibro-Electrotactile Stimulation for Sensory Feedback and Substitution in Rehabilitation. IEEE Transactions on Neural Systems and Rehabilitation Engineering. 22(2), 290–301 (2014). https://doi.org/10.1109/TNSRE.2013.2266482.
55. Kilchenman, R., Goldfarb, M.: Force saturation, system bandwidth, information transfer, and surface quality in haptic interfaces. In: Proceedings 2001 ICRA. IEEE International Conference on Robotics and Automation (Cat. No.01CH37164, Vol. 2). pp. 1382–1387 (2001). https://doi.org/10.1109/ROBOT.2001.932803.
56. Salazar, J., Okabe, K., Hirata, Y.: Path-Following Guidance Using Phantom Sensation Based Vibrotactile Cues Around the Wrist. IEEE Robotics and Automation Letters. 3(3), 2485–2492 (2018). https://doi.org/10.1109/LRA.2018.2810939.
57. Jafari, N., Adams, K.D., Tavakoli, M.: Haptics to improve task performance in people with disabilities: A review of previous studies and a guide to future research with children with disabilities. Journal of Rehabilitation and Assistive Technologies Engineering. 3, 1–13 (2016). https://doi.org/10.1177/2055668316668147.
58. Bark, K., Hyman, E., Tan, F., Cha, E., Jax, S.A., Buxbaum, L.J., Kuchenbecker, K.J.: Effects of Vibrotactile Feedback on Human Learning of Arm Motions. IEEE Transactions on Neural Systems and Rehabilitation Engineering. 23(1), 51–63 (2015). https://doi.org/10.1109/TNSRE.2014.2327229.
59. Sigrist, R., Rauter, G., Riener, R., Wolf, P.: Augmented visual, auditory, haptic, and multimodal feedback in motor learning: A review. Psychonomic bulletin & review. 20(1), 21–53 (2013).
60. Sigrist, R., Rauter, G., Riener, R., Wolf, P.: Terminal feedback outperforms concurrent visual, auditory, and haptic feedback in learning a complex rowing-type task. Journal of motor behavior. 45(6), 455–472 (2013).
61. Duarte, B., McDaniel, T., Tadayon, R., Devkota, S., Strasser, G., Ramey, C., Panchanathan, S.: Haptic Vision: Augmenting Non-visual Travel and Accessing Environmental Information at a Distance. In: Basu, A. and Berretti, S. (eds.) Smart Multimedia. pp. 90–101. Springer International Publishing (2018).

Chapter 10
Intelligent Robotics and Immersive Displays for Enhancing Haptic Interaction in Physical Rehabilitation Environments

Jason Fong, Renz Ocampo, and Mahdi Tavakoli

Abstract Disabling events such as stroke affect millions of people worldwide, causing a need for efficient and functional rehabilitation therapies in order for patients to regain motor function for reintegration back into their normal lives. Rehabilitation regimes often involve performing exercises that mimic the movements used in activities of daily living and are intended to promote recovery from the physical aspects of an injury. However, alongside physical disability, some patients (e.g., stroke patients) develop cognitive deficiencies that affect their ability to think, plan, and carry out tasks. It is a necessity then to consider rehabilitation techniques that can also accommodate patients with cognitive deficiencies alongside those without. Rehabilitation systems that provide haptic interaction to patients practicing therapy tasks work towards both of these ends; physical interaction can provide strength and coordination training for improving physical condition, and providing additional tactile sensory feedback can make tasks more intuitive for patients with cognitive deficiencies. This chapter introduces novel techniques to incorporate robotics, machine learning, and augmented reality for the purposes of enhancing the haptic interactions provided by therapists to assist patients in their rehabilitation process.

1 Rehabilitation for Neurological Injury

As the world's population ages, the demand for rehabilitation medicine services continues to increase. An older population equates to an increase in the prevalence of physical injuries (e.g., from falls) and mobility impairments, but also neurological disabilities such as those experienced by stroke survivors. For rehabilitation-focused

J. Fong · R. Ocampo · M. Tavakoli (✉)
Department of Electrical and Computer Engineering, University of Alberta, Edmonton, AB, Canada
e-mail: jmfong@ualberta.ca; rocampo@ualberta.ca; mahdi.tavakoli@ualberta.ca

© Springer Nature Switzerland AG 2020
T. McDaniel, S. Panchanathan (eds.), *Haptic Interfaces for Accessibility, Health, and Enhanced Quality of Life*, https://doi.org/10.1007/978-3-030-34230-2_10

therapy (as opposed to therapy sessions for teaching compensatory techniques), strength training and neuroplasticity are regarded as key mechanisms of recovery. Neuroplasticity is the ability of the brain to alter and adapt its functionality through the formation of new neuronal connections. In stroke survivors, the repetitive exercise of affected neuromuscular pathways has been found to act as a catalyst for neuroplasticity [87]. As such, there is a strong motivation for rehabilitation services to emphasize repetitive physical exercise for both patients' strength training as well as neuromuscular recovery. However, this means the burden on healthcare providers also increases not only through increased cost expenditures [60] but also through the higher physical demands on therapists who have to support the patients in these repeated exercises. This has resulted in a need for innovation, as evidenced by the growth of research on technology-focused assistance over the last few decades [102]. This chapter presents two branches of innovation that have immense potential in the field of physical rehabilitation. The first is the concept of Learning from Demonstration (LfD) for robot programming and the second is the use of VR and AR displays to enhance serious games. Both of these innovations will be discussed and their applications to enhancing haptic interactions for physical rehabilitation explained in their respective sections.

1.1 Stroke

As the global fifth leading cause of death, stroke causes approximately 6.5 million deaths each year [18]. It is defined as a clinical syndrome of presumed vascular origin, characterized by disturbance of cerebral functions lasting more than 24 h or leading to death [64]. Common symptoms of stroke include a reduction in motor control (i.e., reduced movement, weakness, incoordination), sensory loss or alteration, impaired balance, and impaired tone (i.e., spasticity) [64]. In 2013, the prevalence of stroke for adults between 20 and 64 years of age reached approximately 11 million cases [44]. In Canada alone, stroke and other cardiovascular diseases (CVD) cost the healthcare system $22.2B in 2009 [117], and $20.9B in 2015 [60]. In the United States, the total cost of stroke and other CVD was estimated to be $316.1B in 2013 [18]. Therefore, there is a great incentive for making post-stroke rehabilitation as effective and efficient as possible, not only to lower its economic burden but to improve the quality of life for a significant number of stroke survivors.

1.2 Neuroplasticity and Stroke Rehabilitation

Neuroplasticity can be defined as the ability of the brain to reorganize its structure, function, and connections for the purposes of development and learning or in response to the environment, disease, or injury [34]. For stroke patients, research has found that neuroplasticity can be promoted in patients by actively engaging in repetitive exercises, and has been extensively explored for the purposes of aiding motor recovery [5, 34, 87]. Hemiparesis is one example of a manifestation of post-stroke neuromotor disability where the principles of neuroplasticity are applied. Hemiparesis can be described as the paralysis of one side of the body (and may also be characterized by spasticity and sensory loss) [21] which, in the case of stroke, occurs as a result of a hemispheric lesion in the brain. Degeneration of motor neurons as a result of hemiparesis contributes greatly towards resultant muscle weakness [21, 98], and may be further compounded by hemispatial neglect [114, 150]. Constraint-Induced Movement Therapy (CIMT) is a family of treatments which involves constraint of movement for a patient's unaffected limbs and mass practice using their affected limb and has seen significant success when applied to hemiparetic stroke patients with the objectives of overcoming learned non-use and promoting neuroplasticity [96]. More recently, the facilitation of neuroplasticity has shifted away from having patients practice generic and repetitive single-limb exercises and instead towards bimanual exercises (for upper limb therapy) [22] or task-oriented therapy [79].

1.3   Activities of Daily Living

In the case of post-stroke rehabilitation, task-oriented therapy typically refers to the training and assessment of patients' abilities to perform Activities of Daily Living (ADLs). The most basic of ADLs typically include bathing, personal hygiene and grooming, dressing, toileting, functional mobility (i.e., locomoting and transferring to and from beds and chairs), and self-feeding [142]. The definition, however, can also encompass most actions that allow an individual to live independently. Conventionally, patients will be trained through a combination of strength training and movement strategies, compensatory or otherwise, that allow them to perform ADLs; assessments of their capabilities are then performed by having patients perform the ADLs themselves. Therapists may also perform in-home training of patients for ensuring they can successfully carry out ADLs. However, the use of ADLs as rehabilitation tasks themselves is not standard, although framing the focus of stroke rehabilitation around ADLs has been verified to improve independence and quality of life outcomes [6, 81].

1.4   Patient Motivation

Patient motivation is regarded as an important factor in predicting success rates for rehabilitation [55]. The physical and emotional impacts induced by events such as stroke affect the willingness of the patient to participate. While this motivation to rehabilitate can be influenced by multiple factors in the rehabilitation environment such as family members, staff, and other stroke patients, a significant element lies in rehabilitation exercises itself. Some patients have a lack of information and understanding of the nature of their rehabilitation exercises [94]. Coupled with the long sessions and repetitive motions that are involved [43], patients become uncertain about their rehabilitation outcomes. As a consequence, poor patient participation causes patients to have longer inpatient rehabilitation stay and poorer motor improvement results [84].

2   Haptics-Enabled Rehabilitation Robots

A major source of innovation for rehabilitation therapy is found in the inclusion of robots for facilitating rehabilitation, an area which has seen significant development over the last three decades. The ability of robots to provide repetitive, high-intensity interactions without being subject to fatigue makes them an attractive means of providing the repetitive exercise that is fundamental to expediting a patient's rehabilitation [140]. A significant amount of research in the area has sought to improve the stability of these robots to make them patient-safe, as well as to provide them with the ability to adapt their behaviors, whether it be assisting or resisting a patient during exercise.

2.1   Brief History of Rehabilitation Robots

Initially, most robots used in rehabilitation were for assistive purposes. These robots did not aim to help to regain the lost motor function of the patient, but rather they aimed to assist the patient in performing activities of daily living [147]. These are commonly seen as robots attached to wheelchairs to assist in eating and drinking, grabbing objects, and mobility [61]. It was not until the late 1980s when researchers started to pursue rehabilitation robotics for actual therapy use [86, 140]. Dedicated to assisting and augmenting motor function rehabilitation using robotic devices, research in rehabilitation robotics began as a way to find a solution that alleviates therapists' stress and produces more efficient rehabilitation techniques. In 1988, two double-link planar robots were coupled with a patient's lower limb to provide continuous passive motion for rehabilitation [75]. This was soon followed by an upper-limb rehabilitation device in 1992, the MIT-MANUS, which was used for planar shoulder-and-elbow therapy [63]. Upper-limb rehabilitative devices were further developed after the advent of the MIT-MANUS. These include devices such as the Mirror-Image Movement Enabler (MIME) robotic device, which improved muscle movements through mirror-image training [88], and the Assisted Rehabilitation and Measurement (ARM) Guide, which functions both as an assessment and rehabilitative tool [118]. Robotic rehabilitation that targeted other areas of the body surfaced in the 2000s. These robotic devices allowed rehabilitation for areas such as the wrist [143], hand, and finger [145] for the upper-limb, and gait and ankle training

[29, 40] for the lower limb. More recently, robots designed for training patients to perform ADLs have been developed [58, 100].

2.2 Motivation for Robotic Rehabilitation

The inclusion of robots in therapy to provide therapist-robot-patient interactions presents distinct advantages over conventional therapist-patient interactions:

- Conventional hand-over-hand intervention in which therapists would provide direct supervision and apply direct assistive/resistive forces as needed is highly burdensome on the therapist. As a result, in practice, most therapy sessions are designed to maximize a patient's self-direction, with a therapist designing and supervising interventions while providing mostly verbal guidance [129]. Otherwise, therapy sessions that involve the more preferred hands-on interaction between the therapist and patient must be limited in duration and intensity (where intensity refers to the amount of activity performed), resulting in less practice for patients. On the other hand, robots can perform repetitive movements with superhuman accuracy and reliability, but without suffering from fatigue. This characteristic suits the repetitive nature of strength training and task-oriented therapies alike and can alleviate the physical burden on therapists.
- Assessment in current rehabilitation practice is performed by using standardized assessments such as the Chedoke-McMaster Stroke Assessment [54], Fugl-Meyer Assessment of Sensorimotor Recovery After Stroke [52], and Modified Ashworth Scale [112]. These assessments are driven by therapists' observations and are therefore examined strictly for validity and inter-rater reliability. While specific assessments may rate highly for either criterion, there can never be complete certainty in the results they provide due to the subjective nature of human raters and the resultant coarse resolution of the assessments. On the other hand, sensors in a robotic system can provide numerical measurements that can describe a patient's performance during rehabilitation, which is ideal for supplementing the assessments mentioned.
- The ability of robots to be automated is one of their most important strengths. In the context of facilitating rehabilitation, the automation of rehabilitation robots provides an opportunity to streamline therapy. For example, the ability to time-share a single therapist across multiple patients more efficiently becomes possible. Another example is intelligently automating the amount of assistance or resistance provided during therapy, a concept known as Assistance-as-Needed (AAN), which has received significant interest [20, 115, 130, 144]. AAN makes it possible to introduce robotic assistance on a graduated basis, where the level of automation (from fully manual to fully autonomous) that best suits the needs of the therapist and patient can be automatically chosen.
- Rehabilitative therapy, especially when presented in a "traditional" (i.e., face-to-face) manner, is inherently restricted by distance. Patients must either participate

in rehabilitation sessions at a hospital or other rehabilitation center, or a therapist must visit a patient at their home. In the case where patients are situated in remote or otherwise difficult to access locations, providing rehabilitation may be exceedingly challenging and cost inefficient. Telerehabilitation is the concept of providing rehabilitative support, assessment and intervention over a distance, using internet-based communication as a medium for therapist-patient interaction [120]. This can take the form of purely audio or video communication, audiovisual communication with patient-robot (unilateral) interaction with performance communicated over the internet, or haptic (bilateral) interaction between a therapist side robot and patient side robot, also known as telerobotic therapy [11, 124, 125]. Through telerehabilitation, remote access is inherently addressed and has received significant focus [27, 68]. Another possible advantage comes in the form of early indications from longitudinal studies on telerehabilitation that show small cost savings [71].

2.3   Limitations of the Current State of the Art

Despite these advantages, however, robotic rehabilitation faces limitations. A selection of limitations of considerable importance is given:

1. First and foremost is that analyses of the efficacy of robotic rehabilitation are largely inconclusive as to whether robotic rehabilitation is more or as effective as "conventional" therapy. Improvements in motor function have been shown in studies performed with the MIT-MANUS. Post-stroke patients were recruited for the MIT-MANUS study, and results have shown a statistically significant reduction in shoulder and elbow impairment compared to the group that underwent conventional therapy [141]. Likewise, with the MIME system, post-stroke patients trained with the MIME over an eight-week period showed an improvement in reach and strength of the affected limb [89]. Based on the Fugl-Meyer score, the MIME group produced better results than the conventional therapy group in the two months of therapy. On the other hand, examination of motor improvement using the ARM guide robot was found to be as effective as a control group receiving no assistance [70]. The study hypothesized that the benefits of robotic rehabilitation may be directly tied to the interactions it provides and the robot-generated forces may be providing no additional benefit. As a result, when put in context with the high initial costs of purchasing such robots, acceptance of robotic rehabilitation remains relatively low in clinical settings. The small participant sample sizes associated with each of these works further compounds the uncertainty in the benefits of robotic rehabilitation.
2. The programming of rehabilitation robots has always been done such that the robots provide interactions associated with a specific set of tasks, with no easy method of changing these tasks. As a result, the kinds of interactions a therapist can provide through the robotic medium are limited unless they or another person

(usually a technician) are familiar with computer programming principles and can change the task and/or task-oriented behavior of the robot.

3. Low patient motivation remains an issue in therapy even with the addition of robotics. As robots allow for a reduced therapist intervention, the patients themselves lose motivation due to the lack of encouragement, entertainment, and human interaction [30, 38, 93]. Motivation is seen as an important predictor in successful rehabilitation outcomes.

3   Semi-autonomous Provision of Haptic Interactions for Robot-Assisted Intervention

An important consideration is that the field of rehabilitation robotics should instead focus on the use of robots as supplementary to conventional therapy and as enabling tools in the hands of therapists, instead of as replacements for them [65]. In this frame of mind, the field can be further developed even if rehabilitation robots are as effective as (but not more effective than) conventional therapy methods if improvements are made in other areas (e.g., cost savings or freeing up the therapists' time), addressing *Limitation 1*. Providing semi-autonomy is one way to do so: semi-autonomy keeps the therapist in the loop but allows them to save time and effort through the robot taking a share of the intervention to be done on the patient. Autonomy in robotics implies the existence of machine intelligence, which necessitates the domain of machine learning research.

3.1   Machine Learning in Rehabilitation

The incorporation of machine learning algorithms in rehabilitation (robotic or conventional) has seen a rise in the past two decades. The grand majority of the literature focuses on the use of machine learning algorithms for classifying and recognizing a patient's posture and movement, but not for learning the interventions demonstrated by a therapist. Leightley et al. [82] evaluated the use of support vector machines (SVM) and random forest (RF) algorithms for learning and recognizing general human activities. Li et al. [85] used an SVM and K-nearest neighbors (KNN) classifier to recognize gestures for hand rehabilitation exercises. Giorgino et al. [51] assessed the use of KNN, logistic regression (LR), and decision trees (DT) for identifying upper body posture using a flexible sensor system integrated into the patient's clothes. McLeod et al. [99] compared the use of LR, naive Bayes (NB) classification, and a DT for discriminating between functional upper limb movements and those associated with walking.

However, the power of machine learning models is not limited to only classifying movements. They also have the potential to provide predictions of a patient's condition, which may serve as guides for planning their approach to rehabilitation. Zhu et al. [151] trained an SVM and KNN classifier to predict a patient's rehabilitation potential, both of which provided better predictive abilities than an assessment protocol currently used in the field. Yeh et al. [149] utilized an SVM to classify balance in able-bodied individuals and those with vestibular dysfunction. Begg et al. [17] also used an SVM to classify gait in younger, able-bodied participants and in elderly participants. Lastly, LeMoyne et al. [83] also implemented an SVM for classification of normal and hemiplegic ankle movement.

More recent applications of machine learning build off of these works that classify both a patient's movements and their condition, and now look to address the natural conclusion to this line of research: how rehabilitation systems should adjust either the task or their provided intervention based on these features of a patient. Barzilay et al. [15] train a neural network (NN) to adjust an upper limb rehabilitation task's difficulty based on the upper limb kinematics and electromyography (EMG) signals. Shirzad et al. [127] evaluate the use of KNN, NN, and discriminant analysis (DA) techniques for adjusting task difficulty in relation to a patient's motor-performance and physiological features. Badesa et al. [14] perform a similar evaluation for perceptron learning algorithms, LR, DA, SVMs, NB, KNN, and K-center classifiers. Garate et al. [49] use a fuzzy logic algorithm to relate a patient's joint kinematics to the motor primitive outputs of a Central Pattern Generator (CPG), which effectively provide assistance during gait through the control of an exoskeleton's torques. Gui et al. [57] take a similar approach, this time using EEG measurements as the input to a DA algorithm that provides assistive exoskeleton trajectories through a CPG. It is important to note that in each of these works, the adaptation that is learned by the algorithms is not learned from demonstrations. Rather, these interactions are generated from predetermined models relating patient performance with task difficulty or desired assistance.

3.2 Learning from Demonstration for Haptic Interaction

Learning from Demonstration (LfD) describes a family of machine learning techniques in which a robot observes demonstrations of a task by a human operator (the "demonstration" phase) and learns a policy to describe the desired task-oriented actions, which may or may not be acted upon by the robot in a "reproduction" phase later [13]. The terms "programming by demonstration" or "imitation learning" also refer to the same concept. The policy learned through LfD techniques is a central point to its innovation, and has seen implementation through mapping functions (classification and regression), or through system models (reinforcement learning) [9].

The advantages of using LfD techniques to program robots are clear. After the initial challenge of making the machine intelligent, i.e., teachable, programming the robot can be made as easy as physically holding a robot and moving it through a desired trajectory which is known as kinesthetic teaching. Users themselves do not require knowledge of computer programming. The capabilities of the robot are completely dependent on the level of sophistication of the underlying learning algorithms and the amount of sensors used to characterize a behavior; with highly sophisticated algorithms and sufficient sensors, it is possible to teach more complex aspects of tasks to robots (e.g., understanding a user's intent). The methodology of LfD also requires a human user to be involved in the programming process, meaning the aspect of interacting with an actual human is preserved and conveyed by means of imitation. Lastly, like any other implementation of machine learning for robotics, LfD allows for automation, which translates to time and cost savings.

The concept of semi-autonomous systems and LfD has seen extensive research in the past few decades. Application of LfD principles to human-robot interaction has naturally led to exploration of cooperative tasks. Calinon et al. [26] taught a robot to cooperatively lift a beam in a setup similar to what we propose here. Gribovskaya et al. [56] built upon the same work to ensure global asymptotic stability (GAS) of the system. Peternel et al. [116] created a variant to learn motion and compliance during a highly dynamic cooperative sawing task.

3.3   Learning Haptic Interactions Provided by a Therapist

LfD is an ideal method of introducing semi-autonomy in the field of rehabilitation robotics which our group has investigated since 2015. In addition to the benefits of enabling semi-autonomy as mentioned previously, it is also a plausible method with which therapists with minimum programming experience can easily adjust not only the level of therapeutic assistance or resistance provided to a patient but also set up any number of different therapy tasks, addressing *Limitation 2* (Fig. 10.1). This aspect of mutual adaptation, where users can explore and train robotic aides themselves, is an important step for rehabilitation robotics [16] and is proposed as a viable method of making robotic therapy cost-effective and personalized.

Few groups have applied specifically LfD-based machine learning techniques towards the practice of physical therapy in rehabilitation medicine. Hansen et al. [59] use an adaptive logic network (ALN) to learn a model relating EMG signals and the timing of a patient's activation of an assistive Functional Electrical Stimulation (FES) device during gait. Kostov et al. [77] perform a similar work involving ALNs and inductive learning algorithms, but instead relating foot pressure recordings with FES activation timing. Strazzulla et al. [132] use ridge regression techniques to learn myoelectric prosthetic control during a user's demonstrations, characterized by EMG signals.

Works that use LfD to specifically learn and reproduce the haptic interaction provided by a therapist during interventions represent one branch of the current state of the art in robotic rehabilitation. The merging of these two technologies exploits the hands-on nature of LfD-based robotic systems and addresses some of the shortcomings of robotic rehabilitation as mentioned earlier (i.e., the enabling of cost-savings and ease of programming). Lauretti et al. [80] optimized a system built

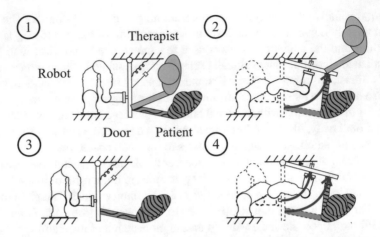

Fig. 10.1 An example of LfD for training a robot to provide haptic interactions that imitate a therapist's intervention. In phases 1 and 2, the therapist will provide haptic interaction for the patient when performing a therapy task while the rehabilitation robot observes the intervention through kinesthetic teaching. The LfD algorithm will then be trained after phase 2. Later in phases 3 and 4, the robot will imitate the haptic interaction demonstrated by the therapist so as to allow the patient to practice in the absence of the therapist while still receiving haptic guidance

on dynamic motor primitives for learning therapist-demonstrated paths for ADLs. Atashzar et al. [10] proposed a framework for both EMG and haptics-based LfD, where the learning of the therapeutic behaviors for an upper limb task was facilitated with an NN.

An extensive amount of research has been performed in the area by our group. Tao et al. [134] utilized a method based on linear least squares regression to provide a simple estimation of the impedance inherent to a therapist's intervention during cooperative performance of upper-limb ADLs with a patient. Maaref et al. [92] described the use of Gaussian Mixture Model (GMM)-based LfD as the underlying mechanism for an assist-as-needed paradigm, evaluating the system for providing haptic interaction for assistance in various upper-limb ADLs. Najafi et al. [107] learned the ideal task trajectory and interaction impedance provided by an able-bodied user with a GMM and provided user experiment evaluations for an upper-limb movement therapy task. Martinez et al. [97] extended the Stable Estimator of Dynamical Systems learning algorithm developed by [76] in order to learn both motion and force-based therapist interventions.

Our group has most recently extended the application of LfD-enhanced robotic rehabilitation in three works. Fong et al. [47] applied kinesthetic teaching principles to a robotic system in order to allow it to first learn and then imitate a therapist's behavior when assisting a patient in a lower limb therapy task (Fig. 10.2). A therapist's assistance in lifting a patient during treadmill-based gait therapy was statistically encoded by the system using a GMM (Fig. 10.3). Later, the therapist's assistance was imitated by the robot, allowing the patient to continue practicing in the absence of the therapist. Preliminary experiments were performed

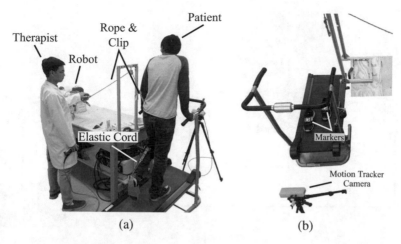

Therapist

Rope & Clip

Patient

Robot

Elastic Cord

Markers

Motion Tracker Camera

(a)   (b)

Fig. 10.2 Photo of the devices used for lifting assistance in [47]. In (**a**) the robot is moved by the therapist by holding and pressing on its end-effector force sensor. This provides a lifting assistance to the patient who walks at their selected pace on the treadmill while harnessed to the robot through the rope and clip. In (**b**) the motion tracker camera is shown placed in front of the patient so as to capture the positions of their toes, which are registered to markers placed on the tops of their shoes

by inexperienced users who took the role of an assisting therapist with healthy participants (wearing an elastic cord to simulate foot drop) playing the role of a patient. The system provided sufficient lifting assistance, but highlighted the importance of learning haptic interactions in the form of the therapist's impedance as opposed to only their movement trajectories.

Our group then applied a similar method of kinesthetic teaching for learning the impedance-based haptic interaction provided by a therapist during intervention in an upper-limb ADL [48]. The kinesthetic teaching process proposed that during performance of the task, the interaction forces exerted on the robot end-effector by each of the agents (task environment, patient, therapist) could be simplified as a set of spring forces, linearized about spatial points of the demonstration (Fig. 10.4). An estimate of the impedance-based interaction provided by the therapist could then be obtained by measuring the "performance differential", i.e., differences in forces along the trajectory, between the patient practicing the task when assisted by the therapist and when attempting the task alone. Experimental validation of the system showed that the interaction impedance was faithfully reproduced, although the resolution of the learnt interaction model briefly produced inaccurate haptic interaction.

The GMM-based LfD system was also applied to a bilateral telerobotic setup to enable telerobotic rehabilitation for home-based delivery. A GMM and GMR-based approach to LfD was implemented with the purpose of learning therapeutic interventions in a collaborative ADL task, where the intervention was dependent on the patient's upper limb position and velocity. By training the GMM with

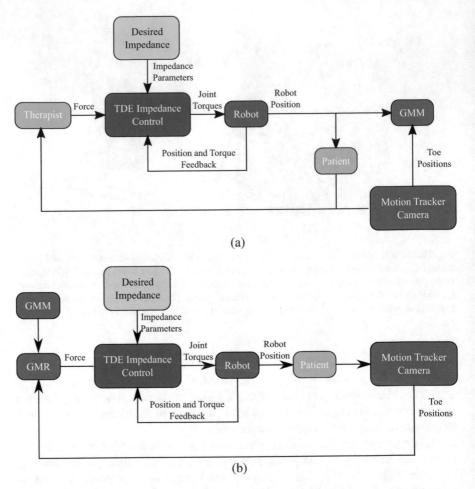

Fig. 10.3 Data flow and interaction between different agents (i.e., the therapist, patient, robot, and motion tracker camera) throughout the LfD process in [47]. (**a**) Shows the process flow when the therapist and patient interact to provide training demonstrations for the GMM algorithm that is used. (**b**) Shows the process flow when the patient is practicing alone with assistance from the robot, which reproduces the learnt haptic interaction using GMR

patient performance (represented by their limb velocity) as a model input, the LfD algorithm inherently learned the adaptive nature of the therapist's intervention with respect to a patient's level of ability (Figs. 10.5 and 10.6).

Lastly, the authors also provided a comparison of the single robot and telerobotic modalities previously implemented, referred to as Robot- and Telerobotic-Mediated Kinesthetic Teaching (RMKT and TMKT), for implementing LfD in robotic rehabilitation. The study provided incentive for rehabilitation-oriented systems to pursue RMKT designs, as the demonstrations provided through the modality were found to be more consistent (Fig. 10.7).

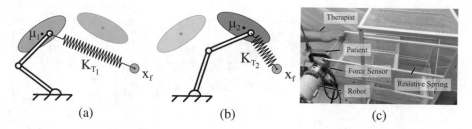

Fig. 10.4 A simplified diagram of position-based impedance retrieval for reproduction of the therapist's behavior in [48]. In (**a**), activation weights for the first Gaussian component of a GMM (colored blue) are highest when the robot is in close proximity to the component. A stiffness constant is retrieved for the corresponding Gaussian and used to generate the forces learned from the therapist. In (**b**), a different stiffness constant is used when the patient progresses into the spatial coordinates associated with a different Gaussian component (colored red). In actual reproduction, the retrieved stiffness constant was composed of a mixture of the learned stiffness constants influenced by multiple components, instead of a single constant from the influence of a single component as shown. (**c**) shows the experimental setup used for validation

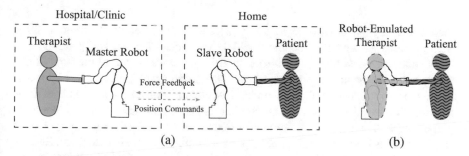

Fig. 10.5 Illustrations of the telerehabilitation system with LfD proposed by our group. The demonstration phase is shown in (**a**) where the patient interacts with the therapist, and the reproduction phase in (**b**) where the patient interacts with a slave robot that emulates the therapist's behavior

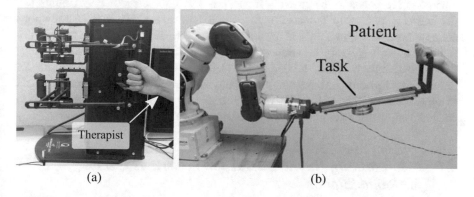

Fig. 10.6 Experimental setup used by our group for bilateral haptics-enabled intervention. (**a**) shows the HD^2 High Definition Haptic Device (Quanser Inc., Markham, Ontario, Canada) used as the master robot by the therapist; (**b**) shows the Motoman SIA-5F (Yaskawa America, Inc., Miamisburg, Ohio, USA) industrial robot that interacts with the patient

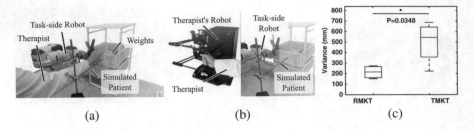

(a)   (b)   (c)

Fig. 10.7 Experimental setups and results for our group's work on comparing RMKT and TMKT. (**a**) shows the RMKT setup. The therapist, patient, and robot force sensor hold and open the drawer together. (**b**) Shows the master robot that is added in TMKT. The therapist holds the master robot and moves the task-side robot through a direct force reflection control loop. (**c**) Shows a comparison of the consistency of user demonstrations (which is proposed to be inversely related to the variance in the trained model's output) for the two modalities

4 Serious Games

First coined by Abt in his 1970 book [3] and popularized by Sawyer in 2002 [123], serious games have become a widely researched field of study garnering a worldwide market worth of 1.5 billion in 2010 [7]. It was not until the early 2000s when fueled by the advancements in hardware, game development, and its success in the commercial market, that researchers worldwide took an interest in novel areas of research that video games can be applied to. Defined as games that serve a main purpose other than pure entertainment, serious games have been expanding in areas such as politics [146], military [126], sports training [53], and health [135]. Serious games can be presented through any form of technology and can be in any genre. For instance, even commercial platforms intended for entertainment, the Nintendo Wii, the PlayStation 2, and the Microsoft Kinect, have been repurposed for research in the rehabilitation field [36, 39, 46]. Various conferences and seminars have emerged in response to the growing interest. In 2010, the Serious Play Conference [2] began a leadership conference for developers of serious games/simulations coming from different fields of expertise from both industry and academia. To promote interdisciplinary research within the field, IEEE launched the first serious games conference in 2009, dubbed as VS-GAMES: Games and Virtual Worlds for Serious Applications. Limitless in potential, games allow for adding entertainment in approaches to education, training, or rehabilitation. In essence, the purpose is to create an enjoyable environment for otherwise tedious activities.

4.1 Incorporating Haptic Interaction in Serious Games for Rehabilitation

Serious games have been shown to increase patient motivation [4, 19, 91], addressing *Limitation 3*. By combining serious games and robot-assisted rehabilitation, the patient becomes engaged in the exercise and may even "forget" that they are in a rehabilitation training session due to the immersion [111]. It has been shown that the combination of the two technologies in rehabilitation training leads to better outcomes than robotics assistance alone [24, 103].

The addition of serious games in the rehabilitation environment to augment physical therapeutic exercises has been shown to produce positive outcomes. Since the games can promote the use of both physical motions and mental processes, the patient becomes actively focused while doing exercises. One of the earlier documented applications of computer games in rehabilitation research was done in 1993 for promoting arm reach using a *Simon* game in which patients are instructed to repeat the sequence of flashing lights by pressing coloured buttons [128]. By adding the game, it structures the exercise such that performance components are actively acquired through goal-directed interaction with the environment instead of acquiring them through random movements or mindless exercise [104].

The availability of various interfaces for serious gaming allows for adaptability and variety for the patient. From keyboards and mice, to body tracking and head-mounted displays, a multitude of devices can be integrated in games. Robots and haptic interfaces can provide haptic feedback to the patients to enable interaction with digital objects. For example, technologies like The Java Therapy System [119] allow for different interfaces (traditional mice, force feedback mice, force feedback joystick) to be used with games.

Serious gaming for rehabilitation does not end outside the doors of the therapy clinics; it plays a key role in home-based rehabilitation. Patients who are discharged from the hospitals and are sent home but are required to do exercises by themselves lack the motivational support from staff and peers in the clinics. This also applies to patients without access to the clinics due to distance or other circumstances. Self-exercise often leads to low motivation and less adherence to the required daily workout dose when lacking motivational support. Engaging patients at home with serious games for rehabilitation promotes adherence to the therapy.

Sources of motivation also come from the environment surrounding the patient. The therapist, the patient's family, and peers in the rehabilitation clinics are all factors that can affect the patient's motivation. Including these elements in a serious game further motivates patients by making the exercise seem more comparable to a social activity. Jadhav et al. [66] developed a system that allowed the therapist to perform an active role in adjusting the complexity of the training regimen while the patient does the exercise from the remote environment. Multiplayer games with peers offer the patients a shared experience and a sense of social acceptance [122]. Being able to cooperate or compete with one another heightens the enjoyment and can potentially produce a more intensive exercise than playing alone [108]. Furthermore, differences in performance levels for an able-bodied person and a post-stroke patient could be equalized such that they can both take part in a rehabilitative game [95].

There is significance in the type of information the patient receives during their exercise. In traditional rehabilitation practices, this may be in the form of distinct changes in the difficulty of an exercise, location of the user's hand with respect to the target position, or general commentary feedback from the therapist. Robotic therapy takes advantage of the ability to record quantitative information to accurately measure even minor improvements in the patient's actions [78]. Coupled with the games, the recorded data could be transformed into an exciting challenge to beat such as using the patient's progress as high scores or achievements to work towards to further engage the patients [41, 135].

Moreover, since the confidence of patients may decrease when recognizing an increased intensity or difficulty in the task, the task difficulty could also be discreetly changed unbeknown to the patient. The game serves to aid in taking away attention from the change, thereby producing a masking effect.

Using the games as a distractor can also be used to alleviate pain. Patients can feel discouraged to continue their rehabilitation exercises when knowing that doing so inflicts pain. The same case is often seen in pediatrics when children become difficult to manage due to the fear of needles. A study on burn patients undergoing burn rehabilitation therapy reported that less pain was experienced when the patients were distracted by VR. The patients spent less time thinking about the pain while distracted [62]. The pain can also be re-attributed as a different sensation (e.g., a needle's sting can be represented as a warm object on the arm) in the VR environment to further draw attention away from it [113, 131].

5   Enhancing Haptic Interaction with Virtual and Augmented Reality

While serious games aid in motivating patients during rehabilitation exercises, they lack the sense of realism found in traditional rehabilitation practices. Real-world tasks involve interaction with real objects such as peg-in-the-hole insertion, block stacking, and pick-and-place operations. Patients touch and see the objects they are interacting with at the exact same location since both the visual and haptic frames are aligned. However, the games used for rehabilitation are typically shown on a 2D screen in front of the user. The disconnect between the patient's arm movement axis and how they see their cursor or avatar move on the screen may unnecessarily impose a mental burden on the patients to match their limb movements with what they visually see on the display. Furthermore, the scaling of movements might need to be accounted for if the workspace and screen are of different sizes. For patients whose cognitive functions are negatively affected by events such a stroke or injury and who require upper-limb neurorehabilitation, the spatial disparity might make it more difficult to perform the task compared to those with no cognitive deficiency.

5.1 Motivation for Visual-Haptic Colocation in Rehabilitation

Visual-haptic colocation is the direct alignment of the virtual environment with the physical environment in terms of visual and physical interaction. In contrast, a typical rehabilitation environment has the users doing rehabilitative tasks on a robot while looking at a screen for feedback of their movements which is often represented by a computer game. However, this spatial disparity between their arm movements and on- screen movements may unintentionally apply an unnecessary mental load on the patient as it requires a mental transformation between the two coordinate frames. This may be most evident in persons affected by mental disability [148]. To alleviate this problem, we propose to use a spatial AR setup in which the virtual and physical environments become aligned to allow the patients to control the rehabilitative tasks more intuitively.

5.2 Virtual Reality and Augmented Reality

Game environments can be shown to users through different display techniques. This can range from a typical 2D computer screen to more state-of-the-art technologies such as the Microsoft Hololens. The environment presented to the user can be described through the Virtuality Continuum (VC) [101] (Fig. 10.8).

The concept of the VC illustrates how the presentation of objects is not limited to a purely virtual or purely real environment. Looking through a cell phone camera and seeing the real world is considered as part of the left end of the continuum, the real environment. On the other hand, games like *Super Mario Brothers* belong purely to the virtual environment (or virtual reality) on the right end of the continuum. The term "Mixed Reality" arises when these two environments are combined within a single display. Depending on the proportion of real and virtual environments represented, the result can be categorized as either Augmented Reality (AR) or Augmented Virtuality (AV). In AR, virtual objects are integrated into the real environment. It involves the augmentation of the real world with virtual (computer generated) objects. At one's viewpoint, these objects are seamlessly integrated as if they physically exist alongside the real-world objects. This potentially allows users to digitally interact with their surrounding

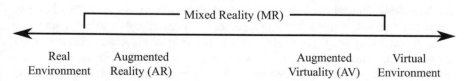

Fig. 10.8 The Virtuality Continuum introduced by Milgram et al. to categorize different mixed reality environments

environment. Games such as *Pokemon Go*, where virtual objects are overlaid on the camera feed, are considered as AR. Conversely, AV adds real objects into the virtual environment. This can be imagined as displaying the user's actual hand in a virtual environment to interact with virtual objects. This is not to be confused with the representation of AR where the entire real environment is overlaid with a virtual one in which only certain real objects are unmasked such that they appear in the virtualized world [50]. It is important to note that these environments are not restricted by a specific type of display method such as a head-mounted display. Even a simple 2D screen would be capable of displaying these environments.

Although the majority of AR and VR development is geared for visual use, there is some research that applies to other senses also [137]. For example, applications for AR in surgery also involve active sensorimotor augmentation techniques to either provide supplementary perceptual information or to augment motor control capabilities of the surgeon during robotic-assisted surgeries [12].

5.2.1 Virtual Reality Displays

There are two main ways that VR is presented: non-immersive VR through a 2D computer screen, or immersive 3D VR using a head-mounted display (HMD). The ReJoyce by Rehabtronics [1] is an example of a non-immersive VR system controlled by a passive robotic interface. A multitude of games can be selected and played with using the various components that are incorporated into the robotic interface. As for immersive VR, the user is brought into a completely virtual 3D environment. The virtual surroundings can be configured in any manner regardless of genre or theme, allowing for creative ways of incorporating it into different applications in the healthcare field. Immersive VR has been used in studies such as treating phobia [33], assessing mild traumatic brain injury [121], gait therapy [67], hand rehabilitation [31], and so on.

5.2.2 Augmented Reality Displays

Video see-through (VST), optical see-through (OST), and spatial AR (i.e. projection-based AR) are the three most common ways of displaying AR [137]. VST takes a video feed of the real world and then virtual objects are directly overlaid on the display on the live video [28]. Registering the virtual objects can be done using fiducial markers attached to real-world surfaces [50]. By digitizing the real-world environment through a camera, it is much easier to manipulate the environment using image processing tools as both virtual objects and the real world are now in the same digital space. This means aspects such as contrast, orientation, and size of the virtual object can be calibrated more easily with the real world. With OST, the virtual and real aspects interact through a semi-transparent mirror [69]. The user can see through the display, and the virtual objects are reflected on the display for the user to see. OSTs let the user experience their environment directly without relying on the screen of a VST and therefore the resolution quality is as good as the user's eyesight. However, it may be bulkier as cameras, monitors, and a mirror are needed to assess the environment and display virtual objects. Finally, projection is another method to implement AR [139]. Projection solves the problem of requiring HMDs or looking at a separate screen to see the computer-generated images. It can be used on both flat and 3D surfaces with cameras or motion trackers for direct interaction. Techniques such as projection mapping are available for the virtual objects to "pop-out" and seem one with the environment. However, projection is limited to low light working environments since it can be easily overpowered by other light sources.

5.3   Virtual and Augmented Reality Rehabilitation Systems

Multiple rehabilitation tasks that incorporate AR and VR have been published in the literature. VR systems present a completely virtual environment to the user, allowing for creative scenarios unbounded by physical limitations found in real environments. AR systems let the user stay within their familiar environment while enabling interaction with the virtual objects that are displayed in the real world. A brief investigation of related work is presented to give insight on current technologies available for rehabilitation in research. These are categorized based on visual technique and the presence or absence of haptics.

5.3.1   Virtual Reality Rehabilitation Systems

There are two main types of displays used in VR systems: non-immersive or immersive. Flat display screens such as a computer monitor fall under non-immersive systems. An example of a product in the market that implements non-immersive VR is the ReJoyce Rehabilitation Workstation, with a multitude of interactive games to simulate a variety of ADL exercises [45]. However, it does not provide haptic feedback during the tasks, only visual and auditory. Immersive VR systems often use HMDs to envelop the patient in a fully virtual world. An Oculus Rift and a non-haptic glove, for example, is utilized by Kaminer et al. [72] for an immersive pick-and-place task. For their upper limb rehabilitation exercises, Andaluz et al. [8] paired the Oculus with the Novint Falcon haptic device. There are, however, fully immersive VR systems that surround the user with a large projection display such as the CAVE [35] or CAREN [42].

5.3.2   Augmented Reality Rehabilitation Systems

AR systems come in three main forms: VST, OST, or projection. The immersion varies depending on the display type as some may use HMDs while others use monitor screens for VST and OST. Examples of research in rehabilitation that utilize VST systems are Burke et al. [25], Correa et al. [32], and Vidrios-Serrano et al. [138]. Burke implemented a reaching task through a game similar to Atari's Breakout using fiducial markers to track real objects and allow for interaction with the virtual environment. Correa developed an AR game involving replication of musical tunes by occluding colored cubes in the proper sequence. Vidrios-Serrano used a haptic device to provide haptic feedback to the users as they viewed the environment through an HMD.

For OST setups, Trojan et al. [136] took a non-haptic approach in developing a mirror training rehabilitation system suitable for home use. Luo et al. [90] created an AR-based hand opening rehabilitation setup using an HMD and a haptic glove; the glove was used to simulate the sensation of holding a real object during their grasp-and-release task.

For projection setups, Hondori et al. [105] created a non-haptic tabletop system for post-stroke hand rehabilitation which incorporated different games. These included interacting with a projected box to play sounds, holding a cup to pour out virtual water, and grasping circles of different sizes. Finally, Khademi et al. [74] implemented a spatial AR setup with a haptic device for monitoring human arm impedance. They also performed a comparison between AR and VR displays for a pick-and-place task [73].

5.4   Dimensionality of Virtual and Augmented Reality Environments

An additional consideration for VR and AR rehabilitation systems lies in the number of dimensions that virtualized rehabilitation tasks are presented in. With the recent surge in developments in VR and AR technologies, the presentation of virtual environments (for rehabilitation or other purposes) in immersive 3D implementations has become more affordable and feasible. Most rehabilitation literature that incorporate serious games involve 2D non-immersive VR displays, or 2D or 3D AR dislays but without colocation of visual and motor axes. Devices such as the ReJoyce Rehabilitation Workstation have additional suites of interactive 2D games to motivate patients and improve upper limb function after stroke [1, 45]. GenVirtual, created by Correa et al. [32], is a musically-oriented spatial 2D AR game where the user replicates a song produced by virtual cubes that light up in a sequence by touching the cubes in the same order as demonstrated. Gama et al. [37] developed MirrARbilitation, a VST 2D non-colocated AR system to encourage and guide users in a shoulder abduction therapy exercise.

In [110], our group integrated spatial AR into robotic rehabilitation to provide colocation between visual and haptic feedback for human users participating in a rehabilitative game (Fig. 10.9). A comparison between the effectiveness of VR vs AR (i.e., non-colocation vs co-location of vision) was performed. Each visualization technique was also compared in the absence and presence of haptic feedback and cognitive loading (CL) for the human user. The system was evaluated by having 10 able-bodied participants play a game targeting upper-limb rehabilitation under all

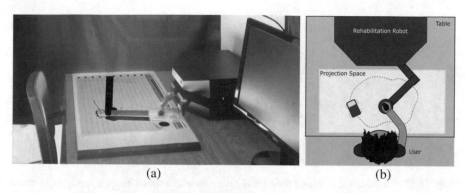

(a)   (b)

Fig. 10.9 Experimental setup used by our group to evaluate the effects of AR technology on performance of 2D rehabilitation tasks. Left: Actual setup. Task is projected onto the desk surface (projector is not in view). Right: Top-down diagram of what the user sees

8 different combinations of conditions lasting approximately 3 min per condition. The results showed that spatial AR (corresponding to colocation of visual frame and hand frame) led to the best user performance when doing the task regardless of the presence or the absence of haptics. It was also observed that for users undergoing cognitive loading, the combination of spatial AR and haptics produced the best result in terms of task completion time.

The use of 3D VR and AR systems with robotics can be seen as the current state of the art for research in visual-haptic colocation. Vidrios-Serrano et al. [138] used a VST 3D non-colocated AR system integrated with a phantom Omni device to interact with the virtual environment in a rehabilitation exercise. Broeren et al. [23] and Murphy et al. [106] used a haptic immersive workbench to test both able-bodied and stroke-impaired persons for rehabilitation and assessment using their OST 3D colocated AR system. Swapp et al. [133] studied the effectiveness of a 3D stereoscopic display AR system for colocated haptic feedback, where the benefits of incorporating visual-haptic colocation as opposed to having no colocation were examined. However, the study did not explore its effects in rehabilitation exercises and did not adjust the display to adapt to the user's head movements.

Our group presented a culmination of these developments in [109]. The authors developed a 3-D spatial augmented reality (AR) display to colocate visual and haptic feedback to the user in three rehabilitative games (Fig. 10.10). A projection-based AR system was implemented with an off-the-shelf projector displaying the virtualized environment on a curved, smooth screen. Head tracking and virtualization of the user's workspace were performed by a Microsoft Kinect sensor, allowing the projection to match the user's perspective for maximizing immersion (Fig. 10.11). The haptic feedback for each task was provided with a Quanser High Definition Haptic Device (HD^2) (Quanser, Inc., Markham, Ontario, Canada) and was intended to both provide the user with guiding cues as well as simulate their physical interactions with the virtualized task. To simulate a rehabilitation scenario, able-bodied participants are put under cognitive load (CL) for simulating disability-induced cognitive deficiencies when performing the tasks. A within-subjects analysis of

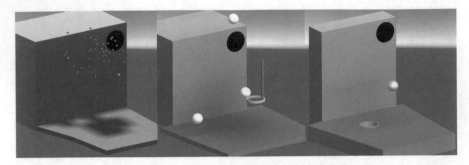

Fig. 10.10 The three virtual games designed in [109]: *Snapping* which involved navigating a cloud of points (left), *Catching* which involved catching falling objects (centre), and *Ball Dropping* which involved accurately dropping objects (right)

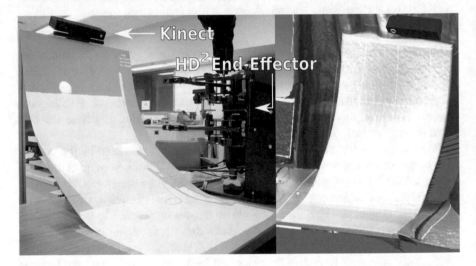

Fig. 10.11 Experimental setup used in [109]. Left: Actual setup with the task projected onto the screen (projector is not in view). Right: Model of the setup created in Unity

ten participants was carried out for the rehabilitative games. Comparison of user task performance for the same games between the AR and VR setups found that AR enabled superior performance with or without cognitive loading. This result is most evident in exercises that require participants to have quick reaction times and movement. Furthermore, even while AR had a significant difference over VR, results for one of the tasks indicated that the recorded performance for AR between non-CL and CL cases was similar, showing the ability of AR to alleviate the negative effects of CL.

Lastly, our group presented a system comprised of a robotic arm for recreating the physical dynamics of functional tasks and a 3D Augmented Reality (AR) display for immersive visualization of the tasks for Functional Capacity Evaluation (FCE) of injured workers. While this system could have been used to simulate a multitude of occupational tasks, one specific functional task was focused on. Participants performed a virtual version of a workplace task using the robot-AR system, and a physical version of the same task without the system. Preliminary results for two able-bodied users indicated that the robot-AR system's haptic interactions resulted in the users' upper-limb kinematics resembling those measured in the real-life equivalent task (Fig. 10.12).

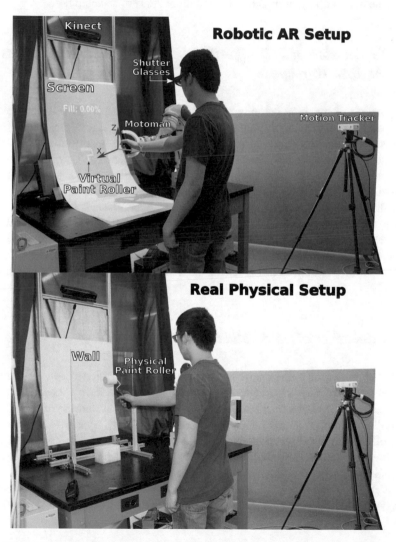

Fig. 10.12 Painting task experimental setups used by our group for robot- and AR-enhanced FCE. Top: the robot-AR condition. Bottom: the real-life equivalent condition. The projector is not shown. Through AR, the paint roller will pop out in 3D from the perspective of the user in a geometrically correct position and orientation relative to the robot end-effector. The Yaskawa Motoman SIA-5F robot provides haptic feedback to the user that match their interactions with the virtual task

6 Future Directions for LfD- and VR/AR-Enhanced Rehabilitation

The innovations presented in this chapter represent parts of the forefront in technological rehabilitation medicine research. As such, there are still many aspects that should be developed before these technologies begin to see clinical adoption. A few of the most important directions for future works are selected here.

6.1 Exploration of LfD Algorithms That Define Models Across the Task Workspace

Incorporation of machine learning techniques is still relatively new in the field of rehabilitation robotics. As such, a wide range of learning algorithms is present in the literature, none of which is presented as a definitive best option. A possible future direction would be to explore and fairly compare LfD algorithms so as to create guidelines for which are optimal for rehabilitation task and interaction learning. Algorithms that generate global models from demonstrations (i.e., that cover the entire task workspace) may represent a good starting point. In these models, desired haptic interactions would be defined for all patient behaviors, which is desirable for safety and ease of programming. This could be performed through simple methods such as surface fitting, but could also be extended to explore more advanced concepts such as fitting Riemannian manifolds, or the SEDS algorithm as seen in [97].

6.2 Clinical Trials and Validation

A common limitation of the majority of the technologies that have been presented in this chapter is that they present proof-of-concept systems, or have not been validated for patient-safe interaction. It is crucial to validate the system by conducting a longitudinal study on actual patients that have a disability. Systems incorporating either of the proposed technologies (LfD and VR/AR) should be compared with a similar traditional rehabilitation setup by analyzing the outcomes of patient neuromuscular and cognitive improvement in order to determine its effectiveness against current methods. Emphasis should also be placed on recruiting large sample sizes, as the majority of studies have shown results for relatively small samples.

References

1. ReJoyce by rehabtronics, https://www.blog.rehabtronics.com/rejoyce
2. Serious Play Conference, https://seriousplayconf.com/
3. Abt, C.C.: Serious games. University press of America (1987)

4. Alankus, G., Proffitt, R., Kelleher, C., Engsberg, J.: Stroke therapy through motion-based games: a case study. ACM Transactions on Accessible Computing (TACCESS) 4(1), 3 (2011)
5. Albert, S.J., Kesselring, J.: Neurorehabilitation of stroke. Journal of neurology 259(5), 817–832 (2012)
6. Alexander, N.B., Galecki, A.T., Grenier, M.L., Nyquist, L.V., Hofmeyer, M.R., Grunawalt, J.C., Medell, J.L., Fry-Welch, D.: Task-specific resistance training to improve the ability of activities of daily living–impaired older adults to rise from a bed and from a chair. Journal of the American Geriatrics Society 49(11), 1418–1427 (2001)
7. Alvarez, J., Alvarez, V., Djaouti, D., Michaud, L.: Serious games: Training & teaching-healthcare-defence & security-information & communication. IDATE, France (2010)
8. Andaluz, V.H., Salazar, P.J., Escudero V., M., Bustamante D., C., Silva S., M., Quevedo, W., Sánchez, J.S., Espinosa, E.G., Rivas, D.: Virtual reality integration with force feedback in upper limb rehabilitation. In: Advances in Visual Computing. pp. 259–268. Springer International Publishing, Cham (2016)
9. Argall, B.D., Chernova, S., Veloso, M., Browning, B.: A survey of robot learning from demonstration. Robotics and Autonomous Systems 57(5), 469–483 (2009), http://www.sciencedirect.com/science/article/pii/S0921889008001772
10. Atashzar, S.F., Shahbazi, M., Tavakoli, M., Patel, R.V.: A computational-model-based study of supervised haptics-enabled therapist-in-the-loop training for upper-limb poststroke robotic rehabilitation. IEEE/ASME Transactions on Mechatronics 23(2), 563–574 (April 2018)
11. Atashzar, S.F., Jafari, N., Shahbazi, M., Janz, H., Tavakoli, M., Patel, R.V., Adams, K.: Telerobotics-assisted platform for enhancing interaction with physical environments for people living with cerebral palsy. Journal of Medical Robotics Research 2(02), 1740001 (2017)
12. Atashzar, S.F., Naish, M., Patel, R.V.: 5 active sensorimotor augmentation in robotics-assisted surgical systems. In: Mixed and Augmented Reality in Medicine, pp. 61–81. CRC Press (2018)
13. Atkeson, C.G., Schaal, S.: Robot learning from demonstration. In: ICML. vol. 97, pp. 12–20. Citeseer (1997)
14. Badesa, F.J., Morales, R., Garcia-Aracil, N., Sabater, J.M., Casals, A., Zollo, L.: Auto-adaptive robot-aided therapy using machine learning techniques. Computer methods and programs in biomedicine 116(2), 123–130 (2014)
15. Barzilay, O., Wolf, A.: Adaptive rehabilitation games. Journal of Electromyography and Kinesiology 23(1), 182–189 (2013)
16. Beckerle, P., Salvietti, G., Unal, R., Prattichizzo, D., Rossi, S., Castellini, C., Hirche, S., Endo, S., Amor, H.B., Ciocarlie, M., Mastrogiovanni, F., Argall, B.D., Bianchi, M.: A human–robot interaction perspective on assistive and rehabilitation robotics. Frontiers in Neurorobotics 11, 24 (2017), https://www.frontiersin.org/article/10.3389/fnbot.2017.00024
17. Begg, R., Kamruzzaman, J.: A machine learning approach for automated recognition of movement patterns using basic, kinetic and kinematic gait data. Journal of Biomechanics 38(3), 401–408 (2005), http://www.sciencedirect.com/science/article/pii/S0021929004002258
18. Benjamin, E.J., Blaha, M.J., Chiuve, S.E., Cushman, M., Das, S.R., Deo, R., de Ferranti, S.D., Floyd, J., Fornage, M., Gillespie, C., Isasi, C.R., Jiménez, M.C., Jordan, L.C., Judd, S.E., Lackland, D., Lichtman, J.H., Lisabeth, L., Liu, S., Longenecker, C.T., Mackey, R.H., Matsushita, K., Mozaffarian, D., Mussolino, M.E., Nasir, K., Neumar, R.W., Palaniappan, L., Pandey, D.K., Thiagarajan, R.R., Reeves, M.J., Ritchey, M., Rodriguez, C.J., Roth, G.A., Rosamond, W.D., Sasson, C., Towfighi, A., Tsao, C.W., Turner, M.B., Virani, S.S., Voeks, J.H., Willey, J.Z., Wilkins, J.T., Wu, J.H., Alger, H.M., Wong, S.S., Muntner, P.: Heart disease and stroke statistics—2017 update: A report from the american heart association. Circulation (2017), http://circ.ahajournals.org/content/early/2017/01/25/CIR.0000000000000485
19. Betker, A.L., Desai, A., Nett, C., Kapadia, N., Szturm, T.: Game-based exercises for dynamic short-sitting balance rehabilitation of people with chronic spinal cord and traumatic brain injuries. Physical therapy 87(10), 1389–1398 (2007)

20. Blank, A.A., French, J.A., Pehlivan, A.U., O'Malley, M.K.: Current trends in robot-assisted upper-limb stroke rehabilitation: promoting patient engagement in therapy. Current physical medicine and rehabilitation reports 2(3), 184–195 (2014)
21. Bourbonnais, D., Noven, S.V.: Weakness in patients with hemiparesis. The American journal of occupational therapy 43(5), 313–319 (1989)
22. Bracewell, R.: Stroke: neuroplasticity and recent approaches to rehabilitation. Journal of Neurology, Neurosurgery & Psychiatry 74(11), 1465–1465 (2003)
23. Broeren, J., Sunnerhagen, K.S., Rydmark, M.: Haptic virtual rehabilitation in stroke: transferring research into clinical practice. Physical Therapy Reviews 14(5), 322–335 (2009)
24. Brütsch, K., Koenig, A., Zimmerli, L., Mérillat-Koeneke, S., Riener, R., Jäncke, L., van Hedel, H.J., Meyer-Heim, A.: Virtual reality for enhancement of robot-assisted gait training in children with neurological gait disorders. Journal of rehabilitation medicine 43(6), 493–499 (2011)
25. Burke, J.W., McNeill, M.D.J., Charles, D.K., Morrow, P.J., Crosbie, J.H., McDonough, S.M.: Augmented reality games for upper-limb stroke rehabilitation. In: 2010 Second International Conference on Games and Virtual Worlds for Serious Applications. pp. 75–78 (March 2010)
26. Calinon, S., Evrard, P., Gribovskaya, E., Billard, A., Kheddar, A.: Learning collaborative manipulation tasks by demonstration using a haptic interface. In: Advanced Robotics, 2009. ICAR 2009. International Conference on. pp. 1–6. IEEE (2009)
27. Carignan, C.R., Krebs, H.I.: Telerehabilitation robotics: Bright lights, big future? Journal of Rehabilitation Research and Development 43(5), 695–710 (August 2006), copyright – Copyright Superintendent of Documents Aug/Sep 2006; Document feature – Diagrams; Photographs; Illustrations; Last updated – 2017-11-09; CODEN – JRRDDB
28. Casas, X., Herrera, G., Coma, I., Fernández, M.: A kinect-based augmented reality system for individuals with autism spectrum disorders. In: Grapp/ivapp. pp. 440–446 (2012)
29. Colombo, G., Joerg, M., Schreier, R., Dietz, V.: Treadmill training of paraplegic patients using a robotic orthosis. Journal of rehabilitation research and development 37(6), 693–700 (2000)
30. Colombo, R., Pisano, F., Mazzone, A., Delconte, C., Micera, S., Carrozza, M.C., Dario, P., Minuco, G.: Design strategies to improve patient motivation during robot-aided rehabilitation. Journal of neuroengineering and rehabilitation 4(1), 3 (2007)
31. Connelly, L., Jia, Y., Toro, M.L., Stoykov, M.E., Kenyon, R.V., Kamper, D.G.: A pneumatic glove and immersive virtual reality environment for hand rehabilitative training after stroke. IEEE Transactions on Neural Systems and Rehabilitation Engineering 18(5), 551–559 (2010)
32. Correa, A.G.D., de Assis, G.A., d. Nascimento, M., Ficheman, I., d. D. Lopes, R.: Genvirtual: An augmented reality musical game for cognitive and motor rehabilitation. In: 2007 Virtual Rehabilitation. pp. 1–6 (Sept 2007)
33. Côté, S., Bouchard, S.: Virtual reality exposure for phobias: A critical review. Journal of CyberTherapy & Rehabilitation 1(1), 75–91 (2008)
34. Cramer, S.C., Sur, M., Dobkin, B.H., O'brien, C., Sanger, T.D., Trojanowski, J.Q., Rumsey, J.M., Hicks, R., Cameron, J., Chen, D., Chen, W.G., Cohen, L.G., Decharms, C., Duffy, C.J., Eden, G.F., Fetz, E.E., Filart, R., Freund, M., Grant, S.J., Haber, S., Kalivas, P.W., Kolb, B., Kramer, A.F., Lynch, M., Mayberg, H.S., McQuillen, P.S., Nitkin, R., Pascual-Leone, A., Reuter-Lorenz, P., Schiff, N., Sharma, A., Shekim, L., Stryker, M., Sullivan, E.V., Vinogradov, S.: Harnessing neuroplasticity for clinical applications. Brain 134(6), 1591–1609 (2011)
35. Cruz-Neira, C., Sandin, D.J., DeFanti, T.A.: Surround-screen projection-based virtual reality: the design and implementation of the cave. In: Proceedings of the 20th annual conference on Computer graphics and interactive techniques. pp. 135–142. ACM (1993)
36. Da Gama, A., Chaves, T., Figueiredo, L., Teichrieb, V.: Poster: improving motor rehabilitation process through a natural interaction based system using kinect sensor. In: 3D User Interfaces (3DUI), 2012 IEEE Symposium on. pp. 145–146. IEEE (2012)
37. Da Gama, A.E.F., Chaves, T.M., Figueiredo, L.S., Baltar, A., Meng, M., Navab, N., Teichrieb, V., Fallavollita, P.: Mirrarbilitation: A clinically-related gesture recognition interactive tool for an ar rehabilitation system. Computer methods and programs in biomedicine 135, 105–114 (2016)

38. Damush, T.M., Plue, L., Bakas, T., Schmid, A., Williams, L.S.: Barriers and facilitators to exercise among stroke survivors. Rehabilitation nursing 32(6), 253–262 (2007)
39. Deutsch, J.E., Borbely, M., Filler, J., Huhn, K., Guarrera-Bowlby, P.: Use of a low-cost, commercially available gaming console (wii) for rehabilitation of an adolescent with cerebral palsy. Physical therapy 88(10), 1196–1207 (2008)
40. Deutsch, J.E., Latonio, J., Burdea, G.C., Boian, R.: Post-stroke rehabilitation with the rutgers ankle system: a case study. Presence: Teleoperators & Virtual Environments 10(4), 416–430 (2001)
41. Djaouti, D., Alvarez, J., Jessel, J.P.: Classifying serious games: the g/p/s model. In: Handbook of research on improving learning and motivation through educational games: Multidisciplinary approaches, pp. 118–136. IGI Global (2011)
42. Van der Eerden, W., Otten, E., May, G., Even-Zohar, O.: Caren-computer assisted rehabilitation environment. Medicine Meets Virtual Reality: The Convergence of Physical & Informational Technologies: Options for a New Era in Healthcare 62, 373–378 (1999)
43. Ekberg, K.: Workplace changes in successful rehabilitation. Journal of Occupational Rehabilitation 5(4), 253–269 (1995)
44. Feigin, V.L., Norrving, B., Mensah, G.A.: Global burden of stroke. Circulation research 120(3), 439–448 (2017)
45. FGTeam: Rejoyce speeds up upper extremity recovery post stroke (January 2015), https://www.fitness-gaming.com/news/health-and-rehab/rejoyce-speeds-up-upper-extremity-recovery-post-stroke.html
46. Flynn, S., Palma, P., Bender, A.: Feasibility of using the sony playstation 2 gaming platform for an individual poststroke: a case report. Journal of neurologic physical therapy 31(4), 180–189 (2007)
47. Fong, J., Rouhani, H., Tavakoli, M.: A therapist-taught robotic system for assistance during gait therapy targeting foot drop. IEEE Robotics and Automation Letters 4(2), 407–413 (2019)
48. Fong, J., Tavakoli, M.: Kinesthetic teaching of a therapist's behavior to a rehabilitation robot. In: 2018 International Symposium on Medical Robotics (ISMR). pp. 1–6 (March 2018)
49. Garate, V.R., Parri, A., Yan, T., Munih, M., Lova, R.M., Vitiello, N., Ronsse, R.: Experimental validation of motor primitive-based control for leg exoskeletons during continuous multilocomotion tasks. Frontiers in Neurorobotics 11, 15 (2017)
50. Garcia, A., Andre, N., Bell Boucher, D., Roberts-South, A., Jog, M., Katchabaw, M.: Immersive Augmented Reality for Parkinson Disease Rehabilitation, pp. 445–469. Springer Berlin Heidelberg, Berlin, Heidelberg (2014), https://doi.org/10.1007/978-3-642-54816-1_22
51. Giorgino, T., Lorussi, F., De Rossi, D., Quaglini, S.: Posture classification via wearable strain sensors for neurological rehabilitation. In: 2006 International Conference of the IEEE Engineering in Medicine and Biology Society. pp. 6273–6276. IEEE (2006)
52. Gladstone, D.J., Danells, C.J., Black, S.E.: The fugl-meyer assessment of motor recovery after stroke: a critical review of its measurement properties. Neurorehabilitation and neural repair 16(3), 232–240 (2002)
53. Göbel, S., Hardy, S., Wendel, V., Mehm, F., Steinmetz, R.: Serious games for health: personalized exergames. In: Proceedings of the 18th ACM international conference on Multimedia. pp. 1663–1666. ACM (2010)
54. Gowland, C., Stratford, P., Ward, M., Moreland, J., Torresin, W., Van Hullenaar, S., Sanford, J., Barreca, S., Vanspall, B., Plews, N.: Measuring physical impairment and disability with the chedoke-mcmaster stroke assessment. Stroke 24(1), 58–63 (1993)
55. Grahn, B., Ekdahl, C., Borgquist, L.: Motivation as a predictor of changes in quality of life and working ability in multidisciplinary rehabilitation. Disability and Rehabilitation 22(15), 639–654 (2000)
56. Gribovskaya, E., Khansari-Zadeh, S.M., Billard, A.: Learning non-linear multivariate dynamics of motion in robotic manipulators. The International Journal of Robotics Research 30(1), 80–117 (2011)

57. Gui, K., Liu, H., Zhang, D.: Toward multimodal human–robot interaction to enhance active participation of users in gait rehabilitation. IEEE Transactions on Neural Systems and Rehabilitation Engineering 25(11), 2054–2066 (2017)

58. Guidali, M., Duschau-Wicke, A., Broggi, S., Klamroth-Marganska, V., Nef, T., Riener, R.: A robotic system to train activities of daily living in a virtual environment. Medical & Biological Engineering & Computing 49(10), 1213 (July 2011), https://doi.org/10.1007/s11517-011-0809-0

59. Hansen, M., Haugland, M., Sinkjær, T., Donaldson, N.: Real time foot drop correction using machine learning and natural sensors. Neuromodulation: Technology at the Neural Interface 5(1), 41–53 (2002), https://onlinelibrary.wiley.com/doi/abs/10.1046/j.1525-1403.2002._2008.x

60. Heart and Stroke Foundation of Canada: Getting to the heart of the matter: solving cardiovascular disease through research (2015), http://www.heartandstroke.ca/-/media/pdf-files/canada/2017-heart-month/heartandstroke-reportonhealth-2015.ashx?la=en

61. Hillman, M.: 2 rehabilitation robotics from past to present–a historical perspective. In: Advances in Rehabilitation Robotics, pp. 25–44. Springer (2004)

62. Hoffman, H.G., Patterson, D.R., Carrougher, G.J.: Use of virtual reality for adjunctive treatment of adult burn pain during physical therapy: a controlled study. The Clinical journal of pain 16(3), 244–250 (2000)

63. Hogan, N., Krebs, H.I., Charnnarong, J., Srikrishna, P., Sharon, A.: Mit-manus: a workstation for manual therapy and training. i. In: Robot and Human Communication, 1992. Proceedings., IEEE International Workshop on. pp. 161–165. IEEE (1992)

64. Intercollegiate Stroke Working Party: National clinical guideline for stroke, vol. 20083. Citeseer (2012)

65. Iosa, M., Morone, G., Cherubini, A., Paolucci, S.: The three laws of neurorobotics: A review on what neurorehabilitation robots should do for patients and clinicians. Journal of Medical and Biological Engineering 36(1), 1–11 (Feb 2016), https://doi.org/10.1007/s40846-016-0115-2

66. Jadhav, C., Krovi, V.: A low-cost framework for individualized interactive telerehabilitation. In: Engineering in Medicine and Biology Society, 2004. IEMBS'04. 26th Annual International Conference of the IEEE. vol. 2, pp. 3297–3300. IEEE (2004)

67. Jaffe, D.L., Brown, D.A., Pierson-Carey, C.D., Buckley, E.L., Lew, H.L.: Stepping over obstacles to improve walking in individuals with poststroke hemiplegia. Journal of Rehabilitation Research & Development 41(3A), 283–292 (2004)

68. Johnson, M.J., Loureiro, R.C., Harwin, W.S.: Collaborative tele-rehabilitation and robot-mediated therapy for stroke rehabilitation at home or clinic. Intelligent Service Robotics 1(2), 109–121 (2008)

69. Juan, M.C., Calatrava, J.: An augmented reality system for the treatment of phobia to small animals viewed via an optical see-through hmd: comparison with a similar system viewed via a video see-through hmd. International Journal of Human-Computer Interaction 27(5), 436–449 (2011)

70. Kahn, L.E., Lum, P.S., Rymer, W.Z., Reinkensmeyer, D.J.: Robot-assisted movement training for the stroke-impaired arm: Does it matter what the robot does? Journal of rehabilitation research and development 43(5) (2014)

71. Kairy, D., Lehoux, P., Vincent, C., Visintin, M.: A systematic review of clinical outcomes, clinical process, healthcare utilization and costs associated with telerehabilitation. Disability and rehabilitation 31(6), 427–447 (2009)

72. Kaminer, C., LeBras, K., McCall, J., Phan, T., Naud, P., Teodorescu, M., Kurniawan, S.: An immersive physical therapy game for stroke survivors. In: Proceedings of the 16th International ACM SIGACCESS Conference on Computers & Accessibility. pp. 299–300. ASSETS '14, ACM, New York, NY, USA (2014), http://doi.acm.org.login.ezproxy.library.ualberta.ca/10.1145/2661334.2661340

73. Khademi, M., Hondori, H.M., Dodakian, L., Cramer, S., Lopes, C.V.: Comparing "pick and place" task in spatial Augmented Reality versus non-immersive Virtual Reality for rehabilitation setting. Conf Proc IEEE Eng Med Biol Soc 2013, 4613–4616 (2013)
74. Khademi, M., Hondori, H.M., Lopes, C.V., Dodakian, L., Cramer, S.C.: Haptic augmented reality to monitor human arm's stiffness in rehabilitation. In: 2012 IEEE-EMBS Conference on Biomedical Engineering and Sciences. pp. 892–895 (Dec 2012)
75. Khalili, D., Zomlefer, M.: An intelligent robotic system for rehabilitation of joints and estimation of body segment parameters. IEEE transactions on biomedical engineering 35(2), 138–146 (1988)
76. Khansari-Zadeh, S.M., Billard, A.: Learning stable nonlinear dynamical systems with gaussian mixture models. IEEE Transactions on Robotics 27(5), 943–957 (Oct 2011)
77. Kostov, A., Andrews, B.J., Popovic, D.B., Stein, R.B., Armstrong, W.W.: Machine learning in control of functional electrical stimulation systems for locomotion. IEEE Transactions on Biomedical Engineering 42(6), 541–551 (1995)
78. Krebs, H.I., Hogan, N., Aisen, M.L., Volpe, B.T.: Robot-aided neurorehabilitation. IEEE transactions on rehabilitation engineering 6(1), 75–87 (1998)
79. Langhammer, B., Stanghelle, J.K.: Bobath or motor relearning programme? a comparison of two different approaches of physiotherapy in stroke rehabilitation: a randomized controlled study. Clinical rehabilitation 14(4), 361–369 (2000)
80. Lauretti, C., Cordella, F., Guglielmelli, E., Zollo, L.: Learning by demonstration for planning activities of daily living in rehabilitation and assistive robotics. IEEE Robotics and Automation Letters 2(3), 1375–1382 (July 2017)
81. Legg, L., Drummond, A., Leonardi-Bee, J., Gladman, J.R., Corr, S., Donkervoort, M., Edmans, J., Gilbertson, L., Jongbloed, L., Logan, P., Sackley, C., Walker, M., Langhorne, P.: Occupational therapy for patients with problems in personal activities of daily living after stroke: systematic review of randomised trials. BMJ (2007), http://www.bmj.com/content/early/2006/12/31/bmj.39343.466863.55
82. Leightley, D., Darby, J., Li, B., McPhee, J.S., Yap, M.H.: Human activity recognition for physical rehabilitation. In: 2013 IEEE International Conference on Systems, Man, and Cybernetics. pp. 261–266. IEEE (2013)
83. LeMoyne, R., Mastroianni, T., Hessel, A., Nishikawa, K.: Ankle rehabilitation system with feedback from a smartphone wireless gyroscope platform and machine learning classification. In: 2015 IEEE 14th International Conference on Machine Learning and Applications (ICMLA). pp. 406–409. IEEE (2015)
84. Lenze, E.J., Munin, M.C., Quear, T., Dew, M.A., Rogers, J.C., Begley, A.E., Reynolds III, C.F.: Significance of poor patient participation in physical and occupational therapy for functional outcome and length of stay. Archives of physical medicine and rehabilitation 85(10), 1599–1601 (2004)
85. Li, W.J., Hsieh, C.Y., Lin, L.F., Chu, W.C.: Hand gesture recognition for post-stroke rehabilitation using leap motion. In: 2017 International Conference on Applied System Innovation (ICASI). pp. 386–388. IEEE (2017)
86. Van der Loos, H.M., Reinkensmeyer, D.J., Guglielmelli, E.: Rehabilitation and Health Care Robotics, pp. 1685–1728. Springer International Publishing, Cham (2016), https://doi.org/10.1007/978-3-319-32552-1_64
87. Lum, P., Reinkensmeyer, D., Mahoney, R., Rymer, W.Z., Burgar, C.: Robotic devices for movement therapy after stroke: Current status and challenges to clinical acceptance. Topics in Stroke Rehabilitation 8(4), 40–53 (2002), https://doi.org/10.1310/9KFM-KF81-P9A4-5WW0, pMID: 14523729
88. Lum, P.S., Burgar, C.G., Shor, P.C.: Evidence for improved muscle activation patterns after retraining of reaching movements with the mime robotic system in subjects with post-stroke hemiparesis. IEEE Transactions on Neural Systems and Rehabilitation Engineering 12(2), 186–194 (2004)

89. Lum, P.S., Burgar, C.G., Shor, P.C., Majmundar, M., Van der Loos, M.: Robot-assisted movement training compared with conventional therapy techniques for the rehabilitation of upper-limb motor function after stroke. Archives of physical medicine and rehabilitation 83(7), 952–959 (2002)
90. Luo, X., Kline, T., Fischer, H.C., Stubblefield, K.A., Kenyon, R.V., Kamper, D.G.: Integration of augmented reality and assistive devices for post-stroke hand opening rehabilitation. In: 2005 IEEE Engineering in Medicine and Biology 27th Annual Conference. pp. 6855–6858 (2005)
91. Ma, M., McNeill, M., Charles, D., McDonough, S., Crosbie, J., Oliver, L., McGoldrick, C.: Adaptive virtual reality games for rehabilitation of motor disorders. In: International Conference on Universal Access in Human-Computer Interaction. pp. 681–690. Springer (2007)
92. Maaref, M., Rezazadeh, A., Shamaei, K., Ocampo, R., Mahdi, T.: A bicycle cranking model for assist-as-needed robotic rehabilitation therapy using learning from demonstration. IEEE Robotics and Automation Letters 1(2), 653–660 (July 2016)
93. Maclean, N., Pound, P., Wolfe, C., Rudd, A.: A critical review of the concept of patient motivation in the literature on physical rehabilitation. Soc Sci Med 50(4), 495–506 (2000)
94. Maclean, N., Pound, P., Wolfe, C., Rudd, A.: Qualitative analysis of stroke patients' motivation for rehabilitation. Bmj 321(7268), 1051–1054 (2000)
95. Maier, M., Ballester, B.R., Duarte, E., Duff, A., Verschure, P.F.: Social integration of stroke patients through the multiplayer rehabilitation gaming system. In: International Conference on Serious Games. pp. 100–114. Springer (2014)
96. Mark, V.W., Taub, E.: Constraint-induced movement therapy for chronic stroke hemiparesis and other disabilities. Restorative neurology and neuroscience 22(3-5), 317–336 (2004)
97. Martínez, C., Tavakoli, M.: Learning and robotic imitation of therapist's motion and force for post-disability rehabilitation. In: Systems, Man, and Cybernetics (SMC), 2017 IEEE International Conference on. pp. 2225–2230. IEEE (2017)
98. McComas, A., Sica, R., Upton, A., Aguilera, N.: Functional changes in motoneurones of hemiparetic patients. Journal of Neurology, Neurosurgery & Psychiatry 36(2), 183–193 (1973)
99. McLeod, A., Bochniewicz, E.M., Lum, P.S., Holley, R.J., Emmer, G., Dromerick, A.W.: Using wearable sensors and machine learning models to separate functional upper extremity use from walking-associated arm movements. Archives of physical medicine and rehabilitation 97(2), 224–231 (2016)
100. Mehrholz, J., Hädrich, A., Platz, T., Kugler, J., Pohl, M.: Electromechanical and robot-assisted arm training for improving generic activities of daily living, arm function, and arm muscle strength after stroke. Cochrane database of systematic reviews (6) (2012)
101. Milgram, P., Kishino, F.: A taxonomy of mixed reality visual displays. IEICE TRANSACTIONS on Information and Systems 77(12), 1321–1329 (1994)
102. Mimouni, M., Cismariu-Potash, K., Ratmansky, M., Shaklai, S., Amir, H., Mimouni-Bloch, A.: Trends in physical medicine and rehabilitation publications over the past 16 years. Archives of Physical Medicine and Rehabilitation 97(6), 1030–1033 (2016), http://www.sciencedirect.com/science/article/pii/S0003999315014100
103. Mirelman, A., Bonato, P., Deutsch, J.E.: Effects of training with a robot-virtual reality system compared with a robot alone on the gait of individuals after stroke. Stroke 40(1), 169–174 (2009)
104. Mosey, A.C.: Psychosocial components of occupational therapy. Lippincott Williams & Wilkins (1986)
105. Mousavi Hondori, H., Khademi, M., Dodakian, L., Cramer, S.C., Lopes, C.V.: A Spatial Augmented Reality rehab system for post-stroke hand rehabilitation. Stud Health Technol Inform 184, 279–285 (2013)
106. Murphy, M.A., Persson, H.C., Danielsson, A., Broeren, J., Lundgren-Nilsson, Å., Sunnerhagen, K.S.: Salgot-s troke a rm l ongitudinal study at the university of got henburg, prospective cohort study protocol. BMC neurology 11(1), 56 (2011)

107. Najafi, M., Adams, K., Tavakoli, M.: Robotic learning from demonstration of therapist's time-varying assistance to a patient in trajectory-following tasks. In: Rehabilitation Robotics (ICORR), 2017 International Conference on. pp. 888–894. IEEE (2017)
108. Novak, D., Nagle, A., Keller, U., Riener, R.: Increasing motivation in robot-aided arm rehabilitation with competitive and cooperative gameplay. Journal of neuroengineering and rehabilitation 11(1), 64 (2014)
109. Ocampo, R., Tavakoli, M.: Improving user performance in haptics-based rehabilitation exercises by colocation of user's visual and motor axes via a three-dimensional augmented-reality display. IEEE Robotics and Automation Letters 4(2), 438–444 (2019)
110. Ocampo, R., Tavakoli, M.: Visual-haptic colocation in robotic rehabilitation exercises using a 2d augmented-reality display. In: 2019 International Symposium on Medical Robotics (ISMR). pp. 1–7 (April 2019)
111. Octavia, J.R., Coninx, K.: Adaptive personalized training games for individual and collaborative rehabilitation of people with multiple sclerosis. BioMed research international 2014 (2014)
112. Pandyan, A., Johnson, G., Price, C., Curless, R., Barnes, M., Rodgers, H.: A review of the properties and limitations of the ashworth and modified ashworth scales as measures of spasticity. Clinical rehabilitation 13(5), 373–383 (1999)
113. Pardini, H.: VR Vaccine, https://www.adforum.com/award-organization/6650183/showcase/2017/ad/34544861
114. Parton, A., Malhotra, P., Husain, M.: Hemispatial neglect. Journal of Neurology, Neurosurgery & Psychiatry 75(1), 13–21 (2004)
115. Pehlivan, A.U., Losey, D.P., O'Malley, M.K.: Minimal assist-as-needed controller for upper limb robotic rehabilitation. IEEE Transactions on Robotics 32(1), 113–124 (2016)
116. Peternel, L., Petrič, T., Oztop, E., Babič, J.: Teaching robots to cooperate with humans in dynamic manipulation tasks based on multi-modal human-in-the-loop approach. Autonomous Robots 36(1-2), 123–136 (January 2014), https://doi.org/10.1007/s10514-013-9361-0
117. Public Health Agency of Canada: Tracking heart disease and stroke in canada 2009, https://www.canada.ca/en/public-health/services/reports-publications/2009-tracking-heart-disease-stroke-canada.html
118. Reinkensmeyer, D.J., Kahn, L.E., Averbuch, M., McKenna-Cole, A., Schmit, B.D., Rymer, W.Z.: Understanding and treating arm movement impairment after chronic brain injury: progress with the arm guide. Journal of Rehabilitation Research and Development 37(6), 653–662 (2014)
119. Reinkensmeyer, D.J., Pang, C.T., Nessler, J.A., Painter, C.C.: Web-based telerehabilitation for the upper extremity after stroke. IEEE transactions on neural systems and rehabilitation engineering 10(2), 102–108 (2002)
120. Ricker, J.H., Rosenthal, M., Garay, E., DeLuca, J., Germain, A., Abraham-Fuchs, K., Schmidt, K.U.: Telerehabilitation needs: a survey of persons with acquired brain injury. The Journal of Head Trauma Rehabilitation 17(3), 242–250 (2002)
121. Robitaille, N., Jackson, P.L., Hébert, L.J., Mercier, C., Bouyer, L.J., Fecteau, S., Richards, C.L., McFadyen, B.J.: A virtual reality avatar interaction (vrai) platform to assess residual executive dysfunction in active military personnel with previous mild traumatic brain injury: proof of concept. Disability and Rehabilitation: Assistive Technology 12(7), 758–764 (2017)
122. Sandlund, M., McDonough, S., Häger-Ross, C.: Interactive computer play in rehabilitation of children with sensorimotor disorders: a systematic review. Developmental Medicine & Child Neurology 51(3), 173–179 (2009)
123. Sawyer, B., Rejeski, D.: Serious games: Improving public policy through game-based learning and simulation (2002)
124. Shahbazi, M., Atashzar, S.F., Tavakoli, M., Patel, R.V.: Position-force domain passivity of the human arm in telerobotic systems. IEEE/ASME Transactions on Mechatronics 23(2), 552–562 (2018)

125. Sharifi, M., Behzadipour, S., Salarieh, H., Tavakoli, M.: Cooperative modalities in robotic tele-rehabilitation using nonlinear bilateral impedance control. Control Engineering Practice 67, 52–63 (2017)

126. Shilling, R., Zyda, M., Wardynski, E.C.: Introducing emotion into military simulation and videogame design: America's army operations and virte. In: Proceedings of the GameOn Conference. vol. 3 (2002)

127. Shirzad, N., Van der Loos, H.M.: Adaptation of task difficulty in rehabilitation exercises based on the user's motor performance and physiological responses. In: 2013 IEEE 13th International Conference on Rehabilitation Robotics (ICORR). pp. 1–6. IEEE (2013)

128. Sietsema, J.M., Nelson, D.L., Mulder, R.M., Mervau-Scheidel, D., White, B.E.: The use of a game to promote arm reach in persons with traumatic brain injury. American Journal of Occupational Therapy 47(1), 19–24 (1993)

129. Söderback, I.: International handbook of occupational therapy interventions. Springer (2009)

130. Squeri, V., Basteris, A., Sanguineti, V.: Adaptive regulation of assistance 'as needed' in robot-assisted motor skill learning and neuro-rehabilitation. In: 2011 IEEE International conference on rehabilitation robotics. pp. 1–6. IEEE (2011)

131. Steele, E., Grimmer, K., Thomas, B., Mulley, B., Fulton, I., Hoffman, H.: Virtual reality as a pediatric pain modulation technique: a case study. Cyberpsychology & Behavior 6(6), 633–638 (2003)

132. Strazzulla, I., Nowak, M., Controzzi, M., Cipriani, C., Castellini, C.: Online bimanual manipulation using surface electromyography and incremental learning. IEEE Transactions on Neural Systems and Rehabilitation Engineering 25(3), 227–234 (2017)

133. Swapp, D., Pawar, V., Loscos, C.: Interaction with co-located haptic feedback in virtual reality. Virtual Reality 10(1), 24–30 (May 2006), https://doi.org/10.1007/s10055-006-0027-5

134. Tao, R.: Haptic teleoperation based rehabilitation systems for task-oriented therapy. Master's thesis, University of Alberta, Edmonton, Canada (2014)

135. Thompson, D., Baranowski, T., Buday, R., Baranowski, J., Thompson, V., Jago, R., Griffith, M.J.: Serious video games for health: How behavioral science guided the development of a serious video game. Simulation & gaming 41(4), 587–606 (2010)

136. Trojan, J., Diers, M., Fuchs, X., Bach, F., Bekrater-Bodmann, R., Foell, J., Kamping, S., Rance, M., Maass, H., Flor, H.: An augmented reality home-training system based on the mirror training and imagery approach. Behav Res Methods 46(3), 634–640 (Sep 2014)

137. Van Krevelen, D., Poelman, R.: A survey of augmented reality technologies, applications and limitations. International journal of virtual reality 9(2), 1 (2010)

138. Vidrios-Serrano, C., Bonilla, I., Vigueras-Gómez, F., Mendoza, M.: Development of a haptic interface for motor rehabilitation therapy using augmented reality. In: 2015 37th Annual International Conference of the IEEE Engineering in Medicine and Biology Society (EMBC). pp. 1156–1159 (Aug 2015)

139. Vieira, J., Sousa, M., Arsénio, A., Jorge, J.: Augmented reality for rehabilitation using multimodal feedback. In: Proceedings of the 3rd 2015 Workshop on ICTs for improving Patients Rehabilitation Research Techniques. pp. 38–41. ACM (2015)

140. Voelker, R.: Rehabilitation medicine welcomes a robotic revolution. JAMA 294(10), 1191–1195 (2005), https://doi.org/10.1001/jama.294.10.1191

141. Volpe, B., Krebs, H., Hogan, N., Edelsteinn, L., Diels, C., Aisen, M.: Robot training enhanced motor outcome in patients with stroke maintained over 3 years. Neurology 53(8), 1874–1874 (1999), http://n.neurology.org/content/53/8/1874

142. Williams, B., Chang, A., Landefeld, C.S., Ahalt, C., Conant, R., Chen, H.: Current diagnosis and treatment: geriatrics 2E. McGraw Hill Professional (2014)

143. Williams, D.J., Krebs, H.I., Hogan, N.: A robot for wrist rehabilitation. In: Engineering in Medicine and Biology Society, 2001. Proceedings of the 23rd Annual International Conference of the IEEE. vol. 2, pp. 1336–1339. IEEE (2001)

144. Wolbrecht, E.T., Chan, V., Reinkensmeyer, D.J., Bobrow, J.E.: Optimizing compliant, model-based robotic assistance to promote neurorehabilitation. IEEE Transactions on Neural Systems and Rehabilitation Engineering 16(3), 286–297 (2008)
145. Worsnopp, T., Peshkin, M., Colgate, J., Kamper, D.: An actuated finger exoskeleton for hand rehabilitation following stroke. In: Rehabilitation Robotics, 2007. ICORR 2007. IEEE 10th International Conference on. pp. 896–901. IEEE (2007)
146. Wright, W., Bogost, I.: Persuasive games: The expressive power of videogames. Mit Press (2007)
147. Yanco, H.A., Haigh, K.Z.: Automation as caregiver: A survey of issues and technologies. Am. Assoc. Artif. Intell 2, 39–53 (2002)
148. Yeh, S.C., Hwang, W.Y., Huang, T.C., Liu, W.K., Chen, Y.T., Hung, Y.P.: A study for the application of body sensing in assisted rehabilitation training. In: 2012 International Symposium on Computer, Consumer and Control. pp. 922–925 (June 2012)
149. Yeh, S.C., Huang, M.C., Wang, P.C., Fang, T.Y., Su, M.C., Tsai, P.Y., Rizzo, A.: Machine learning-based assessment tool for imbalance and vestibular dysfunction with virtual reality rehabilitation system. Computer methods and programs in biomedicine 116(3), 311–318 (2014)
150. Yelnik, A.P., Lebreton, F.O., Bonan, I.V., Colle, F.M., Meurin, F.A., Guichard, J.P., Vicaut, E.: Perception of verticality after recent cerebral hemispheric stroke. Stroke 33(9), 2247–2253 (2002)
151. Zhu, M., Zhang, Z., Hirdes, J.P., Stolee, P.: Using machine learning algorithms to guide rehabilitation planning for home care clients. BMC medical informatics and decision making 7(1), 41 (2007)